Problem-Based Learning for Engineering Education

This book embarks on an in-depth exploration of Problem-Based Learning (PBL) in the context of engineering education. Its overarching goal is to provide a comprehensive understanding of the genesis, potential, challenges, contemporary relevance, and methodological approaches of PBL. The impetus behind this undertaking arises from the acknowledged challenges within engineering education, necessitating a fundamental paradigm shift. The study's justification lies in the imperative adoption of innovative teaching methodologies for student performance improvement and fostering competency among engineering graduates.

The narrative unfolds across four distinct parts: Background and Theoretical Perspectives; Methodological Aspects of PBL; Implications of PBL in Engineering Education and Case Studies; and Plausible Examples and Reflections and Future Direction.

Part I establishes a robust foundation by defining PBL, tracing its historical roots, and weighing its potential against challenges. A critical examination of PBL's relevance underscores the urgent need for a paradigm shift in engineering education. The part weaves a "golden thread" that permeates subsequent explorations. A theoretical perspective of PBL follows, exploring its underpinnings and theoretical frameworks. This is complemented by a comparative analysis, offering insights into PBL's efficacy compared to traditional learning methods.

The second part examines the alignment of PBL with Outcomes-Based Education, lays the groundwork for a coherent and practical integration of these educational approaches within the Engineering curriculum, and also offers methodological approaches to deliver and practice PBL.

Part III explores PBL's application for reinforcing engineering education, emphasising its tangible impact. It probes into PBL as a means to instil entrepreneurial attributes in students and further expands to encompass the decolonisation and internationalisation of engineering education through PBL.

The final part provides case studies, offers plausible examples, reflects on the lessons learned and draws insightful conclusions, synthesising the findings into a cohesive whole.

Overall, this book seeks to make a substantial contribution to the ongoing discourse on innovative teaching methodologies, particularly within the context of engineering education. Through a meticulous examination of theoretical foundations, comparative analyses, and the provision of practical methodologies, it aims to serve as a valuable resource for educators, researchers, and policymakers who seek to enhance the effectiveness and relevance of engineering education in our ever-evolving world.

Dillip K. Das holds a PhD in Urban and Regional Planning, complemented by a background in Civil Engineering and City Planning. He is currently actively involved in teaching, research, and community engagement as a Professor in the Discipline of Civil Engineering at the University of KwaZulu-Natal, Durban, South Africa. He is a co-author of two books, *Planning for Regional Tourism Development* and *Planning for Balanced Regional Development*, and has published over 200 peer-reviewed research articles. He has also led and co-led several research and industry projects, including a project linked to teaching and learning.

Problem-Based Learning for Engineering Education

Developing the Engineers of the Future

Dillip K. Das

Routledge
Taylor & Francis Group
LONDON AND NEW YORK

Designed cover image: Taylor & Francis, template cover

First published 2026
by Routledge
4 Park Square, Milton Park, Abingdon, Oxon OX14 4RN

and by Routledge
605 Third Avenue, New York, NY 10158

Routledge is an imprint of the Taylor & Francis Group, an informa business

British Library Cataloguing-in-Publication Data
A catalogue record for this book is available from the British Library

ISBN: 978-1-032-86366-5 (hbk)
ISBN: 978-1-032-86878-3 (pbk)
ISBN: 978-1-003-52974-3 (ebk)

DOI: 10.1201/9781003529743

Typeset in Times New Roman
by SPi Technologies India Pvt Ltd (Straive)

Contents

Figures

Tables

Preface

The landscape of engineering education across the globe, particularly in the Global South, is undergoing significant transformation. The convolutions of the 21st century continue to transform the professional, social, and environmental roles of engineers. In response to rising demands for competency, improved student performance, and better preparation for the world of work, it has become increasingly evident that traditional models of education dominated by lecture-based instruction and rote learning are no longer adequate.

In this context, Problem-Based Learning (PBL) has emerged as a powerful pedagogical approach. It not only enhances student engagement but also cultivates the critical skills and competencies required for tackling real-world, complex problems. Through PBL, students develop competencies and skills in collaboration, innovation, adaptability, and lifelong learning—attributes that are essential for navigating an increasingly interdisciplinary and rapidly evolving world.

This book, *Problem-Based Learning for Engineering Education: Developing the Engineers of the Future*, is intended to serve as both a theoretical foundation and a practical guide for educators, curriculum designers, researchers, and policymakers seeking to integrate PBL into engineering curricula. Rooted in the constructivist philosophies of educational thinkers such as John Dewey, and informed by contemporary practices across diverse engineering disciplines, the book presents PBL as a dynamic, student-centred approach that can serve as an alternative to—or complement for—traditional models of instruction.

The book is organised into four parts. Part I outlines the background and theoretical underpinnings of PBL, highlighting its historical development, philosophical foundations, and the need for a paradigm shift in engineering education. Part II addresses methodological considerations, including the alignment of PBL with Outcomes-Based Education (OBE), and provides structured models to support effective implementation. Part III transitions from theory to practice, exploring how PBL reinforces core competencies, fosters entrepreneurial thinking, and contributes to both the internationalisation and decolonisation of engineering education. Finally, Part IV offers discipline-specific case studies, plausible examples, and reflective insights that illuminate the application of PBL in Civil, Mechanical, Electrical, Chemical, and Computer/ Software Engineering.

Across these four parts, the chapters emphasise not only the opportunities afforded by PBL but also the practical challenges of embedding it within institutional frameworks. Collectively, they offer thoughtful analysis and constructive recommendations to guide meaningful transformation.

It is hoped that this book will inspire and assist educators seeking new pedagogical approaches, students interested in alternative learning models, and academic leaders engaged in curriculum reform to reimagine engineering education—not as a static transfer of knowledge, but as a vibrant, collaborative, and problem-solving journey.

The author would like to note that Chat GPT-3 and Chat GPT-4 were used to edit and enhance the clarity and readability of the text. SciSpace was used to access literature.

Part I

Background and Theoretical Perspectives

Chapter 1

Problem-Based Learning (PBL)

Background

1.1 Introduction

Contemporary higher education is undergoing a significant transformation, particularly within engineering disciplines. Increasingly, academic and industry stakeholders are advocating for comprehensive reforms in curriculum design, teaching methodologies, and the integration of emerging technologies. These arguments are not merely reactionary; they are driven by fundamental shifts in the nature of work, rapid technological advancement, the demand for interdisciplinary knowledge, and the urgency of addressing global challenges such as climate change, sustainability, and social equity (Gleeson, Klenowski, & Looney, 2020; Howson & Kingsbury, 2021; Wolff, Hattingh & Smith, 2025)

In this evolving landscape, the traditional engineering education model, long centred on mathematical rigour and scientific fundamentals, is being called into question. While these foundations remain essential, there is growing recognition that they are no longer self-sufficient. Educators and employers alike are advocating for more practical, industry-relevant, and student-centred approaches that better reflect the realities of a fast-changing, interconnected world (AACTE, 2010; O'Neill & Short, 2023; UNESCO, 2015). This shift is viewed as critical for preparing graduates to solve complex problems that are often ill-defined, multidisciplinary, and laden with ethical and societal implications.

One of the most intensely debated areas of reform is curriculum design. Engineering programmes are increasingly under pressure to incorporate digital technologies—such as artificial intelligence, the Internet of Things (IoT), and cybersecurity—into their core offerings. These technological advancements are reshaping industries, and, by extension, the expectations placed on engineering graduates (Ahuja & Bala, 2021; Li, 2024). Additionally, some question whether the traditional four-year degree structure remains adequate given the exponential growth of knowledge and the expanding scope of professional responsibilities now required of engineers (Grayson et al., 2014; Zorpette, 1984). There is widespread agreement among stakeholders—students, faculty, alums, and employers—that curricula must not only uphold scientific integrity but also

DOI: 10.1201/9781003529743-2

cultivate interdisciplinary thinking and essential soft skills like leadership, teamwork, and communication (Figueiredo et al., 2023).

Teaching methods are also under critical review. Traditional lecture-based instruction, while still widely practised, is increasingly seen as ineffective for engaging contemporary learners and fostering the higher-order thinking skills needed for modern engineering practice. In response, a variety of interactive pedagogical strategies are being adopted, such as active learning, cooperative learning, and project-based learning, that emphasise student participation, problem-solving, and real-world application (Felder, 1993, 2012; Kim, 2008). These approaches are especially relevant in preparing students to address global challenges and contribute to sustainable development (Mora-Luis & Martín-Gutiérrez, 2020).

Technology has also become a central pillar in educational reform. The rise of digital tools—including virtual labs, simulations, and online, self-paced learning environments—enhances engineering education's flexibility, personalisation, and realism (Ahuja & Bala, 2021; Gati & Kartyas, 2012). These innovations support more accessible hands-on learning experiences, often simulating complex engineering scenarios. However, their implementation also demands a critical lens to ensure that technology not only enhances learning but also fosters ethical reasoning, inclusivity, and reflective thinking (Zorpette, 1984).

Parallel to these pedagogical developments is an increasing alignment between academia and industry. Employers now expect graduates to be competent and "work-ready" from day one, prompting universities to offer internships, co-developed curricula, and applied industry projects (Kim, 2008). While this practical orientation improves employability, it also raises concerns about narrowing education into a vocational pipeline at the expense of broader learning outcomes. Many educators argue that engineering programmes must also prioritize critical thinking, ethical awareness, and civic responsibility—skills essential for engineers to navigate the broader societal impacts of their work.

As engineering education adapts to these demands, universities are embracing greater curriculum flexibility and interdisciplinarity. Students are offered broader options in electives, minors, and dual-degree programmes that blend engineering with disciplines such as business, liberal arts, and social sciences. These interdisciplinary pathways reflect the complex nature of real-world problems, which often require integrated approaches. Alongside technical skills, institutions emphasise developing non-technical competencies such as creativity, empathy, ethical reasoning, and effective communication.

However, despite these efforts, serious challenges persist. Diversity, equity, and inclusion remain unresolved issues, with women and underrepresented minorities still facing significant barriers in engineering fields (Amer et al., 2024; Leong, 2023). In response, universities are experimenting with inclusive curricula, mentorship programmes, and institutional reforms aimed at fostering a more welcoming and equitable learning environment. These initiatives are grounded in a broader shift toward recognising engineering as a sociotechnical discipline—one in which societal and ethical considerations are inseparable from technical innovation.

Structural debates within higher education further complicate the reform agenda. Efficiency-oriented management models, often referred to as “Lean Higher Education,” seek to cut costs and increase performance (Helmold, 2022; Kokkinou & van Kollenburg, 2023). While these models may bring operational benefits, critics warn that they risk reducing education to a commodified product, undermining its academic and social missions. The challenge lies in balancing operational efficiency with the commitment to intellectual rigour, equity, and student development.

Taken together, these debates reveal that engineering education stands at a pivotal crossroads. It faces a unique and urgent challenge: to produce graduates who are not only technically proficient but also socially conscious, globally aware, and capable of addressing the world’s most pressing challenges. Increasingly, engineers are expected to contribute to both economic development and the resolution of complex issues such as environmental degradation, public health, and social inequality. Many of these issues lie outside the traditional scope of engineering. They are often described as “wicked problems”—deeply interconnected problems, lacking clear solutions, and requiring creative, cross-disciplinary responses.

This growing complexity has exposed a gap between the skills students acquire in formal education and the competencies required in the professional world. In many regions, particularly in developing countries, graduates are frequently described as lacking essential competencies, with some deemed unemployable. This troubling disconnect has prompted educators and policymakers to critically examine the limitations of conventional education systems, especially traditional forms of teaching, learning, and assessment.

In this context, Problem-Based Learning (PBL) has gained prominence as a potential solution to many of the challenges facing engineering education today. PBL shifts the focus from passive knowledge acquisition to active, student-driven inquiry. It encourages learners to work collaboratively on real-world problems, drawing on multiple disciplines and engaging in critical reflection. By placing students in the role of problem-solvers, PBL fosters not only technical competence but also creativity, adaptability, and social awareness (Dolmans et al., 2005; Duch et al., 2001; Lalwani, 2024; Mohammed & Ali, 2024; Zumbach & Prescher, 2023).

Advocates argue that PBL offers a transformative framework capable of bridging the gap between education and practice (Rosário & Dias, 2024; Weber, 2014). It prepares students not just to succeed in their first jobs but to thrive in lifelong careers marked by complexity, change, and ethical responsibility. Moreover, PBL aligns with contemporary goals of higher education: to produce graduates who are innovative thinkers, responsible citizens, and effective contributors to sustainable and equitable development.

This book explores the potential of PBL in reimagining engineering education. Drawing on research, case studies, and practical insights, it aims to provide a comprehensive understanding of how PBL can address the competence gap, enhance student engagement, and prepare future engineers for the demands of a rapidly evolving world.

1.2 Defining Problem-Based Learning (PBL)

To probe deeper into the potential of PBL as a transformative pedagogical approach, it is imperative to begin with a precise definition. PBL is an instructional approach that emphasises student-centred, active learning through engagement with real-world problems. Widely adopted in fields such as law, medicine, engineering, and education, PBL is designed to develop not only subject-specific knowledge but also transferable competencies such as critical thinking, collaboration, and communication skills. The learning process is typically conducted in small groups, where students work collaboratively to analyse and solve complex problems, supported by a tutors or facilitators who guides rather than instructs.

As Blumenfeld et al. (1991) described, PBL motivates learners to "pursue solutions to non-trivial problems by asking and refining questions, debating ideas, making predictions, designing plans and/or experiments, collecting and analysing data, drawing conclusions, communicating their ideas and findings to others, asking new questions, and creating artifacts." Similarly, Duch et al. (2001) emphasise that PBL cultivates a wide range of cognitive and interpersonal skills, including critical thinking, effective communication, teamwork, and the ability to navigate complex subject matter. In this context, students assume an active role in their learning journey, shifting from passive recipients of information to autonomous problem-solvers.

According to Thomas (2000), PBL also offers learners the opportunity to engage with long-term, intricate problems that culminate in the creation of real-world solutions or products, blending design, analysis, and decision-making. The overarching aim is to foster lifelong learning and professional readiness.

1.2.1 Core Characteristics of Problem-Based Learning

A review of the literature reveals a set of core characteristics that define PBL as a distinct and impactful pedagogical methodology (Figure 1.1). The core characteristics are:

Real-World Problems as Catalysts: PBL centres learning around complex, authentic problems that reflect real-world challenges, enhancing relevance and student motivation (Hunter et al., 2025; Lalwani, 2024).

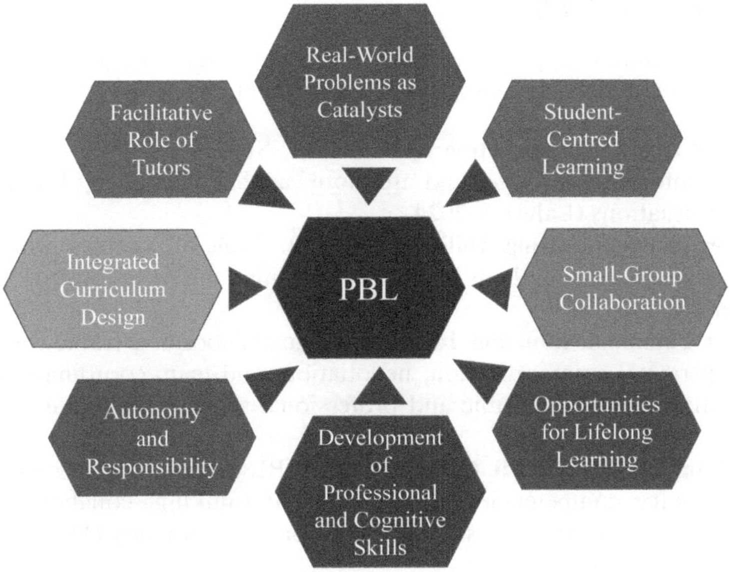

Figure 1.1 Core Characteristics of PBL.

Student-Centred Learning: Students are responsible for identifying what they need to learn, pursuing self-directed study to develop the knowledge and skills required to address the problem (Mohammed & Ali, 2024; Zumbach & Prescher, 2023).

Small-Group Collaboration: Learning occurs in small groups, encouraging communication, negotiation, peer teaching, and collective problem-solving (Prasad & O'Malley, 2022).

Facilitative Role of Tutors: Instructors act as facilitators rather than content deliverers, supporting students in staying focused, thinking critically, and engaging deeply without directly providing answers (Hunter et al., 2025; Prasad & O'Malley, 2022).

Integrated Curriculum Design: PBL is structured to promote interdisciplinary understanding by integrating multiple subjects through the lens of a single, multifaceted problem (Neville & Norman, 2007; Lopes et al., 2020).

Development of Professional and Cognitive Skills: The PBL process promotes the growth of critical thinking, effective communication, teamwork, and research skills (Dolmans et al., 2005; Duch et al., 2001).

Autonomy and Responsibility: PBL empowers students with a high degree of autonomy and accountability, fostering independent, self-regulated learners (Dolmans & Schmidt, 2006; Lopes et al., 2020).

Opportunities for Lifelong Learning: By engaging with the learning process as active participants, students develop inquiry and reflection habits essential for continuous professional development (Duch et al., 2001).

1.2.2 *Efficacy of PBL*

Numerous studies support the efficacy of PBL in achieving educational outcomes aligned with 21st-century skill requirements, such as:

Enhanced Knowledge Retention and Transfer: Students in PBL environments retain information longer and are more capable of applying knowledge in novel situations (Lalwani, 2024).

Improved Problem-Solving Abilities: The central role of complex problem scenarios in PBL strengthens analytical reasoning and decision-making skills (Meifitri, 2020).

Stronger Communication and Teamwork: The collaborative framework fosters interpersonal communication, negotiation, and team coordination—skills essential in both academic and professional settings (Prasad & O'Malley, 2022).

Acquisition of Transversal Competencies: PBL equips learners with cross-disciplinary competencies such as critical thinking, collaboration, and adaptability—attributes essential for global employability (Du et al., 2013; Guerra et al., 2017).

1.2.3 *Critical Challenges*

Despite its strengths, PBL is not without its challenges. Effective implementation demands structural, pedagogical, and institutional support. The critical systemic challenges include:

Resource Intensiveness: PBL requires more time and resources than traditional teaching methods, including smaller class sizes, extensive faculty engagement, and access to appropriate learning materials (Adamyan, 2024; Lalwani, 2024).

Complex Curriculum Integration: Embedding PBL in existing curricula necessitates careful planning and realigning course objectives, assessment methods, and faculty roles (Adamyan, 2024).

Tutor Training and Facilitation: The success of PBL depends heavily on the skill of facilitators, who must be trained to support inquiry without dominating discussion or steering students toward predetermined answers (Prasad & O'Malley, 2022).

Despite its implementation challenges, PBL stands out as a transformative educational model that addresses key limitations of traditional teaching approaches. By engaging students with authentic, interdisciplinary problems and encouraging active participation, PBL not only deepens academic understanding but also prepares learners for real-world professional contexts. It nurtures the development of critical cognitive, social, and emotional competencies essential for navigating a complex and rapidly evolving world. In particular,

PBL supports the acquisition of transversal skills such as collaboration, communication, critical thinking, and problem-solving (Du et al., 2013; Guerra et al., 2017), while simultaneously equipping students with the practical abilities needed for real-life applications. Although its adoption requires careful curriculum design and significant resource investment, the long-term benefits of PBL make it a compelling strategy for reshaping engineering education and other disciplines.

1.3 The Genesis of PBL

Understanding the historical roots and evolution of PBL is essential to appreciating its transformative potential in education. Emerging in the late 1960s, initially within medical education, PBL arose as a response to the inadequacies of traditional didactic teaching, which often failed to equip students with the skills needed to solve complex, real-world problems. Rooted in progressive educational philosophy, PBL marked a significant pedagogical shift toward student-centred, inquiry-driven learning that emphasizes critical thinking, collaboration, and self-direction.

The foundations of PBL are deeply connected to the progressive education movement, particularly the work of John Dewey, who advocated for education that aligns with students' natural tendencies toward exploration and creation. Dewey's assertion that learning should occur through doing, rather than passively absorbing information, laid the groundwork for experiential and active learning models.

1.3.1 Historical Influences and Theoretical Foundations

The intellectual foundations of PBL can be traced to several key educational theorists and movements. Foremost among them was John Dewey, whose progressive educational philosophy emphasised learning by doing. Dewey argued that education should engage learners in active participation and experiential learning, where knowledge is constructed through experience rather than passively received (Schmidt, 2012).

Similarly, Jerome Bruner's theory of discovery learning supported the idea that learning is most effective when students are encouraged to explore, hypothesise, and construct their understanding independently. His work reinforced the notion that problem scenarios could serve as meaningful entry points into complex subject matter, allowing learners to acquire and apply knowledge in authentic contexts (Schmidt, 2012).

Another precursor to PBL was the Case Study Method developed at Harvard University in the 1930s. This method introduced the use of real-world cases as educational tools, requiring students to analyse situations and make decisions based on theoretical knowledge, thereby encouraging practical application and critical thinking (Schmidt, 2012).

1.3.2 Implementation and Development

The practical application of these educational theories was pioneered by Howard Barrows, a physician and educator at McMaster University in Canada. Barrows identified that traditional medical education, focused on memorisation and lectures, did not adequately prepare students for clinical reasoning or complex patient care. In response, he developed an innovative approach to medical instruction centred on solving real-world clinical problems. He recognised the applicability of Dewey's ideas to medical education. His visionary approach sought to develop methods for training physicians that went beyond rote memorisation and instead fostered their capacity for reflection and application of knowledge in real-life scenarios. Barrows aimed to produce doctors who possessed not only comprehensive medical knowledge but also the competency to apply that knowledge effectively in managing their patients' health problems.

Traditionally, medical education relied heavily on memorising vast amounts of information, which students were subsequently expected to apply in clinical settings. However, it became evident that this approach fell short of adequately preparing doctors to confront the complex challenges posed by real-world medical problems. In particular, medical schools noticed that the conventional method of teaching, which relied on lectures and memorisation, had a limited impact on the later performance of medical students during their residencies (Barrows, 1996).

Barrows postulated three pivotal elements for the acquisition of knowledge and competency in medical education:

- An Essential Body of Knowledge: Students should possess a foundational body of knowledge.
- Effective Application of Knowledge: Students must be able to apply this knowledge effectively in evaluating and addressing patients' health problems.
- Knowledge Extension and Improvement: Graduates should be equipped to extend and improve upon this knowledge, enabling them to provide appropriate care for future medical challenges

(Barrows, 1986).

To support these objectives, Barrows introduced a curriculum in which students engaged with authentic patient problems. Rather than receiving all necessary information upfront, students identified their learning needs, researched relevant information, asked pertinent questions, and developed strategies for problem-solving. This method nurtured what Barrows called the "clinical reasoning process" and helped students become self-directed learners—capable of identifying, acquiring, and applying knowledge independently and adaptively.

Barrows and Tamblyn (1980) formally defined this approach as Problem-Based Learning. Their model represented a fundamental shift from teacher-centred instruction to a learner-centred methodology focused on real-life application and continuous knowledge development (Barrows, 1996).

1.3.3 Global Expansion and Adaptation

Following its success at McMaster, PBL was adopted by Case Western Reserve University in the United States and soon spread internationally (Pepper, 2014). Universities such as the University of Limburg in Maastricht, the University of Newcastle in Australia, and the University of New Mexico implemented PBL-based reforms to transform their medical curricula (Barrows, 1986; Boud & Feletti, 1997; De Pinho et al., 2015; McComas, 2014; Pepper, 2014).

In the United Kingdom, PBL became part of broader medical education reforms inspired by the General Medical Council, which called for more integrated and self-directed learning approaches (Burke et al., 2020). These reforms underscored the importance of active engagement and problem-solving in developing competent healthcare professionals.

Today, PBL is practised in more than 60 medical schools worldwide and has expanded well beyond medicine. It is now integrated into fields such as dentistry, nursing, pharmacy, optometry, and is gaining prominence in engineering, science, mathematics, teacher education, and K–12 education (Hsieh & Knight, 2008; Major & Mulvihill, 2018; Mundilarto & Ismoyo, 2017; Pourshafie & Murray-Harvey, 2013; Siew & Mapeala, 2017; Tosun & Şenocak, 2013; Uygun & Tertemiz, 2014; Yadav et al., 2011).

1.4 Theoretical Foundations of PBL

To fully grasp the transformative potential of PBL, it is essential to explore the theoretical frameworks that inform this pedagogical approach. At its core, PBL aligns closely with constructivist learning theory, which emphasises that learners actively build knowledge through experiences, social interaction, and critical reflection (Ferreira & Trudel, 2012). Rather than being passive recipients of information, students in PBL are positioned as active participants who engage collaboratively with complex, real-world problems to generate solutions and construct meaning (Lopes et al., 2020).

While constructivism provides a foundational lens for understanding PBL, it is not the sole theoretical influence. PBL also draws from a variety of complementary perspectives, including cognitive development theory, critical rationalism, and educational psychology, each contributing to the development of learners' higher-order thinking, autonomy, and adaptive expertise. These theoretical underpinnings guide the implementation of PBL and help explain its growing adoption across disciplines.

1.4.1 Constructivist Theory

Constructivist theory serves as the cornerstone of PBL. According to this framework, learning is an active, contextualised process of constructing knowledge rather than acquiring it passively (Ardianti et al., 2022; Loyens et al., 2011). In the PBL classroom, students collaborate to explore and solve open-ended problems, facilitating more profound understanding and long-term retention (Esema et al., 2012). Constructivism also underscores the social dimension of learning—an aspect reflected in PBL's emphasis on group dialogue, peer teaching, and shared inquiry (Loyens et al., 2011).

1.4.2 Cognitive Development Theories

The theories of Jean Piaget and Jerome Bruner further support the pedagogical structure of PBL:

- Piaget's theory emphasises developmental stages of learning and the importance of active exploration. His view that learners construct knowledge through interaction and discovery aligns with PBL's emphasis on engaging students in real-world problem-solving (Ardianti et al., 2022).
- Bruner's Discovery Learning theory reinforces the idea that knowledge is best internalised when learners independently uncover relationships and concepts. This principle is central to PBL, in which learners are guided to identify what they need to know through exploration and reflection (Ardianti et al., 2022).

1.4.3 Critical Rationalism

PBL also draws upon Karl Popper's critical rationalist epistemology, which promotes learning through hypothesis generation, critical inquiry, and revision of ideas. This philosophical stance encourages learners to challenge assumptions, evaluate alternatives, and refine their understanding through evidence-based reasoning (Parton & Bailey, 2008). The alignment between critical rationalism and PBL lies in their mutual emphasis on developing independent thinkers capable of navigating uncertainty and complexity.

1.4.4 Educational and Psychological Theories

Several educational and psychological theories further enrich the theoretical base of PBL:

- Adult Learning Theory (Andragogy) and Transformative Learning Theory underscore the role of self-direction, autonomy, and critical reflection in adult learning. These theories reinforce PBL's focus on learner agency and the transformation of perspectives through meaningful engagement (Storey & Schiavo, 2013).

- Information Processing Theory contributes insights into how learners absorb, organize, and apply information. PBL taps into these cognitive processes by structuring learning around inquiry and problem-solving tasks that demand attention, analysis, and memory retrieval (Storey & Schiavo, 2013).

A more detailed exploration of these theories and their influence on PBL will be undertaken in Chapter 2.

1.5 Potentials and Challenges of PBL

PBL represents a significant departure from traditional didactic models of education. It places students at the centre of the learning process, reconfigures curricula around real-world, complex problems, and redefines the role of the teacher as a facilitator of inquiry rather than a transmitter of knowledge. Through PBL, students progressively develop essential competencies, including critical thinking, collaboration, research literacy, and professional attitudes necessary for success in both academic and real-world settings.

PBL has gained traction across disciplines for its capacity to foster student-centred learning and to improve educational outcomes. However, its practical implementation is not without challenges. This section explores the key potentials of PBL, followed by examining its primary implementation challenges, supported by empirical and theoretical literature.

1.5.1 Potentials of PBL

1.5.1.1 Enhancing Problem-Solving and Cognitive Skills

PBL encourages students to grapple with complex, interdisciplinary problems, promoting the development of cognitive flexibility and problem-solving abilities:

- In fields like environmental biotechnology and chemistry, PBL has been shown to help students apply integrated knowledge to real-world contexts (Garanina, 2020; Reynolds & Hancock, 2010).
- In physics education, PBL fosters the restructuring of cognitive frameworks, enabling students to approach new challenges with adaptive reasoning (Silva, Vidmar & Pastorio, 2024).

1.5.1.2 Promoting Collaborative and Self-Directed Learning

Collaboration and autonomy are integral to PBL, enabling the development of interpersonal and intrapersonal skills:

- PBL nurtures teamwork and communication, which is particularly valuable in social fields like tourism and hospitality, where emotional intelligence and group dynamics are essential.
- It also enhances self-directed learning as students take ownership of their education, supported by tools such as learning analytics, which can tailor learning experiences based on individual behaviours and needs.

(Kilińska & Ryberg, 2019)

1.5.1.3 Increasing Student Engagement and Motivation

Engagement and motivation are significantly boosted when students see relevance in their studies:

- In secondary geography education, PBL has been shown to improve both engagement and content comprehension (Caesar et al., 2016).
- PBL's real-world relevance increases intrinsic motivation, allowing students to connect personally with learning experiences (Garanina, 2020).

1.5.1.4 Facilitating Knowledge Production and Application

Beyond content acquisition, PBL encourages students to generate and apply new knowledge:

- For instance, in geosciences, students have used PBL to co-create a professional code of ethics, demonstrating its role in higher-level knowledge production (Almeida et al., 2014).
- PBL's focus on knowledge transfer prepares learners to apply theoretical insights across novel and dynamic contexts (Malokhat et al., 2019).

1.5.1.5 Broader Educational Benefits

Other core advantages of PBL include:

- Structured discovery: PBL fosters deep learning by guiding students through purposeful exploration.
- Developing critical thinking and creativity: Students learn to formulate hypothsise, evaluate alternatives, and innovate.

- Practical skills acquisition: Through real-life scenarios, students build competencies applicable beyond the classroom.
- Improved motivation: Relevance and autonomy enhance student motivation and engagement.
- Knowledge transferability: PBL equips students to adapt knowledge in diverse, unfamiliar settings—an essential 21st-century skill.

1.5.2 Challenges of PBL Implementation

Despite its strengths, several practical and structural challenges hinder the widespread adoption of PBL. These must be acknowledged to refine their application in educational settings.

1.5.2.1 Resistance to Change

Transitioning from traditional pedagogies to PBL often encounters institutional and individual resistance:

- Educators and students may resist due to discomfort with new roles or unfamiliar learning structures (Eboi et al., 2025; Phage et al., 2023).
- In medical education, deeply entrenched norms make shifting to PBL difficult (Chen et al., 2024).

1.5.2.2 Faculty Training and Role Transformation

Successful PBL depends heavily on facilitator effectiveness:

- Many faculty members lack the training required to guide student-led inquiry (Chen et al., 2024; Mansor et al., 2015).
- Transitioning from the traditional instructor to a facilitator role can be challenging, especially in cultures where teaching is content-driven (Epler & Jacobs, 2022).

1.5.2.3 Resource Limitations

PBL often requires significant institutional support:

- The need for technology, materials, collaborative spaces, and administrative coordination may exceed what some institutions can offer (Eboi et al., 2025; Mansor et al., 2015).
- Sustainability becomes a concern when logistical or financial resources are limited.

1.5.2.4 Assessment and Evaluation Complexities

PBL requires alternative forms of assessment aligned with its learning goals:

- Traditional exams often fail to capture problem-solving, collaboration, and critical thinking (Phage et al., 2023).
- Designing authentic, formative assessment tools that reflect PBL outcomes remains a key challenge (Eboi et al., 2025).

1.5.2.5 Group Dynamics and Communication

Group-based learning presents its own set of interpersonal challenges:

- Unequal participation, conflict, and poor communication can hinder group effectiveness (Maurer & Neuhold, 2012).
- Facilitators must actively manage group processes to ensure equitable engagement.

1.5.2.6 Additional Pedagogical Concerns

Other implementation-related challenges include:

- Uncertainty of prior knowledge: Students may not know what they need to learn or how to approach a problem.
- Diverse learner backgrounds: Facilitators must adapt to a wide range of student knowledge and skills.
- Content coverage: PBL may limit the breadth of curriculum content (Boud & Feletti, 1997).
- High planning demands: PBL requires significant time and effort for curriculum design and facilitation, and increased workload.
- Cost: Resource constraints can impede implementation, especially in underfunded settings.
- Faculty adaptation: Letting go of control and enabling student-led inquiry requires a substantial mindset shift.

1.5.3 Opportunities for Improvement

Despite its challenges, PBL offers a platform for innovation in teaching and learning. Emerging technologies can help mitigate implementation barriers:

- Artificial intelligence (AI) can support personalised learning, dynamic feedback, and group facilitation (Chen et al., 2024).
- Learning analytics offer insights into student behaviour and performance, enabling data-driven instructional adjustments (Kilińska & Ryberg, 2019).

However, these tools also raise ethical concerns and require careful implementation, training, and oversight (Chen et al., 2024).

1.6 Relevance of Problem-Based and Project-Based Learning in the 21st Century

The world is undergoing relentless transformation, marked by increasing complexity and rapid technological advancement. While these innovations open up new opportunities, they also present significant challenges. In this evolving global landscape, the ability to comprehend complex issues and develop the skills to solve them has become essential. To thrive in such a context, today's students—and future generations—must cultivate critical thinking, research, and problem-solving habits.

Although traditional education is expected to foster these competencies, many students still struggle to acquire them. In response to this gap, PBL has emerged as a promising educational approach. This approach requires students not only to demonstrate knowledge but also to actively engage in research, collaborative problem-solving, and reflective learning processes. This dynamic and interactive pedagogy can motivate otherwise passive learners, deepen their understanding, and improve academic performance. Moreover, PBL nurtures creativity, innovation, and a sense of ownership over one's learning, making it particularly relevant for courses and programmes that emphasise active participation and real-world problem-solving.

PBL has become increasingly important in modern education because it focuses on developing essential 21st-century skills. Both approaches emphasise student-centred learning through real-world problems or projects, thereby fostering higher-order thinking, collaboration, and creativity. These methodologies are especially relevant in a world characterised by rapid technological and social change. By aligning educational practices with the evolving demands of contemporary workplaces and societies, they equip students with the cognitive, socio-emotional, and technological skills needed for future success.

1.6.1 Development of 21st-Century Skills

- PBL is particularly effective in fostering the cognitive, socio-emotional, and digital competencies that are crucial for success in today's interconnected world. It promotes autonomy, critical thinking, creativity, and collaboration—skills that are indispensable for navigating complexity and change (Silva, Oliveira & Silva, 2024).
- Through engagement in real-world problem-solving, students strengthen their critical thinking and teamwork abilities, which are increasingly valued by employers across industries (Nurbavliyev, 2023).

1.6.2 Student Engagement and Motivation

- PBL enhances student engagement by providing hands-on, experiential learning opportunities. This active involvement encourages learners to take responsibility for their own educational journey, which significantly boosts motivation and interest (Aprilita & Handican, 2023; Nasution et al., 2025).
- Moreover, the deep, participatory nature of PBL fosters improved conceptual understanding and allows students to apply theoretical knowledge to practical contexts, leading to enhanced academic outcomes (Korkem et al., 2024).

1.6.3 Real-World Application and Lifelong Learning

- One of the most persuasive aspects of PBL is its emphasis on bridging the gap between classroom learning and real-world applications. It trains students to approach problems systematically and to develop the habit of lifelong learning—skills that are critical for career longevity and adaptability (Hawamdeh & Adamu, 2021).
- The integration of digital tools and platforms into PBL further extends its applicability, enabling students to engage meaningfully with current global issues and technological advancements (Tay & Noum, 2020).

1.7 Rethinking Engineering Education: The Case for a Paradigm Shift

The United Nations' 17 Sustainable Development Goals (SDGs) provide a comprehensive global framework to foster a more just, equitable, and sustainable future. These goals address a wide range of interconnected global challenges—including poverty, inequality, climate change, environmental degradation, health, energy, peace, and justice—each with significant local repercussions. Many of these issues are complex, multidimensional, and often described as wicked problems due to their resistance to linear or straightforward solutions. While several of these challenges originate outside traditional engineering domains, they nonetheless demand engineering-informed approaches and solutions.

In this context, engineering education assumes a pivotal role in preparing future engineers to engage meaningfully with sustainability challenges. Today's engineering graduates are expected to possess strong technical knowledge and apply engineering principles across disciplinary boundaries to address real-world, global problems. Accordingly, competence—encompassing both technical skills and broader cognitive abilities—has become a key criterion in assessing engineering graduates' readiness for professional practice.

However, a concerning gap persists: many engineering graduates, particularly in developing countries, do not meet international competency expectations. There is growing consensus that prevailing pedagogical models in engineering education are inadequate for cultivating the wide-ranging skills necessary in modern engineering practice. Beyond technical expertise, graduates must demonstrate creativity, analytical reasoning, design thinking, teamwork, and effective communication. Additionally, entrepreneurial thinking and adaptability are essential in responding to the fast-paced demands of the global economy. However, traditional lecture-based teaching methods frequently fall short in fostering these competencies.

The challenges confronting engineering education today call for transformative change. Long-standing models focused primarily on theoretical instruction and technical training are no longer sufficient to prepare graduates for the dynamic and increasingly complex demands of professional engineering roles. This shift is being driven by a convergence of factors: rapid technological innovation, changing expectations within industry, and mounting evidence from educational research that supports more experiential, student-centred learning approaches.

1.7.1 Technological Advancements and Industry Expectations

- The accelerating pace of technological change renders many traditional engineering skills obsolete. Emerging tools such as digital architecture and computational mechanics are replacing older methods, requiring a strong emphasis on continuous learning and adaptability in engineering education (De Silva, 2008).
- Industry stakeholders now seek engineers who are not only technically competent but also adept at knowledge management, emotional intelligence, and problem-solving skills that are often overlooked in rote-learning environments (De Silva, 2008).

1.7.2 Educational Research and Cognitive Science

- Research in education and cognitive science highlights the shortcomings of conventional pedagogy, including high attrition rates and skills mismatches among graduates. These findings emphasise the need for curriculum restructuring and the adoption of experiential learning approaches (Felder, 2012).
- Integrating Information and Communication Technologies (ICT)—such as computer-aided learning environments—can enhance student engagement and better prepare them for the demands of a digital future (Lee & Sidhu, 2013).

1.7.3 Curriculum and Pedagogical Reforms

- The global movement towards Outcomes-Based Education (OBE) reflects a shift in focus from knowledge acquisition to competence development. This framework shapes the structure and assessment of engineering programmes worldwide (Felder, 2012).

- Innovative teaching strategies—such as contemplative pedagogy and service learning—are being explored to foster technical expertise, ethical awareness, and social responsibility, helping engineers address complex societal issues (Catalano, 2016).

1.7.4 *Institutional and Structural Challenges*

- Despite the recognised need for reform, institutional inertia and financial constraints often hinder progress. Many universities remain resistant to change due to entrenched traditions and the costs of curriculum overhauls (Splitt, 2003).
- Nonetheless, some institutions have successfully implemented reforms, such as revamping introductory engineering courses to incorporate research and hands-on components. These pioneering efforts demonstrate the feasibility and benefits of a more holistic educational model (Lo et al., 2005).

While the call for a paradigm shift in engineering education is gaining momentum, the process of implementing meaningful change remains complex. Institutional inertia, limited funding, and logistical constraints often hinder progress. Nevertheless, the growing success of innovative educational models and a robust body of research advocating for active, competency-based, and interdisciplinary learning approaches present a clear and promising path forward. As the engineering profession evolves within an increasingly interconnected and rapidly transforming world, education systems must adapt, ensuring that graduates are equipped with technical expertise and the mindset, adaptability, and problem-solving capabilities required to address the world's most urgent challenges.

In this context, a fundamental transformation in engineering education is not just desirable but necessary. Valuable insights can be drawn from disciplines such as medicine and health sciences, where PBL has demonstrated its effectiveness in cultivating a wide range of student competencies. Integrating PBL into engineering curricula offers a viable strategy to close the competence gap. This student-centred, inquiry-driven approach fosters deep engagement with real-world, complex problems and nurtures critical thinking, creativity, and collaboration. It aligns closely with the objectives of the SDGs, equipping future engineers to contribute meaningfully to sustainable development at both local and global levels.

1.8 Conclusions

PBL holds significant promise as a transformative force in education, fostering critical competencies such as problem-solving, collaboration, and lifelong learning. Rooted in the progressive philosophies of John Dewey and later shaped by Barrows' innovations in medical education, PBL represents a fundamental shift from passive knowledge transmission to active, inquiry-based learning. It empowers students to take ownership of their learning, preparing them to navigate complex, interdisciplinary challenges in a rapidly evolving world.

In the context of engineering education, this transformation is both timely and necessary. The demands placed on engineering graduates have expanded well beyond technical expertise, requiring creativity, communication, teamwork, and the ability to tackle global challenges such as those outlined in the United Nations' SDGs. Traditional pedagogies have struggled to cultivate these attributes effectively, contributing to a widening competence gap, particularly in developing contexts.

PBL offers a robust framework for bridging this gap. By engaging students in real-world, complex problems and fostering active, student-centred learning, PBL can develop the critical skills and mindset required for modern engineering practice. It supports the cultivation of adaptable, reflective professionals who are equipped not only to solve technical issues but also to contribute meaningfully to sustainable development and broader societal progress.

While implementing PBL presents structural, pedagogical, and technological challenges, a growing body of research and successful case studies demonstrate its viability and impact. Overcoming institutional resistance and resource constraints requires strategic commitment, but the potential benefits far outweigh the difficulties. As engineering and other disciplines continue to evolve within an interconnected, dynamic world, educational systems must also adapt, ensuring graduates are prepared to meet the complex demands of the future.

Exploring the application of PBL in engineering education is therefore not just relevant but essential. By advancing a student-centred, inquiry-driven approach, PBL can play a central role in shaping a more responsive, effective, and future-ready educational paradigm.

PBL holds significant promise for transforming education by fostering critical competencies and promoting lifelong learning. However, educators, institutions, and policymakers must address its structural, pedagogical, and technological challenges to harness its full potential. With thoughtful implementation and continuous refinement, PBL can be a cornerstone of modern, student-centred education.

This study seeks to contribute to the growing body of knowledge on PBL by examining its application within the specific context of engineering education. Although PBL has been extensively researched and successfully implemented in fields such as medicine and health sciences, its adoption in engineering presents unique challenges and opportunities that remain underexplored.

Focusing on engineering education, this study investigates the theoretical foundations, practical applications, and transformative potential of PBL as a pedagogical approach. It aims to address existing gaps by analysing the structural, pedagogical, and technological barriers to implementation, while offering actionable insights and recommendations for educators, institutions, and policymakers.

By extending proven practices from other disciplines to the engineering field, this research supports the development of competent, adaptive, and socially responsible engineers—professionals capable of addressing complex technical problems and contributing to sustainable development and broader societal progress.

References

AACTE. (2010). *21st century knowledge and skills in educator preparation*. American Association of Colleges for Teacher Education. https://files.eric.ed.gov/fulltext/ED519336.pdf

Adamyan, N. (2024). Solutions to problematic situations as an interactive learning tool and for the development of students' thinking. *SHPH Gitakan Teghekagir*, 380–391. https://doi.org/10.54151/27382559-24.1pb-380

Ahuja, K., & Bala, I. (2021). Role of Artificial Intelligence and IoT in next generation education system. In F. Al-Turjman, A. Nayyar, A. Devi, & P. K. Shukla (Eds.), *Intelligence of things: AI-IoT based critical-applications and innovations* (pp. 189–208). Springer. https://doi.org/10.1007/978-3-030-82800-4_8

Almeida, A., Vasconcelos, C., Faria, J., & Pereira, C. (2014). *Problem-based learning at the service of new knowledge production* [Conference presentation]. Repositório Científico do Politécnico de Lisboa. https://repositorio.ipl.pt/handle/10400.21/11632

Amer, A., Sidhu, G., Alvarez, M. I. R., Ramos, J. A. L., & Srinivasan, S. (2024). Equity, diversity, and inclusion strategies in engineering and computer science. *Education Sciences*, *14*(1), 110. https://doi.org/10.3390/educsci14010110

Aprilita, T. D., & Handican, R. (2023). Students' perceptions of the implementation of problem-based learning models in mathematics subjects. *Griya Journal of Mathematics Education and Application*, *3*(3), 291–300. https://doi.org/10.29303/griya.v3i3.353

Ardianti, R., Sujarwanto, E., & Surahman, E. (2022). Problem-based learning: Apa dan bagaimana. *Diffraction: Journal for Physics Education and Applied Physics*, *3*(1), 27–35. https://doi.org/10.37058/diffraction.v3i1.4416

Barrows, H. S. (1986). A taxonomy of problem-based learning methods. *Medical Education*, *20*(6), 481–486. https://doi.org/10.1111/j.1365-2923.1986.tb01386.x

Barrows, H. S. (1996). Problem-based learning in medicine and beyond: A brief overview. *New Directions for Teaching and Learning*, *1996*(68), 3–12.

Barrows, H. S., & Tamblyn, R. M. (1980). *Problem-based learning: An approach to medical education*. Springer.

Blumenfeld, P. C., Soloway, E., Marx, R. W., Krajcik, J. S., Guzdial, M., & Palincsar, A. S. (1991). Motivating project-based learning: Sustaining the doing, supporting the learning. *Educational Psychologist*, *26*(3–4), 369–398.

Boud, D., & Feletti, G. (Eds.). (1997). *The challenge of problem-based learning* (2nd ed.). Kogan Page.

Burke, J., Ditchfield, C., Flynn, M.A., Sneddon, S., & Stapleton, G. (2020). Lessons learned from problem-based learning. *The Clinical Teacher*, *17*(6), 719–722. https://doi.org/10.1111/tct.13147

Caesar, M. I. M., Jawawi, R., Matzin, R., Shahrill, M., Jaidin, J. H., & Mundia, L. (2016). The benefits of adopting a problem-based learning approach on students' learning development in secondary geography lessons. *International Education Studies*, *9*(2), 51–65. https://doi.org/10.5539/ies.v9n2p51

Catalano, G. D. (2016). *Engineering education: Moving toward a contemplative service paradigm*. ASEE PEER. https://peer.asee.org/engineering-education-moving-toward-a-contemplative-service-paradigm

Chen, Y., Qi, H., Qiu, Y., Li, J., Zhu, L., Gao, X., Wang, H., & Jiang, G. (2024). Artificial intelligence in medical problem-based learning: Opportunities and challenges. *Global medical education*. Advance online publication. https://doi.org/10.1515/gme-2024-0015

De Pinho, L. A., Mota, F. B., Conde, M. V. F., Alves, L. A., & Lopes, R. M. (2015). Mapping knowledge produced on problem-based learning between 1945 and 2014: A bibliometric analysis. *Creative Education*, *6*(6), 576–584. http://doi.org/10.4236/ce.2015.66058

De Silva, S. (2008). Shifting the engineering education paradigm – challenges and experience in curriculum development and delivery [Paper presentation]. In *CIB W089 – International Conference on Building Education and Research*, Heritance Kandalama, SL. https://www.irbnet.de/daten/iconda/CIB11693.pdf

Dolmans, D. H. J. M., De Grave, W., Wolfhagen, I. H. A. P., & van der Vleuten, C. P. M. (2005). Problem-based learning: Future challenges for educational practice and research. *Medical Education*, *39*(7), 732–741. https://doi.org/10.1111/j.1365-2929.2005.02205.x

Dolmans, D. H. J. M., & Schmidt, H. G. (2006). What do we know about cognitive and motivational effects of small group tutorials in problem-based learning? *Advances in Health Sciences Education*, *11*(4), 321–336. https://doi.org/10.1007/s10459-006-9012-8

Du, X., Emmersen, J., Toft, E., & Sun, B. (2013). PBL and critical thinking disposition in Chinese medical students—A randomised cross-sectional study. *Journal of Problem-Based Learning in Higher Education*, *1*(1), 72–83. https://doi.org/10.5278/ojs.jpblhe.v1i1.275

Duch, B. J., Groh, S. E., & Allen, D. E. (Eds.). (2001). *The power of problem-based learning*. Stylus Publishing.

Eboi, A. A., Obonyo, L., Kowuor, R. N., Aswani, D. R., & Radoli, L. O. (2025). Advancements and challenges of problem-based learning: A case study of Daystar University. In G. Areba, B. Gisore, & E. Njurai (Eds.), *Creating dynamic space in higher education: Modern shifts in policy, competencies, and governance* (pp. 175–214). IGI Global Scientific Publishing. https://doi.org/10.4018/979-8-3693-6930-2.ch007

Epler, P., & Jacobs, J. (2022). Challenges with designing, implementing, and teaching PBL in higher education. In P. Epler (Ed.), *Guide to integrating problem-based learning programs in higher education classrooms: Design, implementation, and evaluation*, IGI Global Scientific Publishing, (2022, pp. 75–93). https://doi.org/10.4018/978-1-7998-8177-3.ch005

Esema, D., Susari, E., & Kurniawan, D. (2012). Problem-based learning. *Sprachwissenschaft*, *28*(2), 167–174. https://doi.org/10.24246/J.SW.2012.V28.I2.P167-174

Felder, R. M. (1993). American engineering education: Current issues and future directions. *International Journal of Engineering Education*, *9*(4), 286–289.

Felder, R. M. (2012). Engineering education: A tale of two paradigms. North Carolina University. http://www4.ncsu.edu/unity/lockers/users/f/felder/public/Papers/TwoParadigms.pdf

Ferreira, M. M., & Trudel, A. R. (2012). The impact of problem-based learning (PBL) on student attitudes toward science, problem-solving skills, and sense of community in the classroom. *The Journal of Classroom Interaction*, *47*(1), 23–30. [suspicious link removed].

Figueiredo, C., Ferreira, C., Gabriel, B., Valente, R., Oliveira, J. L., Neto, V., Strano, M., Carvalho, T., & Soares, S. C. (2023). Current and future challenges for Engineering Education: Voices from students, alumni, teachers and employers [Paper presentation]. In *2023 International Conference on Information, Communication, and Software Engineering (CISPEE)*, Porto, Portugal. https://doi.org/10.1109/cispee58593.2023.10227652

Garanina, R. M. (2020). Problem-based learning as a means of chemistry education effectiveness improvement at a medical university. *Samara Journal of Science*, *9*(4), 282–289. https://doi.org/10.17816/SNV202094303

Gati, J., & Kartyas, G. (2012). New challenges in engineering higher education. In *2012 IEEE 16th International Conference on Intelligent Engineering Systems* (pp. 279–283). https://doi.org/10.1109/INES.2012.6249845

Gleeson, J., Klenowski, V., & Looney, A. (2020). Curriculum change in Australia and Ireland: A comparative study of recent reforms. *Journal of Curriculum Studies*, *52*(4), 478–497. https://doi.org/10.1080/00220272.2019.1704064

Grayson, D., Collier-Reed, B., Pearce, H., & Shay, S. (2014). A curriculum framework for flexible engineering degrees in South Africa [Paper presentation]. In *Conference on Enhancement and Innovation in Higher Education*, Glasgow, Scotland.

Guerra, A., Ulseth, R., Johnson, B., & Kolmos, A. (2017). Engineering grand challenges and the attributes of the global engineer: A literature review. In J. C. Quadrado, J. Bernardino, & J. Rocha (Eds.), *Proceedings of the 45th SEFI Annual Conference 2017* (pp. 1222–1235). European Association for Engineering Education.

Hawamdeh, M. & Adamu, I. (2021). Problem-Based Learning (PBL): A deep approach to learning in the 21st century. In G. Kurubacak-Meric & S. Sisman-Ugur (Eds.), *Improving scientific communication for lifelong learners* (pp. 38–56). IGI Global Scientific Publishing. https://doi.org/10.4018/978-1-7998-4534-8.ch003

Helmold, M., Küçük Yılmaz, A., Flouris, T., Winner, T., Cvetkoska, V., & Dathe, T. (2022). Lean management in higher education (LHE). In M. Helmold (Ed.), *Lean management, kaizen, kata and keiretsu* (pp. 235–246). Springer. https://doi.org/10.1007/978-3-031-10104-5_17

Howson, C. K., & Kingsbury, M. (2021). Curriculum change as transformational learning. *Teaching in Higher Education*, *28*(8), 1847–1866. https://doi.org/10.1080/13562517.2021.1940923

Hsieh, C., & Knight, L. (2008). Problem-based learning for engineering students: An evidence-based comparative study. *The Journal of Academic Librarianship*, *34*(1), 25–30. https://doi.org/10.1016/j.acalib.2007.11.005

Hunter, C., Archer, S., & Illingworth, C. (2025). Problem-based learning. In R. Papa & S. H. K. Tan (Eds.), *Bloomsbury handbook of educational leadership and management* (pp. 302–305). Bloomsbury Publishing. https://doi.org/10.4337/9781035302932.ch83

Kilińska, D., & Ryberg, T. (2019). Connecting learning analytics and problem-based learning – Potentials and challenges. *Journal of Problem-Based Learning in Higher Education*, *7*(1), 1–24. https://doi.org/10.5278/ojs.jpblhe.v7i1.2545

Kim, K. S. (2008). Current issues of engineering education under a globalised society. *Journal of JSEE*, *56*(6), 36–40. https://doi.org/10.4307/jsee.56.6_36

Kokkinou, A., & van Kollenburg, T. (2023). Critical success factors of lean in higher education: An international perspective. *International Journal of Lean Six Sigma*, *14*(6), 1227–1247. https://doi.org/10.1108/IJLSS-04-2022-0076

Korkem, Y., Gulnara, B., & Alkanova, A. (2024). Unveiling the pedagogical paradigm: Exploring the significance of project-based learning in contemporary education. *International Advanced Research Journal in Science, Engineering and Technology*, *11*(5), 10. https://doi.org/10.17148/iarjset.2024.11510

Lalwani, K. (2024). How to design and moderate an effective problem-based learning discussion. In J. R. Schiller, C. M. Gisondo, & M. B. Goldstein (Eds.), *A practical guide to curriculum and instructional development in medical education* (pp. 301–306). Oxford University Press. https://doi.org/10.1093/med/9780197655979.003.0052

Lee, C. K. & Sidhu, M. S. (2013). Computer-aided engineering education: New learning approaches and technologies. In V. Wang (Ed.), *Handbook of research on teaching and learning in K-20 education* (pp. 317–340). IGI Global Scientific Publishing. https://doi.org/10.4018/978-1-4666-4249-2.ch019

Leong, W. Y. (2023). Advancing diversity and inclusion for women in engineering. In *Proceedings of the 3rd Indian International Conference on Industrial Engineering and Operations Management*, New Delhi, India.

Li, L. (2024). Reskilling and upskilling the future-ready workforce for Industry 4.0 and beyond. *Information Systems Frontiers*, *26*, 1697–1712. https://doi.org/10.1007/s10796-022-10308-y

Lo, J., Goff, R., Lohani, V. K., Walker, T. M., Knott, T., & Griffin, H. (2005). *New paradigm for foundational engineering education*. ASEE PEER. https://peer.asee.org/new-paradigm-for-foundational-engineering-education

Lopes, R. M., Hauser-Davis, R. A., Oliveira, M. M., Pierini, M. F., de Souza, C. A. M., Cavalcante, A. L. M., Santos, C. R. D., Comarú, M. W., & Tinoca, L. A. da F. (2020). Principles of problem-based learning for training and professional practice in ecotoxicology. *Science of the Total Environment*, *702*, Article 134809. https://doi.org/10.1016/j.scitotenv.2019.134809

Loyens, S. M. M., Kirschner, P. A., & Paas, F. (2011). Problem-based Learning. In K. R. Harris, S. Graham & T. Urdan (Eds.), *APA educational psychology handbook*: Vol 2 (p. a). Washington: American Psychological Association. https://ro.uow.edu.au/edu.apers/1235

Major, T., & Mulvihill, T. M. (2018). Problem-based learning pedagogies in teacher education: The case of Botswana. *Interdisciplinary Journal of Problem-Based Learning*, *12*(1). https://doi.org/10.7771/1541-5015.1543

Malokhat, K., Feruzahon, I., Ruzakhon, M., Masuda, H., & Shokhida, R. (2019). The essence and content of problem-based learning. *International Journal of Recent Technology and Engineering*, *8*(4), 5361–5364. https://doi.org/10.35940/ijrte.d5361.118419

Mansor, A. N., Abdullah, N., Wahab, J. A., Rasul, M. S., Nor, M. Y. M., Nor, N. M., & Raof, R. A. (2015). Managing problem-based learning: Challenges and solutions for educational practice. *Asian Social Science*, *11*(4), 259–268. https://doi.org/10.5539/ass.v11n4p259

Maurer, H., & Neuhold, C. (2012). *Problems everywhere? Strengths and challenges of a problem-based learning approach in European studies*. Social Science Research Network. https://papers.ssrn.com/sol3/papers.cfm?abstract_id=1997614

McComas, W. F. (2014). Problem-based learning. In W. F. McComas (Ed.), *The language of science education* (p. 76). Sense Publishers. https://doi.org/10.1007/978-94-6209-497-0_66

Meifitri. (2020). Problem-solving ability and communication skills through a problem-based learning approach. In *Proceedings of the Eighth International Conference on Languages and Arts (ICLA-2019)* (pp. 146–149). Atlantis Press. https://doi.org/10.2991/assehr.k.200819.028

Mohammed, M. A., & Ali, I. (2024). Problem-based learning is helpful in encouraging academic institutions to strive for excellence: Perceptions of Sudanese physiologists as an illustration. *BMC Medical Education*, *24*(1), Article 1353. https://doi.org/10.1186/s12909-024-06285-2

Mora-Luis, C. E., & Martín-Gutiérrez, J. (2020). The change of educational processes, learning and teaching in engineering education. *Eurasia Journal of Mathematics, Science and Technology Education*, *16*(3), em1833. https://doi.org/10.29333/ejmste/116034

Mundilarto, P., & Ismoyo, H. (2017). Effect of problem-based learning on improvement physics achievement and critical thinking of senior high school student. *Journal of Baltic Science Education*, *16*(5), 761–779.

Nasution, K. F., Siahaan, M., & Siregar, N. U. (2025). Penerapan strategi project-based learning di SDS Islam Azizi. *Tsaqofah*, *5*(2), 1215–1227. https://doi.org/10.58578/tsaqofah.v5i2.4769

Neville, A. J., & Norman, G. R. (2007). PBL in the undergraduate MD program at McMaster University: Three iterations in three decades. *Academic Medicine*, *82*(4), 370–374. https://doi.org/10.1097/acm.0b013e318033385d

Nurbavliyev, O. K. (2023). Improving the effectiveness of project-based learning in the process of teaching mathematics at school. *Science and Education*, *4*(1), 81–88. https://doi.org/10.51889/2959-5894.2023.81.1.012

O'Neill, G., & Short, A. (2023). Relevant, practical and connected to the real world: What higher education students say engages them in the curriculum. *Irish Educational Studies*, *42*(4), 693–709. https://doi.org/10.1080/03323315.2023.2221663

Parton, G., & Bailey, R. (2008). Problem-based learning: A critical rationalist perspective. *London Review of Education*, *6*(3), 281–292. https://doi.org/10.1080/14748460802528475

Pepper, C. (2014). Problem-based learning (PBL). In T. S. Rocco, M. C. Smith, R. C. Mizzi, L. R. Merriweather, & J. D. Hawley (Eds.), *The handbook of adult and continuing education* (pp. 1–3). Springer. https://doi.org/10.1007/978-94-007-6165-0_128-2

Phage, R., Molato, B. J., & Matsipane, M. (2023). Challenges regarding transition from case-based learning to problem-based learning: A qualitative study with student nurses. *Nursing Reports*, *13*(1), 389–403. https://doi.org/10.3390/nursrep13010036

Pourshafie, T., & Murray-Harvey, R. (2013). Facilitating problem-based learning in teacher education: Getting the challenge right. *Journal of Education for Teaching*, *39*(2), 169–180. https://doi.org/10.1080/02607476.2013.765190

Prasad, S., & O'Malley, C. B. (2022). An introductory framework of problem-based learning (PBL) and perspectives on enhancing facilitation approaches. *HAPS Educator*, *26*(3), 52–58. https://doi.org/10.21692/haps.2022.016

Reynolds, J. M., & Hancock, D. R. (2010). Problem-based learning in a higher education environmental biotechnology course. *Innovations in Education and Teaching International*, *47*(2), 175–186. https://doi.org/10.1080/14703291003718919

Rosário, A. T., & Dias, J. C. (2024). Implementing problem-based learning in marketing education: A systematic review and analysis. *Education Sciences*, *14*(11), 1139. https://doi.org/10.3390/educsci14111139

Schmidt, H. G. (2012). A brief history of problem-based learning. In H. G. Schmidt, W. M. M. H. van der Molen, W. M. W. te Winkel, & W. H. F. W. Wijnen (Eds.), *Problem-based learning: A subject for educational and psychological research* (pp. 21–40). Springer. https://doi.org/10.1007/978-981-4021-75-3_2

Siew, M. M., & Mapeala, R. (2017). The effects of thinking maps-aided problem-based learning on motivation towards science learning among fifth graders. *Journal of Baltic Science Education*, *16*(3), 379–394.

Silva, B. J. da, Vidmar, M. P., & Pastorio, D. P. (2024). Problem-based learning as a methodological strategy for the development of cognitive flexibility in physics teaching. *Revista Brasileira de Ensino de Física*, *46*, e20230317. https://doi.org/10.1590/1806-9126-rbef-2023-0317

Silva, J. B., Oliveira, M. B. M. de, & Silva, W. D. (2024). Aprendizagem baseada em projetos e problemas (ABP): Uma abordagem para o desenvolvimento de competências no século XXI. *Revista Científica Multidisciplinar O Saber*, *1*(1), 573. https://doi.org/10.51473/rcmos.v1i1.2024.573

Splitt, F. G. (2003). The challenge to change: On realising the new paradigm for engineering education. *Journal of Engineering Education*, *92*(2), 181–187. https://doi.org/10.1002/j.2168-9830.2003.tb00756.x

Storey, V. A., & Schiavo, M. A. (2013). Redefining the higher education landscape through problem-based learning. In V. Wang (Ed.), *Handbook of research on teaching and learning in K-20 education* (pp. 517–534). IGI Global Scientific Publishing. https://doi.org/10.4018/978-1-4666-4249-2.ch030

Tay, P. L., & Noum, S. Y. E. (2020). Digitalising PBL to transition learning into the real world. In P. Kumar, M. Keppell, & C. Lim (Eds.), *Preparing 21st century teachers for teach less, learn more (TLLM) pedagogies* (pp. 279–294). IGI Global Scientific Publishing. https://doi.org/10.4018/978-1-7998-1435-1.ch016

Thomas, J. W. (2000). *A review of the research on project-based learning*. The Autodesk Foundation.

Tosun, C., & Şenocak, E. (2013). The effects of problem-based learning on metacognitive awareness and attitudes toward chemistry of prospective teachers with different academic backgrounds. *Australian Journal of Teacher Education*, *38*(3), 61–73. http://doi.org/10.14221/ajte.2013v38n3.2

UNESCO. (2015). Rethinking education: Towards a global common good? https://unesdoc.unesco.org/ark:/48223/pf0000232555

Uygun, N., & Tertemiz, N. I. (2014). Effects of problem-based learning on student attitudes, achievement and retention of learning in math course. *Education and Science*, *39*(174), 75–90. https://doi.org/10.15390/eb.2014.1975

Weber, J. R. (2014, October 27). *Problem-based learning helps bridge the gap between the classroom and the real world.* Faculty Focus. https://www.facultyfocus.com/articles/teaching-and-learning/problem-based-learning-helps-bridge-gap-classroom-real-world/

Wolff, K., Hattingh, T., & Smith, L. (2025). Global challenges, local responses: Exploring curriculum reform in South African engineering education. *Journal of Engineering Education, 114*, e70006. https://doi.org/10.1002/jee.70006

Yadav, A., Subedi, D., Lundeberg, M. A., & Bunting, C. F. (2011). Problem-based learning: Influence on students' learning in an electrical engineering course. *Journal of Engineering Education, 100*(2), 253–280. https://doi.org/10.1002/j.2168-9830.2011.tb00013.x

Zorpette, G. (1984). The current issues: Formal education: EE programs. *IEEE Spectrum, 21*(11), 44–50. https://doi.org/10.1109/mspec.1984.6370330

Zumbach, J., & Prescher, C. (2023). Problem-based learning and case-based learning. In J. Zumbach, D. A. Bernstein, S. Narciss, & G. Marsico (Eds.), *International handbook of psychology learning and teaching*. Springer. https://doi.org/10.1007/978-3-030-28745-0_58

Chapter 2

Problem-Based Learning

Theoretical Foundations and Pedagogical Approaches

2.1 Introduction

Problem-Based Learning (PBL) is a student-centred pedagogical approach that promotes learning through the investigation and resolution of real-world problems. Rooted in active learning, PBL typically unfolds in collaborative group settings that reflect authentic social and professional environments. Within these settings, interactions among peers and instructors—and the dynamics both within and between groups—play a critical role in shaping the learning experience.

Rather than simply applying existing knowledge, students in PBL actively construct new knowledge through inquiry and collaboration. The approach is inherently interdisciplinary, integrating insights from multiple domains and involving stakeholders across various fields.

A strong theoretical foundation is essential for any pedagogy to be effective and sustainable. PBL is grounded in a diverse range of educational theories, including information processing theory, social constructivism, sociocultural theory, social interdependence theory, and sociocognitive theory. Rather than adhering to a single framework, PBL integrates elements from social, cultural, and systems paradigms. In particular, constructivist theory, cultural theory, and systems theory each inform different aspects of teaching and learning within the PBL model (Hmelo-Silver, 2004; Hmelo-Silver & Eberbach, 2012; Løkke et al., 2023; Lu, Bridges, & Hmelo-Silver, 2014; Storey & Schiavo, 2013; Torre et al., 2016). These perspectives help explain how PBL fosters student learning, development, and the creation of new knowledge.

Before implementing PBL, it is important to understand the nature and variety of problems used in this approach. These problems are often context-specific and can vary significantly across disciplines, meaning there is no universal model suitable for all situations. Therefore, familiarity with the range of PBL approaches practised across higher education institutions globally is crucial for adapting the method effectively to specific educational contexts.

This chapter will first examine the theoretical foundations of PBL. It will then explore the different types of problems commonly used, particularly

DOI: 10.1201/9781003529743-3

within engineering education, and conclude by discussing various PBL models currently in practice, highlighting how they can be tailored to suit diverse learning environments.

2.2 Philosophical Foundations and Theoretical Perspectives of PBL

2.2.1 Philosophical Foundations of PBL

PBL as a pedagogical approach is grounded in key philosophical perspectives, particularly epistemological and ontological. As mentioned previously, it promotes learning through real-world problem-solving, critical thinking, and collaboration, providing a meaningful and reflective educational experience. These philosophical foundations shape how knowledge is perceived, constructed, and applied in PBL, supporting learners in becoming active participants in their learning processes.

2.2.1.1 Epistemological Perspectives

Epistemology deals with the nature, sources, and legitimacy of knowledge. According to Burrell and Morgan (1979), epistemology concerns what is considered acceptable knowledge and how it can be communicated. In the context of PBL, where learning occurs in multidisciplinary environments, a wide variety of knowledge forms—quantitative data, visualisations, interpretive texts, models, and more—are considered valid. This inclusivity of multiple epistemological traditions allows for flexibility in how learning is approached and experienced.

Constructivism is one of the most influential epistemological perspectives underpinning PBL. It posits that knowledge is actively constructed by learners through experience, reflection, and interaction with their environment and peers. In PBL, students build on prior knowledge and collaborate with others to develop new understanding through discussion and inquiry (Hendry et al., 1999; Rani, 2023). Vygotsky's theory (1978) also reinforces this approach, suggesting that cognitive development is mediated by social interaction and the cultural context of learning.

PBL also reflects Karl Popper's theory of critical rationalism, which emphasises knowledge development through conjectures and refutations. This perspective aligns with the iterative learning cycle in PBL, where students hypothesise, test, and revise their ideas through structured dialogue and problem-solving (Parton & Bailey, 2008; Wang et al., 2008). Such an approach promotes deeper critical thinking and intellectual flexibility.

Moreover, PBL draws from pragmatic epistemology, notably Dewey's (1991) idea that thinking is triggered by perplexity and uncertainty. Learners are challenged to solve problems that matter to them, encouraging the integration of prior knowledge with new experiences in authentic contexts (Barrows & Tamblyn, 1980). This experiential and inquiry-based view enhances motivation and engagement.

2.2.1.2 Ontological Perspectives

Ontology concerns the nature of reality and what exists in the world. In PBL, ontology helps explain the types of problems students engage with and the levels at which they analyze these problems. Bhaskar's (1978) critical realism is particularly relevant, offering a stratified view of reality comprising three layers: the Real (causal mechanisms and structures), the Actual (events that occur, whether observed or not), and the Empirical (experiences that are observed or measured). This framework encourages students not only to consider observable data but to think about deeper systemic causes and interrelationships.

For example, a student may analyse empirical evidence in a real-world problem but must also reason about underlying mechanisms that are not immediately visible. This ontological depth is essential for developing higher-order thinking and analytical reasoning within PBL.

In addition, PBL reflects elements of empiricism and logicalism, particularly in how it begins with real-life problems that demand empirical investigation and logical interpretation (Rani, 2023). Learners must collect and interpret evidence to arrive at plausible solutions, reinforcing the importance of observable and verifiable knowledge in problem-solving.

Domain ontologies also support the organisation and retrieval of structured knowledge. These ontologies—conceptual models representing a field's key concepts and relationships—can enhance the problem-solving process by enabling students to map, navigate, and integrate knowledge from various disciplines (Diez-Rodriguez & Olmos-Aguirre, 2008; Fontes et al., 2011).

2.2.1.3 Integration of Philosophical Approaches in PBL

The integration of epistemological and ontological perspectives gives PBL its distinctive philosophical character. The Socratic method, based on dialogic inquiry and questioning, is central to the design and implementation of PBL. This method encourages learners to explore complex ideas, question assumptions, and construct knowledge collaboratively through structured dialogue (Wang et al., 2008).

Moreover, PBL aligns with Freirean ideas of education as empowerment. Freire advocated for learning that is dialogical, critical, and emancipatory, aimed at personal and social transformation. In PBL, students are not passive recipients of knowledge but active agents who co-create meaning through engagement with real-life issues (Barrett, 2001). This critical and participatory stance promotes both individual development and societal awareness.

Pragmatism, especially Deweyan thought, also plays a key role in shaping the learning experience in PBL. Dewey emphasised the need for experiential, contextual learning that bridges theory and practice (Chen, 2008). In PBL, learners confront authentic problems, apply knowledge in meaningful ways, and reflect on their actions—thereby reinforcing the connection between learning and doing.

However, despite the strong philosophical foundation, PBL faces serious criticism in terms of its limitations. The learner-centred, minimally guided structure of PBL can lead to confusion or inefficiency without skilled facilitation and appropriate scaffolding (Wells et al., 2009). Furthermore, reliance on traditional empiricism may fall short in addressing the dynamic and socially constructed aspects of knowledge in PBL, hence the need for alternative or complementary epistemologies such as critical rationalism (Parton & Bailey, 2008). Additionally, the lack of standardised definitions or models of PBL can create inconsistencies in its implementation, reinforcing the importance of developing formal ontologies for clarity and coherence (Fontes et al., 2011).

2.2.2 *Theoretical Constructs*

PBL is grounded in a diverse array of theoretical frameworks that collectively inform its pedagogical design and implementation. While no single theory fully explains PBL, a combination of sociocultural, cognitive, experiential, and systems-based theories provides a strong foundation for its success and sustainability.

As discussed previously (Chapter 1), at its core, PBL draws from major educational theories such as Constructivism, Socio-Cultural Theory, Experiential Learning (Kolb), Cognitive Apprenticeship, and Self-Determination Theory, among others. Influential thinkers, including Piaget, Vygotsky (1978), Lave and Wenger (1991), and Schön (with his theory of reflective practice), have contributed significantly to the development of the PBL approach (Sadlo, 2014).

In addition to these, PBL incorporates insights from Socio-Cognitive Theory, Social Interdependence Theory, Information Processing Theory, Cultural Theory, and Systems Theory. Each of these theoretical perspectives brings essential assumptions about how people learn, emphasising active engagement, social interaction, motivation, contextual learning, and systemic thinking.

Together, these constructs support the key characteristics of PBL: student-centred learning, real-world problem-solving, collaborative inquiry, self-direction, and reflection. In practice, effective PBL design draws upon a tailored mix of these theories to suit specific learning goals and contexts. Each of the theories is discussed below.

2.2.2.1 Constructivism and Social Constructivism

Constructivism posits that learning is an active process in which individuals build new knowledge upon their prior understanding (Jonassen, 1999;

Lopes et al., 2020). In PBL, this theory manifests through problem-solving, hands-on inquiry, and student-driven exploration.

Social constructivism extends this view by emphasising that knowledge is co-constructed through social interaction and collaboration (Hmelo-Silver, 2004; Hmelo-Silver & Eberbach, 2012; Vygotsky, 1978). PBL fosters peer discussions, negotiations, and shared meaning-making, which are central to deep learning and understanding.

2.2.2.2 Socio-Cultural Theory

Vygotsky's Socio-Cultural Theory underscores the importance of cultural context, social tools, and interpersonal interactions in cognitive development. In PBL, students engage in group learning environments where they collaborate using cultural and cognitive tools to co-construct knowledge. This theory supports the notion that learning is mediated by social participation, making PBL particularly effective in diverse educational settings (Løkke et al., 2023).

2.2.2.3 Experiential Learning

Kolb's Experiential Learning Theory (1984) highlights the significance of learning through experience, reflection, and application. PBL directly aligns with this model, as students confront real-world challenges, reflect on their experiences, and iterate solutions. Reflection, a core component emphasised by Schön (1992), enhances students' metacognitive awareness and capacity for self-directed learning.

2.2.2.4 Cognitive Apprenticeship

Cognitive Apprenticeship Theory (Collins, Brown, & Newman, 1989) conceptualizes learning as participation in a community of practice, where novices acquire skills by observing and collaborating with experts. PBL supports this model through team-based learning and mentorship structures, helping students internalise both procedural and conceptual knowledge in authentic contexts.

2.2.2.5 Socio-Cognitive Theory and Networked Learning

Socio-Cognitive Theory focuses on the cognitive and social processes involved in learning. In PBL, this perspective is evident in the creation of knowledge communities where students contribute their diverse skills towards solving shared problems. Concepts such as networked expertise and distributed cognition (Torre et al., 2016) are fundamental to PBL's collaborative structure, allowing for shared understanding and collective learning.

2.2.2.6 *Self-Determination Theory*

Self-Determination Theory (Deci & Ryan, 1985) suggests that autonomy, competence, and relatedness are essential to intrinsic motivation. PBL supports this theory by enabling student autonomy in exploring complex problems, building competence through active inquiry, and fostering relatedness through collaboration and team engagement.

2.2.2.7 *Social Interdependence Theory*

Social Interdependence Theory emphasises the positive effects of cooperative learning environments, where group members depend on one another to achieve mutual goals (Torre et al., 2016). PBL leverages this interdependence through structured group work, shared responsibility, and peer feedback, enhancing both individual and group learning outcomes

2.2.2.8 *Information Processing Theory*

Information Processing Theory examines how learners encode, store, and retrieve knowledge. In PBL, students activate prior knowledge and integrate new information through problem-solving and reflection. This supports the development of flexible problem-solving strategies and enhances long-term knowledge retention (Hmelo-Silver & Eberbach, 2012; Storey & Schiavo, 2013).

2.2.2.9 *Cultural Theory*

Cultural Theory (Beck et al., 2011; Thompson, Ellis, & Wildavsky, 1990; Verweij & Thomson, 2006) argue that problems involve multiple stakeholders with different social solidarities—such as markets, governance, communities, and users—each with unique values and interests. PBL encourages students to consider diverse stakeholder perspectives, evaluate conflicting demands, and navigate trade-offs to propose sustainable, context-sensitive solutions.

2.2.2.10 *Systems Theory*

Systems Theory views a problem as part of a complex, interconnected system comprising multiple subsystems and stakeholders. In PBL, students must identify these subsystems, recognise their interdependencies, and work towards holistic, systemic solutions. This reinforces critical thinking and systems thinking competencies essential for addressing multifaceted real-world issues.

2.2.3 Implications of Learning Theories for PBL

PBL is informed by a combination of learning theories discussed above that together shape its core pedagogical principles. These theories—particularly from constructivist, sociocultural, cognitive, motivational, and systems perspectives—not only provide a solid foundation for the design of PBL but also illuminate the educational value and practical challenges of its implementation. The implications of different theoretical constructs in combinations are discussed below.

2.2.3.1 Constructivism and Experiential Learning

PBL is inherently rooted in constructivist learning theory, which posits that learners actively construct knowledge based on prior experiences. This aligns with experiential learning (Kolb, 1984), where students engage with real-world problems, reflect on their experiences, and develop personal meaning. Research shows that PBL fosters the organisation of new experiences into mental representations or "schemas" (Derry, 1996; Mayer, 1996), enabling deeper understanding of concepts and more effective learning strategies.

PBL tasks, often based on real-life contexts, activate what Dewey (1991) called the "perplexity or doubt" that triggers reflective thinking. As students explore these problems, they use individual and collective prior knowledge to interpret and respond, reinforcing the cyclical nature of learning through experience and reflection.

2.2.3.2 Socio-Cultural and Collaborative Learning

Socio-Cultural Theory (Vygotsky, 1978) underscores the role of social interaction and cultural context in learning. PBL exemplifies this by engaging students in group discussions, peer learning, and co-construction of knowledge through collaboration. This dialogic learning process helps students develop not only cognitive skills but also communication, negotiation, and teamwork capabilities.

Scholars such as Dolmans et al. (2005) emphasise that PBL is grounded in four modern learning insights: constructivist, collaborative, self-directed, and contextual learning. The method relies on the teacher as a facilitator, group learning as a vehicle for interaction, and problems as stimuli for learning, thus supporting Vygotsky's notion of learning within the "zone of proximal development."

2.2.3.3 Cognitive Apprenticeship and Socio-Cognitive Perspectives

Cognitive apprenticeship (Collins et al., 1989) supports learning through guided experience, modelling, and gradual release of responsibility. In PBL, the facilitator plays the role of mentor or coach, scaffolding learning while encouraging increasing independence. This enculturation into disciplinary thinking mirrors real-world professional contexts.

Socio-Cognitive Theory, particularly the idea of networked expertise, also finds relevance in PBL environments. Students collaboratively construct and share knowledge within a community, fostering distributed cognition and leveraging diverse perspectives (Torre et al., 2016).

2.2.3.4 Self-Directed Learning and Motivation

PBL promotes self-directed learning, encouraging students to take responsibility for their learning goals, processes, and outcomes. This aligns with Self-Determination Theory (Deci & Ryan, 1985), which highlights autonomy, competence, and relatedness as key drivers of intrinsic motivation. The structure of PBL tasks allows students to experience ownership over their learning, develop problem-solving competence, and work collaboratively, fostering both personal and group success (Luo, 2019).

2.2.3.5 Cultural Theory and Stakeholder Complexity

Beyond individual and group learning, PBL often addresses complex, real-world problems with multiple stakeholders. Drawing on Cultural Theory (Beck et al., 2011; Thompson, Ellis, & Wildavsky, 1990; Verweij & Thomson, 2006), PBL encourages students to recognise and manage competing narratives and social solidarities, such as market, state, community, and user interests. This involves not just consensus-building but also negotiating trade-offs to develop sustainable solutions. Such tasks enhance students' ability to conduct stakeholder analysis, understand diverse values, and manage complexity—critical competencies in today's interconnected world.

2.2.3.6 Systems Theory and Holistic Thinking

Systems theory adds another dimension to PBL by framing problems as part of larger, interconnected systems. Each issue encompasses various subsystems—social, economic, environmental, and organisational—that interact dynamically. PBL encourages students to adopt a systems-thinking approach, identifying interdependencies, feedback loops, and systemic consequences of proposed solutions.

This holistic perspective enables learners to approach problems strategically, considering both micro- and macro-level impacts while seeking optimal, context-sensitive outcomes.

2.2.3.7 Challenges and Practical Considerations

While the theoretical foundations of PBL are robust, its effective implementation requires careful planning and facilitation. The success of PBL depends heavily on well-designed problem scenarios and skilled instructors who can support inquiry without dominating it (Folomieieva et al., 2024). Transitioning from traditional methods to PBL also involves a shift in mindset for both teachers and learners, requiring the development of new pedagogical strategies and learning habits (Hung, 2015).

Despite these challenges, the integration of diverse learning theories in PBL supports the development of critical thinking, collaboration, autonomy, and real-world problem-solving skills, making it a powerful and adaptable approach across disciplines and educational levels.

2.3 Different Types and Nature of Problems in PBL Across Engineering Education Levels

PBL commonly revolves around meaningful, complex, and real-life problems. These problems are often ill-structured and open-ended, requiring students to engage in inquiry, analysis, and synthesis. However, the structure and complexity of these problems vary depending on the educational level within engineering programmes.

Engineering education generally spans three tiers:

- **Technicians** (Diploma level)
- **Engineering Technologists** (Bachelor of Engineering Technology/Bachelor of Technology (3 years)
- **Engineers** (Bachelor of Engineering or Bachelor of Science in Engineering) (4 years)

Professionals encounter three types of problems: well-defined, broadly defined, and complex. However, the nature of the problems they address varies depending on their roles and levels of expertise. As such, different categories of professionals typically engage with different types of problems. They are discussed below.

- **Technicians** deal with well-defined problems.
- **Technologists** engage with broadly defined problems.
- **Engineers** address complex engineering problems.

The effectiveness of learning in PBL is highly dependent on the structure and design of the problems. Well-structured and thoughtfully implemented problems enhance learner motivation, deepen subject understanding, and foster commitment to problem-solving (Hung, 2006). Table 2.1 presents a summary and comparison of different types of problems.

Table 2.1 Comparison of Well-Defined, Broadly Defined, and Complex Problems

Aspect	*Well-defined Problems*	*Broadly defined Problems*	*Complex Problems*
Definition	Problems with clear parameters and predictable solutions	Problems with some ambiguity, multiple factors, and familiar structures	Ill-posed, unique problems with many unknowns and dynamic elements
Nature of Problem	Structured, clear, predictable	Semi-structured, moderately complex	Unstructured, unpredictable, dynamic
Type of Knowledge Required	Limited theoretical; mostly practical and procedural	Theoretical and procedural knowledge with analytical methods	In-depth, interdisciplinary, abstract, and theoretical knowledge
Approach to Solution	Standardised procedures, codes, and practices	Analytical techniques with flexibility, partially outside standard methods	Creative, iterative, exploratory; often no predefined method
Innovation Needed	Minimal	Moderate	High—requires originality and adaptability
Use of Standards & Codes	Sufficient	May extend beyond standard codes	Often outside the domain of standards and codes
Scope of Problem	Narrow and focused on components	Broader, may span subsystems	Wide-ranging, often across full systems or interconnected systems
Stakeholder Involvement	Few stakeholders with aligned interests	Several groups with potentially conflicting needs	Diverse stakeholders—government, civil society, technical, and societal actors
Examples	Fixing a leaking pipe, setting a road curve, and testing tensile strength	Traffic congestion management, power distribution, and neighbourhood water systems	Water supply for large cities, intelligent transport systems, and climate solutions
Level in Engineering Education	Technician (Diploma level)	Technologist (Bachelor of Engineering Technology)	Engineer (Bachelor of Engineering or Science in Engineering)
Context of Application	Local, specific, limited	Broader local, extendable	Multidisciplinary, regional or global, high-impact
Decision-Making Complexity	Low—clear outcomes and methods	Moderate—requires judgment within known boundaries	High—requires judgment under uncertainty and competing factors
Repeatability of Solutions	High—can be applied to similar problems	Moderate—can be adapted with contextual understanding	Low solutions are often unique and non-transferable

2.3.1 *Well-Defined Problems*

Well-defined problems are structured, standardised, and predictable. They typically involve:

- Standardised methods for resolution
- Minimal theoretical knowledge but significant practical skills
- Use of codes, standards, and best practices
- Limited scope (often discrete components of systems)
- Low innovation requirements
- Few stakeholders with aligned interests
- Local significance

Examples:

- Fixing plumbing issues in a building
- Setting a horizontal road curve
- Measuring the tensile strength of a material

2.3.2 *Broadly Defined Problems*

These problems involve multiple variables, potential conflicts, and constraints. Characteristics include:

- Belonging to a familiar family of problems solvable using known strategies
- Resolution through analytical techniques and principles
- Requiring broader procedural and methodological knowledge
- Limited reliance on standards and codes
- Involving components of complex systems
- Engagement with multiple stakeholders with differing interests
- Significant locally, with potential for broader impact

Examples:

- Alleviating traffic congestion on a specific road
- Developing a local power distribution grid
- Managing a control system for electronic appliances
- Designing a neighbourhood water reticulation system

2.3.3 *Complex Engineering Problems*

Complex problems are ill-posed and multifaceted, often requiring refinement and deep expertise. They typically:

- Are ill-posed, over- or under-specified and in need of refinement
- Cannot be solved without in-depth engineering knowledge

- Involve conflicting technical, social, economic, environmental, and managerial considerations
- Lack obvious solutions, requiring abstraction and innovative thinking
- Demand application of fundamental principles and first-principles analysis
- Are unique and fall outside the scope of standard codes or practices
- Involve a diverse range of stakeholders (e.g., government, civil society, NGOs)
- Consists of multiple interacting subproblems
- Require decision-making under uncertainty
- Carry wide-ranging consequences across multiple contexts

Examples:

- Planning irrigation systems in arid regions
- Designing urban water supply systems in water-scarce cities
- Developing Integrated Rapid Transit systems for metropolitan areas
- Creating smart infrastructure systems for transportation
- Developing flood-resistant construction materials or road pavements

2.3.4 Levels of Complex Problems (Illustrative Hierarchy)

Level 1: Complex Challenge

- *Example*: Achieving sustainable transportation in large cities (socio-economic, environmental)

Level 2: Complex Problem

- *Example*: Developing public transportation systems (socio-economic, environmental)

Level 3: Engineering-Level Problem

- *Example*: Planning an Integrated Rapid Transit system in a South African city

Level 4: Technical Engineering Problem

- Examples:
  - Planning an integrated BRT (Bus Rapid Transit) network
  - Designing non-motorised transportation systems
  - Developing intelligent instrumentation for urban cycling systems

Students, particularly at higher levels, can undertake Level 3 or 4 problems. They can also align their project work with the UN Sustainable Development Goals (SDGs) by breaking down socio-economic-environmental challenges into smaller, actionable, complex engineering problems.

2.3.5 *Evidence of a Complex Problem*

Students must demonstrate the complexity of a problem through attributes such as:

- Broad scope involving full systems or subsystems
- Multidisciplinary, dynamic contexts
- Unpredictability and absence of obvious solutions
- Significant and diverse resources
- Interactions among technical, economic, environmental, and social factors
- Real-world constraints (financial, legal, infrastructural, etc.)
- High-impact consequences in multiple domains

2.3.6 *Activities and Expected Outcomes for Addressing Complex Problems*

To effectively solve complex problems, students should perform and document many (or all) of the following activities:

1. **Problem Formulation**: Articulating interconnectedness, stakeholder roles, risks, and broader significance
2. **Project or Design Brief**: Planning the entire problem-solving approach
3. **Literature Review**: Engaging critically with existing research and similar case studies
4. **Research/Design Planning**: Structuring experiments, data collection tools, or investigation frameworks
5. **Data Collection/Investigation**: Conducting relevant experiments using appropriate scientific methods
6. **Analysis**: Employing high-level mathematics, modelling tools, and engineering principles
7. **Innovation and Ideation**: Generating novel ideas or alternate solutions
8. **Evaluation**: Assessing alternatives using scientific criteria
9. **Solution Design**: Providing substantiated conclusions and solutions that may be unique or adaptable across contexts

2.3.7 *Difference between Complex and Complicated Problems*

A complex problem is not the same as a complicated problem, though they are often mistakenly used interchangeably. The key differences lie in the nature of the problem, how it is analysed, and the kind of solutions it requires. Table 2.2 presents the critical difference between these two types of problems.

Table 2.2 Difference between Complicated and Complex Problems

Aspect	*Complicated Problem*	*Complex Problem*
Nature	Predictable, analysable	Unpredictable, emergent
Causality	Linear and identifiable	Non-linear, often unclear
Expertise required	Subject matter experts	Cross-disciplinary, adaptive thinking
Repeatability of solution	High-standard methods apply	Low solutions are context-specific
Examples	ERP implementation, bridge design	Organisational change, climate policy planning

2.3.7.1 Complicated Problems

Complicated problems are challenging but knowable. They involve multiple interconnected parts, but their relationships follow a predictable chain of causation that can be understood with sufficient expertise.

- These problems can be deconstructed and analysed systematically.
- Subject matter experts, research, and logical analysis can lead to effective solutions.
- Once solved, solutions can be repeated and standardised across similar situations.
- There is typically a clear process or methodology for addressing the problem.

Example: Implementing an Enterprise Resource Planning (ERP) system. While complex in terms of components (finance, supply chain, HR, etc.), experts can analyze and integrate these modules in a logical sequence using best practices. The same process can then be repeated in other organisations with adjustments.

Analogy: Imagine a diagram with dots connected in an intricate but logical pattern (Figure 2.1). In the figure, the dots are connected in a complicated pattern, but have a chain of causation, and experts can understand it, analyse the pattern and develop an equation that can be repeatedly applied to similar cases. In other words, an expert can study the pattern, identify the rules governing the connections, and derive an equation or system that applies to other, similar configurations.

Figure 2.1 Example of a Complicated Problem.

2.3.7.2 Complex Problems

Complex problems, on the other hand, are dynamic, unpredictable, and not easily solvable through existing expertise or processes. They:

- Involve many interdependent and interacting variables, some unknown or poorly understood.
- Lack a clear chain of causation or linear relationship between cause and effect.
- Have no single, repeatable solution; what works in one situation may not work in another.
- Require adaptive, iterative, and often creative approaches.
- Are heavily influenced by context, stakeholders, and evolving factors.

Example: Streamlining operations in a large organisation. This involves navigating informal structures, varying human behaviours, changing market dynamics, and cultural aspects. A successful solution in one organisation might fail completely in another.

Analogy: The same diagram now has dots connected randomly (Figure 2.2), with no clear pattern or rule. Not even experts can easily decipher the structure. Solutions are context-specific and non-transferable.

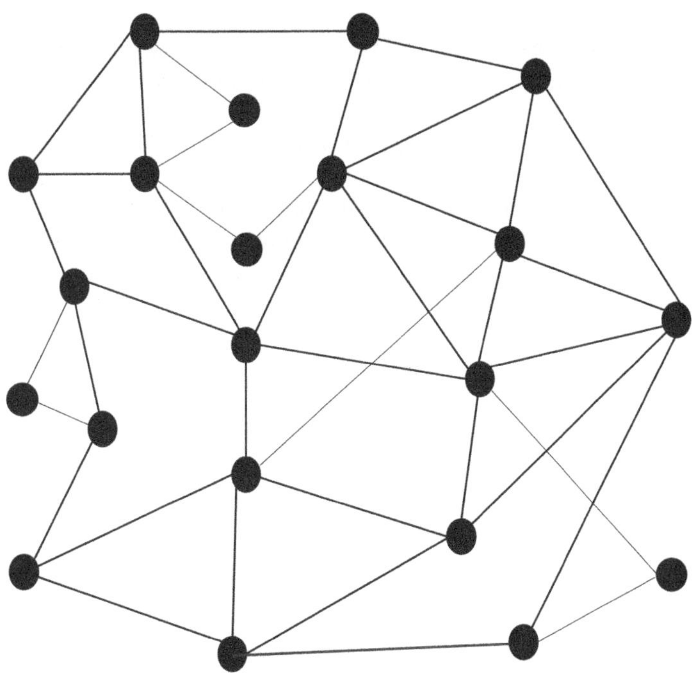

Figure 2.2 Example of a Complex Problem.

2.4 PBL Models in Engineering

PBL has been implemented across a wide variety of educational contexts, yielding diverse outcomes depending on the model, discipline, and implementation strategy. This section explains three prominent PBL models in engineering education: the Aalborg University model, the University College London model, and a South African adaptation. Each model reflects unique pedagogical approaches tailored to different educational settings and objectives.

2.4.1 The Aalborg University Model

Aalborg University (AAU) in Denmark is widely recognised as a pioneer in integrating PBL into higher education curricula, particularly in engineering disciplines. AAU's PBL model is grounded in a comprehensive, mandatory introductory course designed to familiarise students with the theoretical underpinnings, methodological frameworks, and practical applications of PBL (Askehave et al., 2015; Kolmos, Holgaard & Clausen, 2021; Servant-Miklos & Kolmos, 2022; Servant-Miklos, 2020). This course serves as a foundation for students' understanding of how PBL is conducted, aligning its principles with the specific demands and nuances of various scientific and technical disciplines.

The core principles of the AAU PBL model include:

- **Problem-Centric Learning**: At AAU, learning begins with a real-world problem. This problem serves as the starting point and driving force for all learning activities.
- **Project Organisation as the Structural Framework**: The entire educational experience is structured around project work, with projects serving as the central organising principle for both students and faculty.
- **Course Support for Project Work**: Academic courses are designed to complement and support students' project work, providing essential knowledge and skills that can be immediately applied.
- **Emphasis on Cooperation**: Successful problem-based project work at AAU hinges on close collaboration among students, fostering teamwork, communication, and collective problem-solving.
- **High Standards for Project Groups**: Groups engaged in PBL projects are expected to produce high-quality work that sets an example for others, reflecting both academic rigour and practical relevance.
- **Student Responsibility**: Students are encouraged to take ownership of their learning process and outcomes, developing autonomy and self-directed learning skills.

Students typically engage in group-based projects for extended periods—usually three to four months—working independently but with guidance and support from teachers or supervisors. These projects emphasise the learning process and the final deliverables, allowing students to experience the full cycle of problem-solving, from problem definition to solution implementation (Dahl, 2018; Davidsen, Andersen, & Christiansen, 2019; Kolmos et al., 2019).

A distinguishing feature of the AAU model is integrating learning factories into the PBL framework. Learning factories are simulated environments replicating real industrial value chains, providing students with an authentic context to apply theoretical knowledge. This hands-on learning environment facilitates interdisciplinary collaboration and action-oriented problem-solving, preparing students to address complex challenges in real-world settings (Kolmos et al., 2019; Servant-Miklos, 2020).

Within this environment, after an initial phase of mandatory laboratory training, students gain significant autonomy in deciding their project approaches, selecting methods, and designing experiments. This autonomy empowers students to apply knowledge from multiple course modules to practical industrial scenarios, fostering deeper learning and transferable skills.

AAU's PBL approach equips students with competencies that align closely with PBL principles and supports their ongoing professional development throughout their academic journey.

2.4.2 The UCL Model: How to Change the World (HtCtW)

University College London (UCL) implements PBL in its engineering and management programmes through the innovative How to Change the World (HtCtW) initiative. This programme serves as a culminating experience following the first two years of the Integrated Engineering Programme (IEP), drawing students from a diverse range of departments to work collaboratively on complex, interdisciplinary challenges.

The HtCtW programme is designed with several key objectives:

- **Equipping Students with Practical Problem-Solving Methods**: The programme aims to develop students' abilities to tackle complex, real-world problems using structured, practical approaches.
- **Fostering Interdisciplinary Teamwork**: Students gain hands-on experience working in diverse teams, combining different disciplinary perspectives to create innovative solutions.
- **Consolidating Prior Learning**: HtCtW consolidates the skills and knowledge students have acquired in the first two years of the IEP, enabling them to apply these in authentic problem-solving contexts.

The programme spans an intensive two-week period, during which students progress through a clearly defined sequence of stages:

- **Team Building**: Students form diverse interdisciplinary groups and establish effective working relationships.

- **Problem Identification**: Teams analyse and define the problem they will address, working closely with external partners who bring real-world challenges.
- **Conceptualisation and Design**: Teams develop conceptual models and design solutions, applying engineering principles alongside creative thinking.
- **Problem-Solving and Prototyping**: Students implement and test their solutions, iterating as necessary.
- **Showcasing Solutions**: Teams present their work to a panel of assessors, stakeholders, and industry partners for evaluation and feedback.

The teaching and facilitation team in HtCtW is diverse, including Cohort Mentors (CM), Cohort Leads (CL), Postgraduate Teaching Assistants (PGTAs), and Undergraduate Teaching Assistants (UGTAs). This structure ensures robust student support throughout the programme, providing guidance, mentorship, and feedback.

PBL sessions are highly interactive and occur in phases, featuring direct teaching by mentors, collaborative student activities, formative assessments at critical milestones, and consultations with technical and challenge partners (experts and problem owners). Additionally, students have access to rich online and classroom resources. The final assessment process involves a jury evaluation that includes internal and external experts (Johansson & Svensson, 2019).

A distinctive feature of the HtCtW programme is the deep engagement with external technical and challenge partners—real experts and problem owners—who actively mentor and assess students. This connection to real-world problems and stakeholders greatly enriches the learning experience, preparing students for professional engineering practice.

2.4.3 The South African Model (Test Case): Mini How to Change the World

The South African adaptation of the UCL model, implemented at the Central University of Technology (CUT) in the Free State, offers a condensed version of the HtCtW programme, often referred to as the Mini How to Change the World. This model is designed to introduce the foundational principles of PBL within a limited timeframe, typically one week or several short sessions.

Despite its brevity, the programme emphasises active student engagement and collaborative learning around authentic, complex problems. However, due to time constraints, the programme may not cover all aspects of comprehensive PBL models.

The Mini HtCtW programme is multidisciplinary, addressing a broad range of socio-economic and environmental challenges framed within the United Nations Sustainable Development Goals (UN-SDGs). Student teams focus on solving a single problem throughout the programme, applying engineering principles and methodologies.

Participants primarily include undergraduate students enrolled in engineering, built environment, and information technology programmes, specialising in Civil Engineering, Mechanical Engineering, Electrical Engineering, and the Built Environment.

The programme is structured as an intensive 3-to-4-day initiative, offered as a standalone experience outside the standard curriculum, targeting Bachelor of Engineering Technology and Diploma students.

Facilitation is provided by a team of university lecturers drawn from engineering, built environment, and Information Technology (IT) departments, supplemented by industry professionals and consultants. Additionally, a team of academics and professionals from UCL supports the programme, ensuring alignment with international PBL standards.

The programme follows a systematic progression:

- Team building and group formation.
- Problem identification and framing.
- Conceptualisation and design development.
- Problem-solving and solution implementation.
- Presentation of solutions to assessors and stakeholders.

A hallmark of this programme is its strong interaction with technical and challenge partners—external experts and problem owners—who contribute real-world insights and evaluation.

This model provides students with a valuable introduction to PBL methodology and collaborative problem-solving, preparing them to engage with complex challenges despite the shorter timeframe.

2.5 Approaches of PBL Practised in Different Disciplines

PBL has been implemented across a wide range of disciplines and contexts. This section discusses PBL approaches used in four fields: Mathematics, Physical Education, Physics, and Environmental Science. The purpose is to explore how PBL has been adapted to suit different subject areas and to evaluate the outcomes of these implementations. A comparative analysis of these approaches is summarised in Table 2.3.

2.5.1 Problem-Solving Approach in Mathematics

This model focuses on enhancing students' problem-solving skills, particularly targeting those with weaker abilities in this area. The learning process follows several stages—problem understanding, planning, execution, and reflection—designed to foster heuristic learning.

Table 2.3 PBL Approaches and Key Features in Different Disciplines

Discipline	*Key Features*	*Implementation Highlights*	*Outcomes and Benefits*
Mathematics	Focus on heuristic problem-solving	Stages: understanding, planning, execution, reflection	Improved analytical skills and problem-solving ability
Physical Education	Active group work with practical skill assessment	14-week intervention comparing PBL and traditional	Enhanced skills, increased interaction, reduced freeloading
Physics	Integration of Augmented Reality for abstract concepts	AR-supported problem-solving cycles	Better comprehension and engagement
Environmental Science	Emphasis on individual reflection within collaborative learning	IRPs and peer-assessed tutorials	Increased metacognition, accountability, and collaboration

Objective: The primary aim in mathematics education is to improve students' problem-solving abilities, especially for those with initially lower proficiency.Implementation Stages:

- **Problem Understanding**: Students are presented with a problem that requires acquiring new knowledge.
- **Planning**: Students interpret the problem and develop a plan to approach it.
- **Execution**: Solutions are developed through analysis, calculation, and application of mathematical concepts.
- **Reflection**: Students review their problem-solving process and outcomes, promoting heuristic and self-directed learning.

2.5.1.1 Learning Process

Students actively engage with problems that challenge them to explore new concepts and methods, encouraging critical thinking and independent learning.

2.5.1.2 Outcomes and Benefits

Research, including a case study on vector analysis, demonstrates that PBL significantly improves mathematical skills. Students develop stronger analytical abilities and greater confidence in tackling complex problems (Arends, 2008; Mushlihuddin, Nurafifah, & Irvan, 2018).

2.5.2 PBL in Physical Education (Ming Chuan University)

A study at Ming Chuan University investigated the implementation of PBL in physical education, specifically examining its impact on badminton skill development. The study compared outcomes between uncontrolled and controlled learning environments over 14 weeks, with weekly 100-minute sessions.

2.5.2.1 Study Design

- **Uncontrolled Group**: Students worked in small groups, engaging in unstructured questioning, problem-solving, group discussions, and peer assessments.
- **Controlled Group**: Students received traditional classroom-based instruction.

2.5.2.2 Assessment

Student performance was evaluated using badminton skill tests, which included tasks such as picking up and serving a shuttlecock with a racket.

2.5.2.3 Findings

The PBL approach in the uncontrolled group significantly enhanced skill development by fostering interaction, discussion, and mutual support among students. Additionally, it helped mitigate the "free rider problem," where some group members contribute less in collaborative settings (Luo, 2019).

2.5.3 PBL in Physics Education (Integration of Augmented Reality)

To address challenges such as abstract physics concepts, equipment shortages, and common misconceptions, Augmented Reality (AR) was integrated into PBL activities in physics education (Cai et al., 2017). The PBL process followed stages of problem presentation, problem definition, data gathering, solution generation, reflection, and evaluation. AR tools were incorporated into select stages to provide interactive and immersive experiences that enhanced students' understanding.

2.5.3.1 Rationale for AR Integration

Physics often involves complex, abstract phenomena that are difficult to visualise. Additionally, limited access to laboratory equipment and prevalent misconceptions further hinder student learning.

2.5.3.2 *Implementation Process*

- Students engaged in the standard PBL stages: problem presentation, problem definition, data gathering, solution generation, reflection, and evaluation.
- AR technology was used during key stages to clarify abstract concepts through interactive visualisation.

2.5.3.3 *Assessment*

Student comprehension was evaluated using multiple tests focused on physics problem-solving and conceptual understanding.

2.5.3.4 *Impact*

Integrating AR within the PBL framework improved students' grasp of challenging physics concepts, increased engagement, and enhanced overall learning outcomes (Fidan, Mustafa, & Tuncel Meric, 2019).

2.5.4 *PBL Model with Individual Reflection Paper (Environmental Science Programme)*

In an environmental science programme, PBL was adapted to emphasise individual reflection through the use of Individual Reflection Papers (IRPs). Students were presented with real-world scenarios and tasked with identifying their learning needs and formulating questions for self-directed study. Based on collaboratively developed questions, they compiled IRPs reflecting their individual insights. These papers were then discussed during tutorial meetings, where students also evaluated each other's contributions. Assessments focus on students' preparation and active, constructive participation in tutorials, using a three-level grading scale: fail, sufficient, and reasonable (Johansson & Svensson, 2019).

2.5.4.1 *Purpose*

To cultivate deeper individual reflection and critical thinking within a collaborative PBL environment.

2.5.4.2 *Methodology*

- Students analysed real-world environmental scenarios.
- They identified personal learning needs and formulated questions to guide their self-study.
- IRPs were written based on group-developed questions.

2.5.4.3 *Tutorial Meetings*

- IRPs were discussed in small-group tutorials where students provided peer assessments.
- The process emphasised constructive feedback and collaborative learning.

2.5.4.4 *Assessment Criteria*

Students were evaluated on their preparation and the quality of their participation, rated on a three-tier scale: fail, sufficient, and reasonable.

2.5.4.5 *Outcomes*

This approach fostered metacognitive skills, enhanced accountability, and improved learning through reflection and peer interaction (Johansson & Svensson, 2019).

These diverse PBL models demonstrate the versatility of this educational approach across various disciplines and contexts, providing opportunities for active learning, interdisciplinary collaboration, and a strong focus on problem-solving and reflection.

2.6 Conclusion

PBL is a pedagogically and philosophically rich approach that draws on diverse epistemological and ontological perspectives—including constructivism, critical rationalism, critical realism, pragmatism, and Freirean empowerment—to facilitate meaningful, deep learning. By fostering critical thinking, learner autonomy, and active engagement with authentic, real-world problems, PBL cultivates thoughtful, competent, and socially responsible individuals.

While practical challenges in implementation exist, the strong philosophical foundations of PBL make it a powerful and transformative educational strategy. This chapter has emphasised the necessity of grounding PBL within a robust theoretical framework, providing an in-depth examination of the various theories and principles that underpin it. It also discusses various types, the nature of the problems dealt with, and their nuances by the professionals at different levels. Moreover, it has highlighted the broad applicability of PBL across different institutions and disciplines globally.

The evidence shows that PBL is highly adaptable, effectively serving as both an integrated component within curricula and a standalone educational approach. Importantly, PBL holds particular significance in engineering education, where its emphasis on problem-solving, collaboration, and reflection aligns closely with the demands of the profession. More relevantly, PBL represents a vital and versatile pedagogical model capable of preparing graduates to meet the complex challenges of today's world.

References

Arends, R. I. (2008). *Learning to teach*, The McGraw-Hill Companies.

Askehave, I., Prehn, H. L., Pedersen, J., & Pedersen, M. T. (Eds.). (2015). *PBL: Problem-based learning*. Aalborg University. https://www.aau.dk/digitalAssets/148/148025_pbl-aalborg-model_uk.pdf

Barrett, T. (2001). Philosophical principles for problem-based learning: Freire's concepts of personal development and social empowerment. http://eprints.teachingandlearning.ie/1741/

Barrows, H.S. and Tamblyn, R. M. (1980). *Problem-based learning: An approach to medical education*. Springer Publishing Company.

Beck, M. B., Thompson, M., Ney, S., Gyawali, D., & Jeffrey, P. (2011). On governance for re-engineering city infrastructure. *Engineering Sustainability: Proceedings of the Institution of Civil Engineers*, *164*(2), 129–142. https://doi.org/10.1680/ensu.2011.164.2.129

Bhaskar, R. (1978). *A realist theory of science*. Harvester Press.

Burrell, G., & Morgan, G. (1979). *Sociological paradigms and organisational analysis: Elements of the sociology of corporate life*. Heinemann.

Cai, S., Chiang, F. K., Sun, Y., Lin, C., & Lee, J. J. (2017). Applications of augmented reality-based natural interactive learning in magnetic field instruction. *Interactive Learning Environments*, *25*(6), 778–791. https://doi.org/10.1080/10494820.2016.1181094

Chen, N. C. (2008). An educational approach to problem-based learning. *Kaohsiung Journal of Medical Sciences*, *24*(3), S23–S30. https://doi.org/10.1016/S1607-551X(08)70090-1

Collins, A., Brown, J. S., & Newman, S. E. (1989). Cognitive apprenticeship: Teaching the crafts of reading, writing, and mathematics. In L. B. Resnick (Ed.), *Knowing, learning, and instruction: Essays in honour of Robert Glaser* (pp. 453–494). Lawrence Erlbaum Associates.

Dahl, B. (2018). What is the problem with problem-based learning in higher education mathematics? *European Journal of Engineering Education*, *43*(1), 112–125. https://doi.org/10.1080/03043797.2017.1320354

Davidsen, J., Andersen, P. V. K., & Christiansen, E. (2019). Problem-based learning in a box: Lessons learned from an educational design experiment. *Journal of Problem-Based Learning in Higher Education*, *7*(1), 120–132. https://doi.org/10.5278/ojs.jpblhe.v7i1.2375

Deci, E. L., & Ryan, R. M. (1985). *Intrinsic motivation and self-determination in human behaviour*. Plenum Press.

Derry, S. J. (1996). Cognitive schema theory in the constructivist debate. *Educational Psychologist*, *31*(3-4), 163–174. https://doi.org/10.1080/00461520.1996.9653263

Dewey, J. (1991). *How we think*. Prometheus Books.

Diez-Rodriguez, H., & Olmos-Aguirre, J. O. (2008, March 16–18). *Ontology-guided problem-based learning* [Paper presentation]. *6th IASTED International Conference on Web-Based Education*, Innsbruck, Austria.

Dolmans, D. H. J. M., de Grave, W. S., Wolfhagen, I. H. A. P., & van der Vleuten, C. P. M. (2005). Problem-based learning: Future challenges for educational practice and research. *Medical Education*, *39*(7), 732–741. https://doi.org/10.1111/j.1365-2929.2005.02205.x

Fidan, M., & Tuncel, M. (2019). Integrating augmented reality into problem-based learning: The effects on learning achievement and attitude in physics education. *Computers & Education*, 142, Article 103635. https://doi.org/10.1016/j.compedu.2019.103635

Folomieieva, N., Koriakin, O., Matsenko, L., Siaska, I., & Fed, I. (2024). Problem-based learning in higher education: A path towards training innovative and competent specialists. *Amazonia Investiga, 13*(77), 201–217. https://doi.org/10.34069/AI/2024.77.05.15

Fontes, L. M. de O., Neto, F. M. M., & Pontes, A. Á. A. (2011). OntoPBL: Uma ontologia de domínio sobre aprendizagem baseada em problema. In *Anais do XXII Simpósio Brasileiro de Informática na Educação* (pp. 838–847). Sociedade Brasileira de Computação.

Hendry, G. D., Frommer, M., & Walker, R. (1999). Constructivism and problem-based learning. *Journal of Further and Higher Education, 23*(3), 369–371. https://doi.org/10.1080/0309877990230306

Hmelo-Silver, C. E. (2004). Problem-based learning: What and how do students learn? *Educational Psychology Review, 16*(3), 235–266. https://doi.org/10.1023/B:EDPR.0000034022.16470.f3

Hmelo-Silver, C. E., & Eberbach, C. (2012). Learning theories and problem-based learning. In O. S. Tan (Ed.), *New perspectives on problem-based learning* (pp. 3–17). Springer. https://doi.org/10.1007/978-94-007-2515-7_1

Hung, W. (2006). The 3C3R model: A conceptual framework for designing problems in PBL. *Interdisciplinary Journal of Problem-Based Learning, 1*(1), 55–77. https://doi.org/10.7771/1541-5015.1006

Hung, W. (2015). Problem-based learning: Conception, practice, and future. In W. Hung, M. J. Jacobson, & O. S. Tan (Eds.), *Problem-based learning in the 21st century: New directions and challenges* (pp. 75–92). Springer. https://doi.org/10.1007/978-981-287-521-1_5

Johansson, M., & Svensson, T. (2019). Individual reflection paper: Supporting students' learning in the critical phase of self-directed learning in PBL. *Journal of Problem-Based Learning in Higher Education, 7*(1), 97–106. https://doi.org/10.5278/ojs.jpblhe.v7i1.2418

Jonassen, D. H. (1999). Designing constructivist learning environments. In C. M. Reigeluth (Ed.), *Instructional-design theories and models: A new paradigm of instructional theory* (Vol. 2, pp. 215–239). Lawrence Erlbaum Associates.

Kolb, D. A. (1984). *Experiential learning: Experience as the source of learning and development*. Prentice-Hall.

Kolmos, A., Bøgelund, P., & Spliid, C. (2019). Learning and assessing problem-based learning at Aalborg University. In M. Moallem, W. Hung, & N. Dabbagh (Eds.), *The Wiley handbook of problem-based learning* (pp. 437–458). Wiley Blackwell. https://doi.org/10.1002/9781119173243.ch19

Kolmos, A., Holgaard, J. E., & Clausen, N. R. (2021). Progression of student self-assessed learning outcomes in systemic PBL. *European Journal of Engineering Education, 46*(1), 67–89. https://doi.org/10.1080/03043797.2020.1789070

Lave, J., & Wenger, E. (1991). *Situated learning: Legitimate peripheral participation*. Cambridge University Press.

Løkke, S., Nielsen, H. N., & Holgaard, J. E. (2023). Problem-based learning approach facilitating sustainable waste management. *Journal of Problem-Based Learning in Higher Education, 11*(3). https://doi.org/10.54337/ojs.jpblhe.v11i3.7828

Lopes, R. M., Hauser-Davis, R. A., Oliveira, M. M., Pierini, M. F., de Souza, C. A. M., Cavalcante, A. L. M., Santos, C. R. D., Comarú, M. W., & da Fonseca Tinoca, L. A. (2020). Principles of problem-based learning for training and professional practice in ecotoxicology. *Science of the Total Environment*, *702*, Article 134809. https://doi.org/10.1016/j.scitotenv.2019.134809

Lu, J., Bridges, S., & Hmelo-Silver, C. E. (2014). Problem-based learning. In R. K. Sawyer (Ed.), *Cambridge handbook of the learning sciences* (2nd ed., pp. 298–318). Cambridge University Press.

Luo, Y. J. (2019). The influence of problem-based learning on learning effectiveness in students' of varying learning abilities within physical education. *Innovations in Education and Teaching International*, *56*(1), 3–13. https://doi.org/10.1080/14703297.2017.1389289

Mayer, R. E. (1996). Learning strategies for making sense out of expository text: The SOI model for guiding three cognitive processes in knowledge construction. *Educational Psychology Review*, *8*(4), 357–371. https://doi.org/10.1007/BF01463939

Mushlihuddin, R., Afifah, N., & Irvan, I. (2018). The effectiveness of problem-based learning on students' problem-solving ability in the vector analysis course. *Journal of Physics: Conference Series*, *948*, Article 012028. https://doi.org/10.1088/1742-6596/948/1/012028

Parton, G., & Bailey, R. (2008). Problem-based learning: A critical rationalist perspective. *London Review of Education*, *6*(3), 281–292. https://doi.org/10.1080/14748460802528475

Rani, T. P. (2023). Penerapan model project-based learning dalam perspektif ontologi dan epistemologi filsafat pendidikan matematika [Application of the project-based learning model in the ontological and epistemological perspective of the philosophy of mathematics education]. *Strategi: Jurnal Inovasi Strategi dan Model Pembelajaran*, *3*(1), 1–8. https://doi.org/10.51878/strategi.v3i1.1956

Sadlo, G. (2014). Using problem-based learning during student placements to embed theory in practice. *International Journal of Practice-based Learning in Health and Social Care*, *2*(1), 6–19. https://doi.org/10.11120/pblh.2014.00029

Schön, D. A. (1992). *The reflective practitioner: How professionals think in action*. Ashgate. (Original work published 1983)

Servant-Miklos, V. F. C. (2020). Problem-oriented project work and problem-based learning. *Interdisciplinary Journal of Problem-Based Learning*, *14*(1). https://doi.org/10.14434/ijpbl.v14i1.28596

Servant-Miklos, V. F. C., & Kolmos, A. (2022). Student conceptions of problem and project-based learning in engineering education: A phenomenographic investigation. *Journal of Engineering Education*, *111*(4), 792–812. https://doi.org/10.1002/jee.20478

Storey, V. A., & Schiavo, M. A. (2013). Redefining the higher education landscape through problem-based learning. In V. A. Storey (Ed.), *Redefining student success: Building a new path to college and career* (pp. 517–534). IGI Global. https://doi.org/10.4018/978-1-4666-4249-2.ch030

Thompson, M., Ellis, R., & Wildavsky, A. (1990). *Cultural theory*. Westview Press.

Torre, D. M., van der Vleuten, C. P. M., & Dolmans, D. H. J. M. (2016). Theoretical perspectives and applications of group learning in PBL. *Medical Teacher*, *38*(2), 189–195. https://doi.org/10.3109/0142159X.2015.1009429

Verweij, M. & Thomson, M. (2006). *Clumsy solution for a complex world: Governance, politics perception*. Springer.

Vygotsky, L. S. (1978). *Mind in society: The development of higher psychological processes*. Harvard University Press.

Wang, S. Y., Tsai, J. C., Chiang, H. C., Lai, C. S., & Lin, H. J. (2008). Socrates, problem-based learning and critical thinking- a philosophic point of view. *Kaohsiung Journal of Medical Sciences*, *24*(3 Suppl), S6–S13. https://doi.org/10.1016/S1607-551X(08)70088-3

Wells, S. H., Warelow, P., & Jackson, K. L. (2009). Problem-based learning (PBL): A conundrum. *Contemporary Nurse*, *33*(2), 191–201. https://doi.org/10.5172/conu.2009.33.2.191

Chapter 3

A Comparative Analysis of Problem-Based Learning and Traditional Methods of Learning

3.1 Introduction

Teaching and learning in higher education play a crucial role, particularly in developing countries. In recent years, the objectives of higher education have expanded beyond traditional academic pursuits to include contributions to socio-economic development and the promotion of social justice and equity. As Waghid (2014) emphasises, it is essential that the knowledge students acquire and produce addresses challenges related to equity, sustainability, and justice, while also advancing broader socio-economic progress.

Educational approaches in higher education have undergone considerable transformation, reflecting evolving pedagogical philosophies and the complex demands of contemporary societies. Broadly, these approaches can be categorised into traditional teaching methods and innovative pedagogies, each with distinct characteristics and educational implications (Chandraja, Ajayan, Shabi, & George, 2024; Renz & Vogel, 2024).

Traditional teaching methods are grounded in teacher-centred pedagogy, where the instructor is the primary source of knowledge, and students assume a passive learning role (Tularam & Machisella, 2018). Typically characterised by lecture-based instruction, this model emphasises transmitting theoretical content in a structured format. Students are often expected to memorise and reproduce facts for standardised assessments. Interaction between students and instructors is generally minimal, limiting opportunities for dialogue, critical inquiry, or collaborative engagement (Saira, Zafar, Hafeez, 2021; Tularam & Machisella, 2018). Although this method is effective for delivering foundational content, it frequently fails to cultivate higher-order thinking skills such as critical analysis, creativity, and problem-solving. Additionally, it often overlooks diverse learning styles and does not sufficiently prepare learners for the complexities of real-world scenarios (Saira, Zafar, Hafeez, 2021; Tularam & Machisella, 2018).

In contrast, innovative pedagogies prioritise active, student-centred learning, aiming to address the limitations inherent in traditional methods

DOI: 10.1201/9781003529743-4

(Baig & Yadegaridehkordi, 2023; Baquero & Escortell, 2022; Dziuban, Graham, Moskal, et al., 2018; Kolb & Kolb, 2011; Smiderle, Rigo, Marques et al., 2020. One such approach, PBL, has emerged as a powerful, transformative educational model. PBL situates learning within the context of real-world problems, encouraging students to collaborate in groups to identify solutions. This approach fosters critical thinking, self-directed learning, and applying theoretical knowledge to practical contexts. Unlike traditional methods, where the teacher directs the flow of information, PBL repositions the instructor as a facilitator who supports and guides students throughout the learning process (Hunter et al., 2025; Prasad & O'Malley, 2022). While PBL promotes deeper engagement and skill development, it demands thorough planning, sustained student participation, and careful evaluation of individual and group contributions (Renz & Vogel, 2024).

Beyond PBL, several other innovative approaches continue to redefine the landscape of modern higher education. These include the flipped classroom, experiential learning, gamification, and blended learning (Baig & Yadegaridehkordi, 2023; Baquero & Escortell, 2022; Dziuban, Graham, Moskal, et al., 2018; Kolb & Kolb, 2011; Smiderle, Rigo, Marques et al., 2020). The flipped classroom model allows students to engage with instructional materials outside of class, thereby enabling more interactive and application-oriented activities during class sessions (Baig & Yadegaridehkordi, 2023; Kong, 2021). Experiential learning emphasises learning through direct experience—such as internships, simulations, or fieldwork—to bridge the gap between theory and practice (Kolb & Kolb, 2011; Kong, 2021). Gamification incorporates game-like elements into the learning process to enhance student motivation and engagement (Smiderle, Rigo, Marques et al., 2020). Blended learning, meanwhile, combines face-to-face instruction with digital resources, offering increased flexibility and accessibility to learners across diverse contexts (Dziuban, Graham, Moskal, et al., 2018). The details of these methods are discussed in the following sections.

3.2 Traditional Teaching and Learning

3.2.1 Characteristics of Traditional Teaching and Learning

Traditional teaching and learning have long served as the backbone of formal education systems worldwide. Rooted in teacher-centred instruction, these methods emphasise the transmission of established knowledge through structured, hierarchical approaches. While foundational in shaping educational practices, traditional methods have faced criticism for their limitations in fostering creativity, engagement, and real-world problem-solving. Nonetheless, they remain relevant in specific educational contexts, particularly where discipline, structure, and standardised knowledge acquisition are valued. The key characteristics of traditional teaching and learning are discussed below.

3.2.1.1 Teacher-Centred Instruction

A hallmark of traditional education is its teacher-centred nature, where the teacher assumes the central role as the primary source of knowledge and authority in the classroom (Sengupta-Irving & Enyedy, 2014; Woods & Copur-Gencturk, 2024). Students are typically passive recipients of information, expected to absorb content delivered through lectures with limited interaction (Azzalis et al., 2009). Large class sizes often limit the possibility of individual attention and personalised feedback (Xiu-Hong, 2006). For example, in a civil engineering course on transportation engineering, the instructor may use a whiteboard to derive formulas related to highway design. At the same time, students take notes without actively engaging in problem-solving.

3.2.1.2 Emphasis on Knowledge Transmission and Memorisation

Traditional pedagogy prioritises transmitting objective knowledge, often through direct instruction and rote memorisation (Hanspal et al., 2024; Zhang, 2011). Students are expected to recall facts and concepts during standardised assessments, which serve as the primary measure of academic success. In engineering education, for instance, students may memorise formulas for stress–strain relationships or fluid dynamics without applying them to practical scenarios, limiting their problem-solving capacity.

In language education, the grammar-translation method exemplifies this approach, emphasising explicit grammar instruction and vocabulary memorization over communicative competence (Jin & Cortazzi, 2011).

3.2.1.3 Structured and Formalised Learning Environment

Traditional classrooms are often rigid and formal, with fixed schedules, standardised curricula, and hierarchical teacher–student dynamics (Ozola, 2012). Students are usually seated in rows facing the teacher, who controls the flow of information. Learning is compartmentalised into subject areas (Azzalis et al., 2009).

In engineering programmes, this is evident in structured courses such as soil mechanics or circuit analysis, where students follow a uniform progression and are evaluated through standardised testing.

3.2.1.4 Standardised Curriculum and Assessments

Traditional education follows a predefined curriculum with a strong emphasis on uniformity. The curriculum is designed to cover a broad range of topics, ensuring that all students learn the same content. Assessments like written exams and quizzes are standardised to evaluate students' knowledge. In the context of engineering education, the curriculum is rigidly structured, covering a predefined set of subjects such as mathematics, physics, and core engineering disciplines. Assessments are primarily conducted through written exams and quizzes. For example, in an electrical engineering programme, students might be tested on circuit analysis through a standardised exam that requires solving equations for voltage, current, and resistance using Ohm's law and Kirchhoff's rules.

3.2.1.5 Limited Student Interaction and Engagement

Another defining feature of traditional learning is the limited opportunity for student interaction, dialogue, or collaboration (Azzalis et al., 2009; Zhang, 2011). Teaching tends to focus on content delivery rather than student inquiry or discussion, which can hinder the development of critical thinking and communication skills. In a traditional fluid mechanics class, for example, students may be taught Bernoulli's equation via board work without conducting related experiments or engaging in group problem-solving activities.

3.2.1.6 Discipline, Uniformity, and Cultural Preservation

Traditional education emphasises discipline and conformity, promoting a uniform learning pace for all students regardless of individual needs or abilities (Zhang, 2011). This uniformity can disadvantage students who learn at different speeds or require differentiated instruction. Moreover, these methods are often embedded in historical and cultural contexts. For example, indigenous African education integrates storytelling, myths, and proverbs as educational tools to preserve cultural values and communal identity (Hanspal et al., 2024; Omolewa, 2007).

3.2.1.7 Focus on Theoretical Knowledge Over Practical Application

Traditional education often stresses theoretical understanding over practical experience, with limited exposure to hands-on or real-world applications (Zhang, 2011). Engineering students, for instance, may understand the theoretical underpinnings of structural analysis but lack opportunities to apply them in design or simulation environments. Where they exist, lab sessions often reinforce rather than challenge existing theoretical instruction.

3.2.1.8 Minimal Use of Technology

In traditional settings, textbooks, chalkboards, and printed materials serve as primary instructional resources, with minimal technology integration (Ozola, 2012). In engineering classrooms, this may result in outdated learning practices that fail to prepare students for the digital competencies required in the modern workplace. While some institutions have recently adopted digital platforms like Moodle or Blackboard, the use of such tools remains inconsistent.

3.2.1.9 Uniformity in Learning Pace

In traditional classrooms, all students are expected to learn at the same pace, regardless of their individual abilities or learning styles. In traditional engineering classrooms, this uniform approach is common, where students are expected to progress through the material at the same rate, irrespective of their prior knowledge or learning needs. For example, in a chemical engineering course on heat transfer, the instructor might allocate a fixed number of lectures to cover conduction, convection, and radiation, moving on to the next topic even if some students have not fully grasped the basics. This rigid approach can leave slower learners behind while failing to adequately challenge more advanced students. The lack of flexibility in such environments can hinder personalised learning and may not effectively address the diverse needs of students.

3.2.2 Examples of Traditional Learning in Practice

- ***Lecture-Based Instruction in Thermodynamics***
  A professor delivers a series of lectures on entropy and enthalpy, providing theoretical explanations and derivations. Students are expected to memorise these concepts and solve related problems during exams.
- ***Standardised Exams in Structural Analysis***
  Students in a civil engineering programme take a midterm exam focused on calculating shear force and bending moment diagrams for beams. The exam assesses their ability to apply formulas but does not involve practical design or real-world problem-solving.
- ***Rote Memorisation in Electrical Engineering***
  Students memorise formulas for calculating impedance, reactance, and resonance in AC circuits. These formulas are used in exams but are rarely connected to practical scenarios, such as designing circuits for specific applications.
- ***Lab Sessions as Demonstrations***
  In traditional engineering education, lab sessions often serve as demonstrations rather than interactive experiences. For example, in a mechanical engineering lab, students might observe an instructor operating a wind tunnel to study aerodynamics without conducting their own experiments or analysing data.
- ***Textbook-Centric Learning in Fluid Mechanics***
  Students rely heavily on textbooks to study concepts like laminar and turbulent flow. They solve problems from the textbook, but may not use computational tools like Computational Fluid Dynamics (CFD) software to simulate fluid behaviour.

3.2.3 Teaching and its Core Elements

Teaching is a complex and dynamic process that involves guiding and supporting learners through the acquisition of knowledge, skills, and attitudes. As defined by Abbatt and McMahon (1993), teaching is fundamentally about helping others to learn. It is not merely the transmission of information, but a purposeful activity designed to create conditions that foster effective learning (Clawson & Haskins, 2006). Effective teaching comprises several interrelated elements, each playing a critical role in ensuring that learning is meaningful, inclusive, and impactful (Figure 3.1).

3.2.3.1 Defining Learning Objectives

One of the foundational elements of teaching is the clear articulation of learning objectives. These objectives outline what students are expected to learn and are often shaped by the roles or tasks they are preparing to undertake (Abbatt & McMahon, 1993). Clearly defined objectives help in aligning teaching strategies, materials, and assessments to ensure students gain the necessary knowledge, skills, and attitudes relevant to their academic or professional fields.

3.2.3.2 Facilitating Learning

Facilitation of learning involves thoughtful planning of lessons, designing supportive environments, and providing appropriate resources. This process must be learner-centred, taking into account the diverse backgrounds, learning

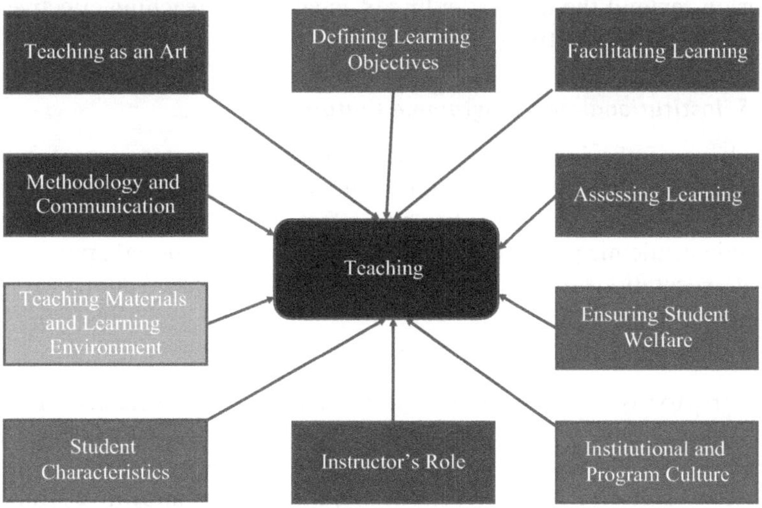

Figure 3.1 Core Elements of Teaching.

styles, and language proficiencies of students (Abbatt & McMahon, 1993). Instructors are not just conveyors of information but play the role of facilitators who encourage active participation, adapt methodologies, and nurture critical engagement. Teaching is most effective when it is inclusive and responsive to learners' needs.

3.2.3.3 Assessing Learning

Assessment is essential to determine whether students have successfully achieved the intended learning outcomes. Teachers must use varied and effective assessment strategies to evaluate students' knowledge and competencies, guide remedial actions, and inform further instruction (Abbatt & McMahon, 1993). These evaluations are not only a measure of academic performance but also a tool for ensuring students are prepared to apply what they have learned in real-world scenarios.

3.2.3.4 Ensuring Student Welfare

Beyond academic instruction, effective teaching involves caring for students' overall well-being. This includes fostering a safe and supportive learning environment, building trust, and offering emotional or psychological support when students encounter challenges (Abbatt & McMahon, 1993). Promoting student welfare is integral to holistic education, ensuring that learners are equipped to thrive both academically and personally.

Teaching extends beyond the core instructional responsibilities and is influenced by broader contextual and interpersonal factors. These additional dimensions extend the understanding of what makes teaching effective across various educational settings.

3.2.3.5 Institutional and Programme Culture

The culture of the institution and its specific programmes can significantly shape teaching practices and learning outcomes. A supportive, inclusive institutional environment enhances teaching effectiveness, while the norms and values within academic programmes influence both instructional approaches and student expectations (Clawson, 2008; Clawson & Haskins, 2006).

3.2.3.6 Instructor's Role

The instructor is at the heart of the teaching process. Effective instructors demonstrate enthusiasm, empathy, and the capacity to inspire students (Banner & Cannon, 2017). They must be adaptable, capable of modifying their teaching styles to suit diverse learner needs and integrating innovative strategies to maintain student engagement (Wang, 2005).

3.2.3.7 Student Characteristics

Students are not passive recipients of knowledge but active participants in their education. Understanding their unique backgrounds, learning preferences, and motivations enables educators to tailor their approaches and foster deeper engagement. Encouraging student agency promotes a more interactive and meaningful learning experience (Banner & Cannon, 2008).

3.2.3.8 Teaching Materials and Learning Environment

The materials used in instruction—including textbooks, digital resources, and visual aids—must be current, relevant, and aligned with the curriculum (Clawson, 2008; Molenda & Subramony, 2020). Additionally, the learning environment—whether physical or virtual—should be conducive to concentration, collaboration, and participation (Clawson, 2008).

3.2.3.9 Methodology and Communication

Teaching methodologies encompass the strategies and techniques employed to facilitate learning. Effective approaches are evidence-based, flexible, and often combine traditional and modern practices (Clawson, 2008; Molenda & Subramony, 2020). Clear communication—both verbal and non-verbal—is essential for successful instruction and includes the appropriate use of media to enhance understanding.

3.2.3.10 Teaching as an Art

Teaching is not solely a technical endeavour but also an art. It requires creativity, emotional intelligence, and the ability to connect with students meaningfully. Teachers must integrate tried-and-tested methods with innovative approaches to create a dynamic and engaging learning experience (Banner & Cannon, 2017; Wang, 2005). Continuous professional development and reflective practice are essential for maintaining relevance in a rapidly changing educational landscape (Banner & Cannon, 2017).

Teaching is a multidimensional practice that blends structure with adaptability, content knowledge with emotional intelligence, and tradition with innovation. The effectiveness of teaching depends on the harmonious integration of its various elements—defining objectives, facilitating learning, assessing progress, and supporting student well-being—alongside institutional influences, instructor attributes, student diversity, and pedagogical methodologies. Recognising teaching as both a science and an art underscores the importance of flexibility, empathy, and lifelong learning in addressing the evolving demands of education.

3.2.4 Learning

Learning is a continuous and dynamic process that results in changes in behaviour, perceptions, or actions through the acquisition of knowledge, skills, values, and attitudes (Ambrose et al., 2010). It occurs through both formal means, such as structured educational programmes, and informal settings, including everyday life experiences. At its core, learning enhances performance and prepares individuals for future tasks. According to Abbatt and McMahon (1993), learning is a lifelong endeavour, not limited to the classroom but present in various forms and contexts throughout one's life.

Several key aspects shape how individuals learn (Figure 3.2). One important dimension is the variety of learning styles. People have unique preferences for how they absorb and process information. While some learners may rely on memorisation techniques, others may thrive through exploration, inquiry, or hands-on experiences. Another crucial factor is the depth of learning. Superficial learning often involves rote memorisation with little understanding, whereas deep learning emphasises connecting new information to prior knowledge, resulting in meaningful comprehension and better application. Motivation is also central to the learning process, varying between intrinsic drivers such as curiosity or personal growth, and extrinsic incentives like academic success or professional advancement.

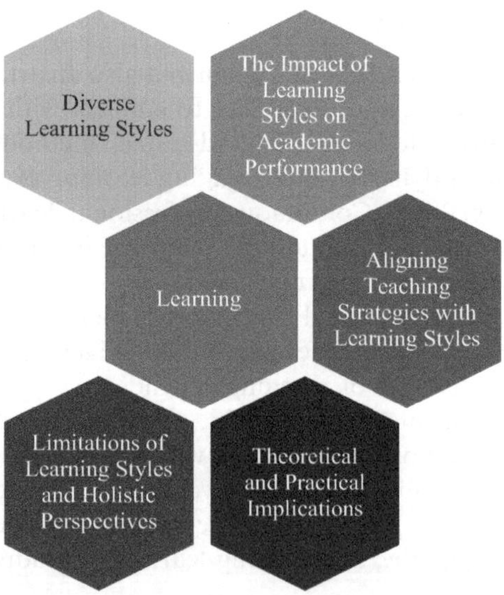

Figure 3.2 Key Aspects That Shape Learning.

3.2.4.1 Diverse Learning Styles

Learning styles refer to the preferred ways individuals engage with and retain information, and they play a significant role in educational success. Traditionally, learning styles have been categorised into visual, auditory, and kinesthetic modalities. However, recent research has expanded this view. De Paz Lazaro et al. (2022) identified nine distinct learning styles that go beyond the conventional three, addressing a broader spectrum of cognitive and sensory preferences.

Several theoretical models further elaborate on these styles. Felder and Silverman's learning style model classifies learners along four dimensions: active versus reflective, sensing versus intuitive, visual versus verbal, and sequential versus global (Villalba & Henao, 2012). This model helps educators understand how students prefer to receive and process information. Similarly, Herrmann's Brain Dominance Instrument (HBDI) categorises learners based on cerebral dominance. It offers insight into individual cognitive strengths by identifying which part of the brain—logical, analytical, emotional, or intuitive—a person predominantly uses (Villalba & Henao, 2012).

3.2.4.2 The Impact of Learning Styles on Academic Performance

Learning styles have been shown to significantly influence academic achievement. Students whose learning experiences align with their preferred styles often perform better academically. For example, Vaishnav and Bajaj (2013) and García (2018) found that matching instruction to students' learning styles can enhance comprehension and retention. Among various styles, the reflective learning style—where learners prefer thoughtful analysis before engaging—has been found to be particularly dominant in some student populations, further underscoring the need to understand individual differences in learning preferences (García, 2018).

3.2.4.3 Aligning Teaching Strategies with Learning Styles

Effective teaching requires aligning instructional methods with students' learning preferences. Putri and Marsofiyati (2024) emphasised the value of incorporating diverse strategies, such as using visual aids for visual learners, promoting discussion-based learning for auditory learners, and involving hands-on activities for kinesthetic learners. Active and collaborative learning strategies—like group projects, peer teaching, and real-life case studies—also support multiple learning styles simultaneously, fostering greater student engagement and deeper understanding (De Paz Lazaro et al., 2022).

However, when there is a mismatch between teaching and learning styles, students may experience frustration or disengagement. Gilakjani (2012) notes that such a mismatch can hinder learning outcomes, emphasising the importance of flexibility in teaching. Educators are therefore encouraged to adopt a balanced, inclusive approach that accommodates a range of learner needs (Peacock, 2001).

3.2.4.4 Theoretical and Practical Implications

Learning is shaped not only by style preferences but also by cognitive, affective, social, and physiological factors. Vaishnav and Bajaj (2013) and Li (2005) highlight that understanding these diverse influences can help educators design more inclusive and effective learning environments. Additionally, cultural and biological differences play a crucial role in shaping how individuals learn, reinforcing the importance of culturally responsive teaching practices (Vaishnav & Bajaj, 2013).

3.2.4.5 Limitations of Learning Styles and Holistic Perspectives

While learning styles offer a valuable lens through which to understand learner diversity, they are not the sole determinants of educational success. Overreliance on learning styles may oversimplify the complexity of learning processes. Other factors—such as motivation, prior knowledge, teaching quality, and the learning environment—are equally significant. Some scholars argue for a more holistic approach to education that integrates learning styles with broader pedagogical and contextual considerations. Ultimately, effective education requires recognising the unique needs of each learner while designing adaptable and inclusive teaching strategies.

3.3 Approaches to Learning

3.3.1 Surface Approach to Learning

Surface learning is an approach characterised by minimal engagement with learning activities and a tendency to focus on rote memorisation rather than understanding. It often results in superficial retention of information disconnected from meaningful application (Biggs, 1999; Biggs & Tang, 2007). Learners who adopt this approach tend to uncritically accept new facts and ideas, resulting in fragmented knowledge that lacks depth and is easily forgotten over time (Ramsden, 1992). This approach is frequently associated with low intrinsic motivation and limited comprehension, as learners prioritise short-term outcomes over long-term understanding (Tagg, 2003).

Surface learners are primarily driven by extrinsic motivators, such as the desire to obtain high grades or meet external expectations. Their learning strategies are typically centred on memorisation and the reproduction of content rather than genuine comprehension (Hooijer, 2010; Shamuratova, 2024). These students often perceive academic tasks as obligatory and focus on fulfilling minimum requirements, such as recalling key points for assessments, instead of engaging in critical or reflective thinking (Kieser et al., 2006). This task-oriented mindset leads them to prioritise performance over mastery, making their learning process reactive rather than proactive.

The educational environment plays a crucial role in shaping whether students adopt a surface or deep learning approach. Teaching practices that emphasise content coverage over comprehension, and assessment methods that reward memorisation rather than analytical thinking, tend to encourage surface learning behaviours (Tang & Robinson, 2010). High-stakes testing environments, in particular, can pressure students into adopting surface strategies to meet immediate performance goals.

Additionally, disciplinary differences influence the prevalence and impact of surface learning. In some fields—such as business or early-stage coursework across disciplines—strategic surface learning may lead to academic success, especially when assessments are aligned with factual recall (Hooijer, 2010; Svedin et al., 2013). However, in disciplines that require integrative thinking, such as engineering or the sciences, surface approaches may undermine the development of essential problem-solving skills.

While surface learning is generally viewed as a less effective approach, it may yield short-term academic success under certain conditions. Studies have shown that in some contexts—particularly in early education or content-heavy programmes where memorisation is rewarded—a surface approach can correlate with satisfactory academic performance (Donnison & Penn-Edwards, 2012; Svedin et al., 2013). However, these benefits are often limited to specific settings and do not extend to complex or higher-order learning tasks.

In the long term, reliance on surface learning is associated with poorer educational outcomes. It limits students' ability to synthesise information, apply knowledge in new situations, and develop critical thinking skills—abilities that are essential for academic progression and professional competence (Lee et al., 2020). Consequently, educators are encouraged to design curricula and assessments that discourage rote learning and instead promote deeper cognitive engagement.

3.3.2 Deep Approach to Learning

The deep approach to learning emphasises critical analysis, meaningful engagement, and the integration of new ideas with existing knowledge. Unlike surface learning, which is often driven by external motivations and characterised by rote memorisation, deep learning is rooted in intrinsic motivation and a desire to understand (Biggs, 1999; Biggs & Tang, 2007). Students who adopt a deep approach seek to make sense of what they learn, applying knowledge to real-world problems and fostering long-term retention. This method promotes intellectual curiosity, mindfulness, and personal growth, making learning an active and enjoyable process (Tagg, 2003).

One of the core features of deep learning is intrinsic motivation—learners are genuinely interested in the subject matter and driven by a desire to understand rather than achieve high grades (Hooijer, 2010; Shamuratova, 2024). They tend to engage with content by connecting it to prior knowledge and real-life experiences, which enhances critical thinking, synthesis, and application across contexts (Chen & Dhillon, 2012; Zakarya, 2020). Although this approach often results in a more comprehensive understanding of material, it does not always translate directly to higher academic performance, particularly when assessments are designed to reward factual recall rather than conceptual insight (McLoone & Oluwadun, 2014).

Several factors contribute to the adoption of a deep approach. A supportive educational environment encouraging exploration, innovation, and appropriately balanced workloads can significantly enhance deep learning (Peng & Tuan, 2018; Phanudulkitti et al., 2018). Additionally, teaching methods such as problem-based learning, inquiry-based learning, and reflective practice are particularly effective in fostering deep engagement with academic content (McLoone & Oluwadun, 2014). Student characteristics—including prior knowledge, personal learning goals, and cognitive engagement—also influence the likelihood of adopting a deep approach (Zhang, 2024).

Despite its many advantages, deep learning faces several challenges. Traditional assessment methods often fail to capture the depth of student understanding, instead favouring memorisation and factual recall, which can discourage students from adopting a deeper approach (McLoone & Oluwadun, 2014). Additionally, the effectiveness of deep learning may vary by discipline. For instance, in business studies, strategic learning approaches—focused on optimising performance outcomes—may align more closely with academic success than deep approaches (Hooijer, 2010).

It is also important to recognise that learning is not a 'one-size-fits-all' process. While the deep approach is widely regarded as effective, it may not always be sufficient or appropriate in isolation. Integrating deep learning with other approaches, such as strategic or surface strategies when contextually appropriate, can yield more balanced and comprehensive educational outcomes (Mathias & Newton, 2016). By aligning teaching methods and assessment practices with intended learning outcomes, educators can better support students in developing a deep understanding and academic success (Biggs & Tang, 2007).

3.4 Outcomes-Based Education (OBE)

Outcomes-Based Education (OBE) is an educational framework emphasising clearly defined learning outcomes. Instead of focusing solely on content delivery or instructional methods, OBE prioritises what students are expected to know, do, and value by the end of a course or programme. This learner-centred approach has gained widespread adoption, particularly in engineering education, where ensuring students' competence is critical to addressing complex, real-world challenges (Barsoum, 2017; OECD, 1996). Despite its popularity and reported benefits, questions remain about its effectiveness in consistently achieving these outcomes across diverse educational contexts.

A foundational element of OBE is its alignment of teaching, learning activities, and assessment with Intended Learning Outcomes (ILOs) (Figure 3.3). This principle, known as constructive alignment, ensures that all aspects of the educational process support the achievement of desired competencies (Biggs, 1999; Biggs & Tang, 2007). The focus on predefined outcomes allows educators to design curricula and instructional strategies that are purpose-driven and measurable (Aggarwal, 2024).

OBE adopts a student-centred approach, promoting active engagement by making students aware of the goals they are expected to meet. This fosters a sense of ownership, motivation, and responsibility in learning (Cantika & Alhayat, 2023; Hamidi et al., 2024). Moreover, the framework provides flexibility in teaching methods, enabling instructors to tailor their strategies according to their students' needs and learning styles (Qadir et al., 2020).

OBE is implemented through carefully designed curricula and assessment tools that align with specific outcomes. OBE assessment includes formative and summative methods, structured to evaluate whether students have attained the desired competencies (Ohatkar & Deshpande, 2022; Sharma, 2023). This approach is particularly relevant in engineering education, where knowledge application and problem-solving are essential for professional readiness.

Studies support the positive impact of OBE on student skills and behaviour, particularly in developing critical thinking, accountability, and a deeper

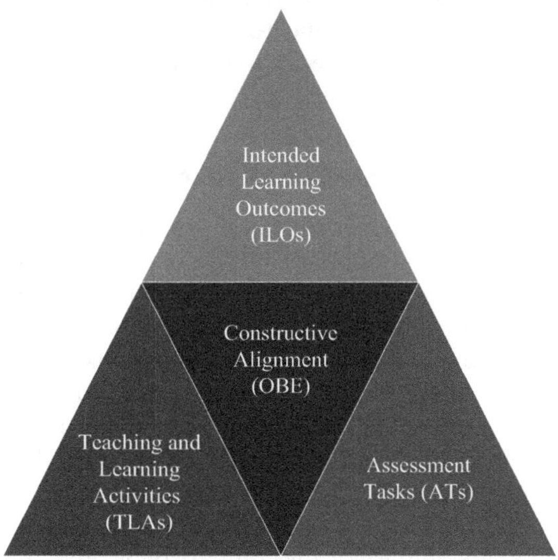

Figure 3.3 Constructive Alignment for Outcomes Based Education.

engagement with learning (Hamidi et al., 2024). Empirical evidence from case studies has demonstrated improvements in academic performance and outcome attainment following the implementation of OBE frameworks (Aggarwal, 2024; Cantika & Alhayat, 2023).

Despite its advantages, OBE is not without criticism. A primary concern involves the complexity of implementing constructive alignment. In practice, perfect alignment between learning activities, assessments, and ILOs can be challenging to achieve, especially in theoretical or abstract subjects where assessing competencies is less straightforward (Brady, 1996; Donnelly, 2007). Furthermore, critics argue that the emphasis on predefined outcomes may restrict students' cognitive exploration, limiting creative or divergent thinking opportunities.

Another challenge is resistance from educators, some of whom view OBE as burdensome due to its extensive planning and documentation requirements. Additionally, misconceptions persist that OBE is overly rigid or incompatible with traditional teaching approaches, despite its inherent flexibility (Qadir et al., 2020). Successful implementation also hinges on adequate professional development, as educators must be well-equipped to design and deliver outcome-aligned learning experiences (Chan, 2023).

However, OBE represents a shift towards purposeful and accountable teaching and learning, with a strong emphasis on preparing students for real-world challenges. When effectively implemented, it enhances student engagement, fosters critical competencies, and aligns academic programmes with societal and professional needs. However, to maximize its potential, it is essential to address the structural and pedagogical challenges that accompany its adoption. A balanced approach that incorporates innovative teaching methods, flexible assessments, and ongoing educator training will be key to ensuring that OBE supports academic excellence and contributes to sustainable development and lifelong learning.

3.5 PBL Compared to Traditional Teaching and Learning

In the evolving landscape of higher education—particularly in engineering—the effectiveness of teaching methodologies has become a central focus. Traditional approaches, characterised by lecture-based instruction and passive knowledge transmission, have long dominated academic settings. However, the growing demand for real-world problem-solving, critical thinking, and collaborative skills has driven increased interest in PBL. As education shifts towards more active and student-centred models, PBL has emerged as a compelling alternative to traditional methods. The comparison below explores the distinctions between PBL and traditional teaching across critical dimensions, including student engagement, learning outcomes, assessment practices, motivation, and professional preparedness. A summary of the comparison is presented in Table 3.1.

Table 3.1 Comparative Analysis of Traditional Learning and PBL

Dimension	*Traditional Teaching and Learning*	*Problem-Based Learning (PBL)*
Teaching Approach	Teacher-centred; knowledge is transmitted via lectures and textbooks in a classroom environment.	Student-centred; students learn through solving real-world problems. Learning occurs in a PBL or project environment. The problem is at the core of the learning process.
Student Engagement	Passive learning: limited interaction or application. The individual learner is responsible for his or her learning. The teacher leads the classroom and the teaching and learning process.	Active participation encourages collaboration, inquiry, and ownership of learning. Although the individual learner is responsible for learning, it is conducted in groups and teams. Students lead the learning instead of the teacher.
Learning Outcomes	Focus on theoretical knowledge and memorisation.	Emphasises critical thinking, knowledge integration, and practical application
Knowledge Retention	Often short-term, based on rote memorisation	Long-term retention through contextual and experiential learning
Assessment Practices	Exams, quizzes, and written assignments testing recall and understanding	Holistic assessments: projects, presentations, and reports evaluating problem-solving and skills
Motivation	Often extrinsically driven (grades, exams)	Intrinsically driven by relevance, autonomy, and engagement
Professional Preparation	Emphasis on foundational theory; limited real-world exposure	Develops teamwork, communication, and problem-solving skills relevant to the industry
Skill Development	Prioritises technical knowledge; soft skills are underdeveloped	Equips students with critical soft skills: collaboration, communication, and adaptability
Flexibility	Fixed content and structure	Adaptive to students' interests, learning paths, and real-world relevance
Implementation Needs	Requires fewer resources and less instructional redesign	Requires trained facilitators, resource investment, and curriculum redesign

3.5.1 Teaching Approaches and Student Engagement

Traditional teaching and learning methods in engineering education are typically lecture-centric, where the instructor is the primary information source. Students often assume a passive role, receiving content through lectures, textbooks, and standardised materials. This model emphasises theoretical understanding, with limited opportunities for hands-on learning or engagement with real-world problems. For instance, in a traditional fluid mechanics course, students may study Bernoulli's equation through lectures with minimal application or collaborative exploration (Biggs & Tang, 2007).

In contrast, Problem-Based Learning (PBL) adopts a student-centred approach, emphasising active engagement with complex, real-world problems. Students work collaboratively, apply existing knowledge, and seek new information to develop viable solutions. This method promotes critical thinking, teamwork, and problem-solving. For example, students in a civil engineering course might be challenged to design a sustainable water management system, requiring interdisciplinary integration across engineering and socio-economic domains (Barrows, 1996). As a result, students are more engaged, motivated, and take greater ownership of their learning.

3.5.2 Learning Outcomes and Knowledge Retention

Traditional methods often prioritise rote memorisation and theoretical mastery. While this approach may support strong exam performance, it often fails to promote long-term knowledge retention or the ability to apply concepts in practice. For instance, engineering students might grasp the principles of thermodynamics but struggle to utilise them in real-world design problems (Biggs, 1999).

PBL, however, facilitates deep learning by encouraging students to analyse problems, integrate concepts, and apply knowledge in context. Research has shown that PBL improves knowledge retention and transferability, equipping students with the skills to navigate novel challenges (Hmelo-Silver, 2004). In an electrical engineering course, for example, troubleshooting a faulty circuit requires theoretical recall and practical problem-solving skills, promoting deeper comprehension.

3.5.3 Assessment and Evaluation

Traditional assessment methods—such as written exams, quizzes, and individual assignments—primarily evaluate students' ability to recall theoretical knowledge. These methods may not adequately assess critical thinking, collaboration, or real-world problem-solving abilities. For instance, in a materials science course, students might be tested on the properties of metals through written exams that fail to assess practical application skills (Biggs & Tang, 2007).

PBL employs holistic and performance-based assessments, including group projects, presentations, and reflective reports. These evaluations consider technical competence, collaboration, communication, and innovation. For example, in a mechanical engineering course, students might be assessed on designing a prototype, with criteria covering technical soundness, teamwork, and presentation quality (Barrows, 1996).

3.5.4 Motivation and Student Autonomy

Traditional teaching methods may result in student disengagement, particularly for learners who thrive on interaction and autonomy. The passive structure can alienate students who fail to see the relevance of abstract theories to real-world challenges (Tagg, 2003).

Conversely, PBL nurtures intrinsic motivation by situating students at the centre of the learning process. Through engagement with authentic, discipline-relevant problems, students are more likely to feel motivated and invested. They take responsibility for their learning, navigating how to approach problems, seek solutions, and evaluate outcomes (Hmelo-Silver, 2004). This increased autonomy contributes to a richer and more meaningful educational experience.

3.5.5 Skill Development and Professional Preparation

Traditional approaches often produce students with strong theoretical foundations but may fall short in preparing them for professional engineering practice's complex, interdisciplinary, and collaborative nature (Biggs, 1999). The lack of exposure to practical challenges can leave graduates underprepared for real-world roles.

PBL is intentionally designed to cultivate professional competencies such as critical thinking, teamwork, communication, and applied problem-solving. For instance, students working on an urban infrastructure project in a civil engineering course must collaborate across disciplines, communicate effectively, and apply technical knowledge in a real-world context (Barrows, 1996). These skills are directly transferable to professional environments and essential for career success.

3.5.6 Comparative Effectiveness in Academic Performance

Studies show that PBL enhances academic performance in areas requiring higher-order thinking. Students in PBL environments often outperform their peers in tasks involving procedural understanding and application, though traditional methods may still be more effective for conceptual knowledge acquisition (Masek & Yamin, 2012; Rajuli & Baharun, 2017).

In medical education, PBL is widely regarded as superior for developing competencies, although research findings are mixed. Some studies report significant advantages of PBL in fostering deep learning, while others find no significant differences compared to traditional approaches (Vélez Silva et al., 2024).

3.5.7 *Student Engagement and Learning Attitudes*

PBL fosters greater engagement and positive attitudes towards learning by promoting active participation and collaboration. It enables students to take ownership of their education, improving motivation and satisfaction (Leng et al., 2022; Movafegh & Azimaraghi, 2015). In contrast, while traditional methods may be criticised as passive, they can offer a structured learning environment that some students find helpful for mastering dense, theoretical material (Marcinauskas et al., 2024).

3.5.8 *Implementation Challenges and Institutional Considerations*

Despite its benefits, PBL presents implementation challenges, such as the need for trained facilitators, resource-intensive planning, and supportive infrastructure. The success of PBL depends heavily on facilitators' ability to guide student inquiry effectively and the availability of suitable problems and learning materials (Koriťáková et al., 2023; Saif & Khan, 2015).

Institutions considering a transition to PBL must address these logistical and cultural barriers and ensure that faculty and students are adequately prepared for the shift (Nepal, 2013).

3.5.9 *Disciplinary Applicability and Hybrid Models*

PBL has shown effectiveness in developing critical thinking and teamwork skills in engineering and computing disciplines. However, traditional methods may still be preferred for establishing foundational theoretical knowledge (Rajuli & Baharun, 2017; Saavedra, 2023). In health and medical education, PBL is increasingly adopted due to its alignment with practical and diagnostic learning, though more empirical support is needed to confirm its superiority (Movafegh & Azimaraghi, 2015; Vélez Silva et al., 2024).

Given the strengths and limitations of both approaches, a hybrid model—blending the structure and efficiency of traditional teaching with the engagement and skill development of PBL—may provide the most effective strategy for diverse educational contexts.

3.6 Conclusion

Both traditional teaching and PBL offer distinct advantages and limitations within engineering education. Traditional methods provide a structured and theoretical foundation, which remains essential for understanding core engineering principles. However, they often fall short in preparing students for professional practice's dynamic, real-world challenges.

PBL, by contrast, promotes active learning, critical thinking, and practical problem-solving. It fosters essential skills such as collaboration, adaptability, and innovation, which are increasingly demanded in engineering. Integrating PBL into engineering curricula can enhance students' readiness to tackle complex, interdisciplinary problems.

Despite its historical value, traditional teaching has been criticised for emphasising authority, standardisation, and passive learning, which can suppress student autonomy and creativity. These limitations raise concerns about its effectiveness in developing the applied skills and mindset required in modern, innovation-driven environments.

While traditional methods contribute to academic discipline and conceptual clarity, they are often insufficient. A balanced pedagogical approach—combining the structure of traditional teaching with the engagement and real-world focus of PBL—may provide the most effective framework for equipping students to meet the demands of contemporary engineering practice.

References

Abbatt, F., & McMahon, R. (1993). *Teaching health care workers* (2nd ed.). Macmillan.

Aggarwal, T. (2024). An empirical analysis of outcome-based education research. *International Journal for Multidisciplinary Research*, *6*(5). https://doi.org/10.36948/ijfmr.2024.v06i05.28001

Ambrose, S. A., Bridges, M. W., DiPietro, M., Lovett, M. C., & Norman, M. K. (2010). *How learning works: Seven research-based principles for smart teaching*. Jossey-Bass.

Azzalis, L. A., Sato, S. N., de Mattos, M. M., Fonseca, F. L. A., & Giavarotti, L. (2009). Active learning versus traditional teaching. *Journal of Biochemistry Education*, *7*(2), 2.

Baig, M. I., & Yadegaridehkordi, E. (2023). Flipped classroom in higher education: A systematic literature review and research challenges. *International Journal of Educational Technology in Higher Education*, *20*, Article 61. https://doi.org/10.1186/s41239-023-00430-5

Banner, J. M., & Cannon, H. C. (2008). *The elements of learning* (Unabridged ed.). Yale University Press.

Banner, J. M., & Cannon, H. C. (2017). *The elements of teaching*. Yale University Press.

Baquero, A., & Escortell, R. (2022). Blended learning: A new trend in education. In *Proceedings of the 16th International Technology, Education and Development Conference*, 4484–4491. https://doi.org/10.21125/iceri.2022.1080

Barrows, H. S. (1996). Problem-based learning in medicine and beyond: A brief overview. *New Directions for Teaching and Learning*, *1996*(68), 3–12. https://doi.org/10.1002/tl.37219966804

Barsoum, G. (2017). Engineering education and the development of professional competencies. *Engineering Education*, *12*(1), 15–25.

Biggs, J. (1999). *Teaching for quality learning at university*. Society for Research into Higher Education & Open University Press.

Biggs, J., & Tang, C. (2007). *Teaching for quality learning at university* (3rd ed.). Open University Press.

Brady, L. (1996). Outcome-based education: A critique. *The Curriculum Journal*, *7*(1), 5–16. https://doi.org/10.1080/0958517960070102

Cantika, V. M., & Alhayat, A. (2023). Implementation of outcome-based education in a leading high school (a case study at SMA Taruna Bakti Bandung). *Proceedings Series on Social Sciences & Humanities*, *12*, 52–59. https://doi.org/10.30595/pssh.v12i.773

Chan, J. Y. (2023). Law teachers' adoption and acceptance of OBE: A case study of Hong Kong. *The Law Teacher*, *57*(4), 530–547. https://doi.org/10.1080/03069400.2023.2258023

Chandraja, C. V., Ajayan, T., Shabi, R., & George, M. (2024). Innovative pedagogies: Adapting teaching strategies for modern learning environments. In *21st century teaching and learning in classrooms* (pp. 63–74). Nitya Publications. https://doi.org/10.58532/nbennurctch9

Chen, L., & Dhillon, J. (2012). Deep approaches to learning in improving reading skills: A case study from Yunnan Agricultural University. *Theory and Practice in Language Studies*, *2*(8), 1603–1613. https://doi.org/10.4304/TPLS.2.8.1603-1613

Clawson, J. G. (2008). *Fundamental elements in teaching* (Darden Case No. UVA-PHA-0046). Darden Business Publishing. https://doi.org/10.2139/ssrn.911821

Clawson, J. G., & Haskins, M. E. (2006). Teaching management: Fundamental elements in teaching. In J. G. Clawson & M. E. Haskins (Eds.), *Teaching management* (pp. 12–25). Cambridge University Press. https://doi.org/10.1017/CBO9780511617850.002

De Paz Lazaro, A. C., Damian Espinoza, Y. E., & Verde Lujan, H. E. (2022). Styles and forms of learning in university education. *Minerva*, *3*(8), 84–92. https://doi.org/10.47460/minerva.v3i8.67

Donnelly, K. (2007). Australia's adoption of outcomes-based education: A critique. *Issues in Educational Research*, *17*(2), 123–145.

Donnison, S., & Penn-Edwards, S. (2012). Focusing on first-year assessment: Surface or deep approaches to learning? *The International Journal of the First Year in Higher Education*, *3*(2), 9–20. https://doi.org/10.5204/INTJFYHE.V3I2.127

Dziuban, C., Graham, C. R., Moskal, P. D., Norberg, A., & Sicilia, C. (2018). Blended learning: The new normal and emerging technologies. *International Journal of Educational Technology in Higher Education*, *15*, Article 3. https://doi.org/10.1186/s41239-017-0087-5

García, A. E. (2018). Estilos de aprendizaje y rendimiento académico [Learning styles and academic performance]. *Revista Boletín Redipe*, *7*(7), 218–228. https://revista.redipe.org/index.php/1/article/download/536/509

Gilakjani, A. P. (2012). A match or mismatch between learning styles of the learners and teaching styles of the teachers. *International Journal of Modern Education and Computer Science*, *4*(11), 51–60. https://doi.org/10.5815/ijmecs.2012.11.05

Hamidi, H., Hejran, A. B., Sarwari, A., & Edigeevna, S. G. (2024). The effect of outcome-based education on the behaviour of students. *European Journal of Theoretical and Applied Sciences*, *2*(2), 764–773. https://doi.org/10.59324/ejtas.2024.2(2).68

Hanspal, G., Singh, H., & Goswami, P. K. (2024). Bridging tradition and innovation: A balanced approach to teaching and learning. *African Journal of Biomedical Research*, *27*(3S). https://doi.org/10.53555/AJBR.v27i3S.6112

Hmelo-Silver, C. E. (2004). Problem-based learning: Its history, philosophy, and future. *Interdisciplinary Journal of Problem-Based Learning*, *1*(1), 6–21. https://doi.org/10.7771/1541-5015.1006

Hooijer, J. G. (2010). *In the business of learning: Approaches to learning of undergraduate students in business* [Doctoral dissertation, Erasmus University Rotterdam]. Erasmus University Rotterdam Repository. https://doi.org/10.6100/IR684850

Hunter, C., Archer, S., & Illingworth, C. (2025). Problem-based learning. In R. Papa & S. H. K. Tan (Eds.), *Bloomsbury handbook of educational leadership and management* (pp. 302–305). Bloomsbury Publishing. https://doi.org/10.4337/9781035302932.ch83

Jin, L., & Cortazzi, M. (2011). Re-evaluating traditional approaches to second language teaching and learning. In S. Gass & A. Mackey (Eds.), *The Routledge handbook of second language acquisition* (pp. 576–593). Routledge. https://doi.org/10.4324/9780203836507.CH34

Kieser, J. A., Herbison, P., Waddell, N., Kardos, T. B., & Innes, P. (2006). Learning in oral biology: A comparison between deep and surface approaches. *The New Zealand Dental Journal*, *102*(3), 64.

Kolb, A. Y., & Kolb, D. A. (2011). Experiential learning theory: A dynamic, holistic approach to management learning, education and development. In S. J. Armstrong & C. Fukami (Eds.), *Handbook of management learning, education and development* (pp. 68–87). Sage.

Kong, Y. (2021). The role of experiential learning on students' motivation and classroom engagement. *Frontiers in Psychology*, *12*, Article 771272. https://doi.org/10.3389/fpsyg.2021.771272

Koriťáková, G., Madleňák, A., Kopecký, M., Malcho, M., & Caban, J. (2023). Comparison of problem-based and team-based learning strategies: A multi-institutional investigation. *Frontiers in Education*, *8*. https://doi.org/10.3389/feduc.2023.1301269

Lee, J., Kim, R. J. Y., & Choi, H. (2020). Most surface learning in the third year: Dental student learning approaches and implications for curriculum and assessment. *Journal of Dental Education*, *84*(4), 464–472. https://doi.org/10.1002/JDD.12043

Leng, B., Ding, X. X., & Wang, S. (2022). A comparative analysis of traditional teaching and PBL model. In *Proceedings of the 2022 5th International Conference on Humanities Education and Social Sciences (ICHESS 2022)*, 306. https://doi.org/10.2991/assehr.k.220504.306

Li, J. (2005). Mind or Virtue: Western and Chinese Beliefs About Learning. *Current Directions in Psychological Science*, *14*(4), 190–194. https://doi.org/10.1111/j.0963-7214.2005.00362.x

Marcinauskas, L., Iljinas, A., Čyvienė, J., & Stankus, V. (2024). Problem-based learning versus traditional learning in physics education for engineering program students. *Education Sciences*, *14*(2), Article 154. https://doi.org/10.3390/educsci14020154

Masek, A., & Yamin, S. (2012). A comparative study of the effect of problem-based learning and traditional learning approaches on students' knowledge acquisition. *International Journal of Engineering Education*, *28*(5), 1161–1167.

Mathias, J., & Newton, D. P. (2016). *Deep and shallow approaches to learning mathematics are not mutually exclusive*. British Society for Research into Learning Mathematics. https://dro.dur.ac.uk/20319/

McLoone, P., & Oluwadun, A. (2014). Approaches to learning in higher education: A review. *African Educational Research Journal*, *2*(3), 110–115.

Molenda, M., & Subramony, D. P. (2020). *The elements of instruction: A framework for the age of emerging technologies*. Routledge.

Movafegh, A., & Azimaraghi, O. (2015). Problem-based learning. *Advanced Journal of Science*, *2*(1), 1–4.

Nepal, K. P. (2013). Comparative evaluation of PBL and traditional lecture-based teaching in undergraduate engineering courses: Evidence from controlled learning environment. *International Journal of Engineering Education*, *29*(1), 17–22.

OECD (1996). *The knowledge-based economy*. Organisation for Economic Co-operation and Development. OECD Publishing.

Ohatkar, S. N., & Deshpande, A. M. (2022). Outcome-based education: A learner-centric pedagogical framework with case studies in digital communication and signal processing courses. *Journal of Engineering Education Transformations*, *36*(S1), 38–42. https://doi.org/10.16920/jeet/2022/v36is1/22172

Omolewa, M. (2007). Traditional African modes of education: Their relevance in the modern world. *International Review of Education*, *53*(5), 593–612. https://doi.org/10.1007/S11159-007-9060-1

Ozola, S. (2012). Student-centred learning: A dream or reality. In *Conference proceedings. Vol. 10* (pp. 235–241). Kyustendil, Bulgaria, Bulgarian Comparative Education Society.

Peacock, M. (2001). Match or mismatch? Learning styles and teaching styles in EFL. *International Journal of Applied Linguistics*, *11*(1), 1–20. https://doi.org/10.1111/1473-4192.00001

Peng, M. Y.-P., & Tuan, S.-H. (2018). Disentangling the antecedents of relationship between deep approach to learning and student learning outcomes. In *Proceedings of the International Conference on Information Technology*, 148–151. https://doi.org/10.2991/ICITME-18.2018.29

Phanudulkitti, C., Farris, K. B., Makmee, P., & Kittisopee, T. (2018). Deep approach to learning of pharmacy students: A multilevel analysis. *Pharmacy Education*, *18*(1), 99–109.

Prasad, S., & O'Malley, C. B. (2022). An introductory framework of problem-based learning (PBL) and perspectives on enhancing facilitation approaches. *HAPS Educator*, *26*(3), 52–58. https://doi.org/10.21692/haps.2022.016

Putri, R., & Marsofiyati, M. (2024). Pengaruh gaya belajar terhadap prestasi peserta didik [The influence of learning styles on student achievement]. *Jurnal Inovasi Pendidikan dan Pembelajaran*, *1*(3), 146–161. https://doi.org/10.62383/edukasi.v1i3.322

Qadir, J., Shafi, A., Al-Fuqaha, A., Taha, A.-E. M., Yau, K.-L. A., Ponciano, J., Hussain, S., Sheikh Muhammad, S., Rais, R. N. B., Rashid, M., & Tan, B. L. (2020). Outcome-based engineering education: A global report of international OBE accreditation and assessment practices. In *2020 ASEE Virtual Annual Conference Content Access*. https://doi.org/10.35542/osf.io/rde62

Rajuli, S., & Baharun, N. (2017). A comparative study of problem-based and traditional teaching in computing subjects. In N. Zainuddin, A. A. A. Ghani, A. M. Nor, & M. S. M. Rahim (Eds.), *Proceedings of the 6th International Conference on Computing and Informatics* (pp. 275–284). Springer. https://doi.org/10.1007/978-981-10-2772-7_28

Ramsden, P. (1992). *Learning to teach in higher education*. Routledge.

Renz, F. M., & Vogel, J. U. (2024). Be the change you want to see: Problem-based learning to promote diversity, justice, equity, inclusion, belonging, and sustainability in the classroom and workplace. *Merits*, *4*(1), 79–94. https://doi.org/10.3390/merits4010006

Saavedra, P. G. R. (2023). Comparative analysis between a traditional design curriculum and a problem and project-based learning (PBL) aligned curriculum applied to electromagnetism under systematic evaluation. In *2023 IEEE International Conference on Automation, Control and Instrumentation Technology (ICACIT)*, 1–5. https://doi.org/10.1109/icacit59946.2023.10403701

Saif, Q., & Khan, R. (2015). Problem based learning in comparison to traditional teaching as perceived by the MBBS students of a North Indian teaching hospital. *IOSR Journal of Research & Method in Education (IOSR-JRME)*, *5*(2), 9–14.

Saira, Z. N., & Hafeez, M. (2021). A critical review on discussion and traditional teaching methods. *Psychology and Education*, *58*(1), 1871–1886.

Sengupta-Irving, T., & Enyedy, N. (2014). Why engaging in mathematical practices may explain stronger outcomes in affect and engagement: Comparing student-driven with highly guided inquiry. *Journal of the Learning Sciences*, *24*(4), 550–592. https://doi.org/10.1080/10508406.2014.928214

Shamuratova, D. (2024). The surface and deep approaches to learning and teaching the English language in higher education. *Universum: Metallurgiya va Mashinasozlik*, *1*(1.2), 263–266. https://doi.org/10.69617/uzmu.v1i1.2.1004

Sharma, M. (2023). Outcome-based education pyramid: A comprehensive framework for enhancing educational outcomes. *Thiagarajar College of Preceptors Edu Spectra*, *5*(S1), 67–73. https://doi.org/10.34293/eduspectra.v5is1-may23.012

Smiderle, R., Rigo, S. J., Marques, L. B., Peçanha de Miranda Coelho, J. A., & Jaques, P. A. (2020). The impact of gamification on students' learning, engagement and behaviour based on their personality traits. *Smart Learning Environments*, *7*, Article 3. https://doi.org/10.1186/s40561-019-0098-x

Svedin, M., Bälter, O., Scheja, M., & Pettersson, K. (2013). A surface approach to learning rewards first-year engineering students. In *2013 IEEE Global Engineering Education Conference (EDUCON)*, 143–147. https://doi.org/10.1109/LATICE.2013.40

Tagg, J. (2003). *The learning paradigm college*. Anker Publishing Company.

Tang, T., & Robinson, T. (2010). Learning approach and perception of learning context in economics education: A causality test. *The International Journal of Learning: Annual Review*, *17*(2), 21–40. https://doi.org/10.18848/1447-9494/CGP/V17I02/46863

Tularam, G. A., & Machisella, P. (2018). Traditional vs non-traditional teaching and learning strategies: The case of e-learning! *International Journal for Mathematics Teaching and Learning*, *19*(1), 129–158.

Vaishnav, R., & Bajaj, C. K. C. (2013). Learning style and academic achievement of secondary school students. *Voice of Research*, *1*(4), 1–4.

Vélez Silva, V. N., Palma Zambrano, G. A., & Borroto Cruz, E. R. (2024). Problem-based learning vs traditional method in the development of competencies in students in the health area: A systematic review. *Salud, Ciencia y Tecnología*, *4*, 601. https://doi.org/10.56294/saludcyt2024.601

Villalba, N. L., & Henao, A. G. (2012). Learning styles: A pedagogical proposal to optimise the teaching of foreign languages in the Bachelor's program in Modern Languages at the University of Quindío. *Revista de Investigaciones Universidad del Quindío*, *23*(2), 24–35. https://doi.org/10.33975/RIUQ.VOL23N2.402

Waghid, Z. W. (2014). Education for social justice through sustainable development, economic development and equity. *South African Journal of Higher Education*, *28*(4), 1448–1463.

Wang, X. (2005). On the art of teaching. *Journal of Weifang University*, *5*(5), 116–118. https://doi.org/10.3969/j.issn.1671-4288.2005.05.038

Woods, P. J., & Copur-Gencturk, Y. (2024). Examining the role of student-centred versus teacher-centred pedagogical approaches to self-directed learning through teaching. *Teaching and Teacher Education*, *138*, 104415. https://doi.org/10.1016/j.tate.2023.104415

Xiu-Hong, G. (2006). A review of traditional higher education teaching model. *Journal of Shandong Administrative Institute and Shandong Provincial Institute of Economic Management*, *3*, 118–120. https://doi.org/10.3969/j.issn.1008-3154.2006.03.037

Zakarya, S. R. M. (2020). Utilising the deep approach to learning for developing EFL higher-order reading skills among engineering students. *Journal of the Faculty of Education*, *31*(1), 43–86. https://doi.org/10.21608/JFEB.2020.135289

Zhang, T. (2011). *Basic characteristics of traditional classroom teaching and plight it is confronted*. Theory and Practice of Education.

Zhang, X. (2024). Exploring factors affecting students' deep approach to learning in higher education: An empirical study. *Journal of College Student Development*, *65*(2), 213–228. https://doi.org/10.1353/csd.2024.a919352

Chapter 4

Challenges in Engineering Education and the Prospective Role of Problem-Based Learning

4.1 Introduction

The world today is confronted with a vast spectrum of challenges, ranging from routine technical tasks to highly complex and persistent problems often described as "wicked." These wicked problems, a term introduced by Rittel and Webber (1973), are characterised by ambiguity, dynamic variables, and the involvement of multiple stakeholders with conflicting values and priorities. These problems span technical, social, economic, and environmental domains and often resist straightforward solutions. Engineers today are increasingly called upon to engage with such types of challenges which are not purely technical, such as climate change, energy transitions, urbanisation, and equitable infrastructure development. Addressing these issues requires a thorough understanding of sociopolitical dynamics, environmental systems, and economic constraints.

Engineers, trained in analytical thinking and problem-solving, are uniquely equipped to address such multifaceted challenges. However, succeeding in this role requires more than technical competency; it demands the ability to navigate ambiguity, engage diverse perspectives, and act with ethical and social responsibility. Engineering education plays a pivotal role in preparing professionals capable of responding to these evolving demands. It must move beyond the traditional scope of imparting technical knowledge and focus on equipping students with the skills and attributes needed to engage with broader societal issues. This includes fostering interdisciplinary collaboration, encouraging innovation, and cultivating the ability to develop sustainable and inclusive solutions. As Biggs and Tang (2007) emphasise, effective education depends on aligning teaching methods, learning activities, and assessment strategies with intended learning outcomes, ensuring that students acquire competencies relevant to the complexity of real-world problems.

Engineering education, therefore, must integrate sustainability principles, systems thinking, and interdisciplinary learning approaches. This aligns with the vision set forth by the Brundtland Commission (1987), which defined sustainable development as meeting the needs of the present without compromising

DOI: 10.1201/9781003529743-5

the ability of future generations to meet theirs. Engineers are central to this vision, as they are responsible for designing technologies, systems, and infrastructure that uphold environmental integrity, social equity, and economic viability (OECD, 1996).

To prepare engineers for this responsibility, education must embrace sustainable pedagogy—approaches that promote deep learning, critical thinking, and the practical application of knowledge to complex, real-world scenarios. Deep learning, as described by Biggs (1999), occurs when students meaningfully engage with subject matter, connecting new concepts with prior knowledge and applying them creatively to solve problems. Active learning methods such as problem-based learning (PBL) have proven particularly effective in encouraging this level of engagement and understanding (Hmelo-Silver, 2004).

While traditional engineering challenges, such as optimising a process or designing a physical structure, often involve well-defined parameters and solutions, most real-world problems are far more complex. For instance, the development of renewable energy systems requires not only technical design but also a sound understanding of public policy, community needs, and environmental impacts (Barsoum, 2017). Thus, engineering education must be intentionally structured to help students integrate both technical and non-technical knowledge domains.

Wicked problems, in particular, present unique challenges due to their ill-structured nature and their resistance to final solutions. Addressing them requires a deep understanding of the problem context, engagement with a wide range of stakeholders, and the ability to reconcile competing interests (Graves & Garrett, 2014; Schuelke-Leech, 2020). Engineers must not only recognise the interconnectedness of these issues but also consider the broader ethical and societal implications of their decisions. As Krumdieck (2024) argues, engineers must embrace a sense of social responsibility, ensuring that technological interventions are both ethically sound and socially acceptable.

Current models of engineering education, however, often fall short in preparing graduates for such complexity. With a predominant focus on structured problem-solving, students are frequently offered limited opportunities to grapple with ill-defined, interdisciplinary issues. This disconnect hinders their readiness to engage with societal needs. Bridging this gap requires rethinking curricula to include design thinking, systems thinking, and stakeholder-centred approaches that promote flexibility, creativity, and critical reflection (Schuelke-Leech, 2020). Equally important is the ability to leverage data and digital technologies to understand complex systems, identify trends, and assess the potential impacts of proposed solutions. The analysis of large datasets allows for better-informed decision-making and scenario testing, though engineers must remain critical of computational tools and cognisant of their limitations and ethical implications (Graves & Garrett, 2014).

Effectively addressing complex and wicked problems also requires engineers to collaborate beyond disciplinary boundaries. Solutions must be developed in partnership with scientists, policymakers, industry experts, and local communities. Such transdisciplinary collaboration fosters inclusive, robust outcomes that reflect a balance of technological potential and social realities (Harrison, 1986). Frameworks like Transition Engineering and Human-Centric Functional Modelling (HCFM) are valuable in facilitating such collaborations, providing methods for integrating stakeholder perspectives and enabling system-level change (Ahrens et al., 2024; Bhogal, 2022).

Ultimately, to meet the needs of a rapidly changing world, engineers must adopt adaptive, holistic, and reflective mindsets. The limitations of traditional engineering approaches become evident when facing problems shaped by diverse and often competing interests. As Hipólito and Khanduja (2024) suggest, incorporating insights from complex systems theory, ethics, and participatory processes allows for the development of more nuanced and impactful responses to global challenges. By aligning engineering education with these imperatives, future engineers can be prepared not just as problem-solvers but also as thoughtful leaders and change agents committed to creating a more sustainable and equitable world.

However, engineering education and pedagogy face several challenges that must be critically addressed to remain relevant and effective. This chapter examines the context, purpose, and key characteristics of engineering education, highlighting its evolving challenges and the role of PBL. It underscores the importance of sustainable pedagogy and the need to align educational practices with the complex and changing needs of society.

4.2 Context and Purpose of Engineering Education

The context and purpose of engineering education are multifaceted, shaped by cultural, societal, and practical dimensions that influence how knowledge and skills are developed and applied. At its core, engineering education aims to produce graduates who are not only technically competent but also capable of addressing complex, real-world challenges. This requires a shift from traditional content delivery to more contextual and experiential learning that reflects the environments in which engineering solutions are implemented (Crawley et al., 2008).

The Conceive–Design–Implement–Operate (CDIO) framework exemplifies this approach by embedding the lifecycle of engineering projects within the educational process, providing students with a structured yet practical understanding of engineering in context (Crawley et al., 2008). Educational approaches vary across countries—such as China, Sweden, and the United States—reflecting local needs, industrial contexts, and societal priorities. Despite these differences, there is a shared recognition of the need for curricular reform to improve relevance, practical competence, and societal impact (Jørgensen, 2007).

Engineering education thus extends beyond technical training to focus on developing engineers who are lifelong learners, effective communicators, critical problem-solvers, and socially adaptable professionals (Yin et al., 2012). This is supported by student-centred learning environments, continuous curricular improvement, and strong alignment with industry and societal needs. As Rolstadås and Paci (2013) argue, engineering education is central to national economic and social progress, particularly in advanced economies where innovation is key to competitiveness.

A crucial aspect of this education is contextualisation—embedding engineering problems within broader societal, environmental, and ethical dimensions. Doing so enhances students' ability to apply technical knowledge in meaningful ways, deepening their understanding of the real-world consequences of engineering decisions (Moriarty, 1993). Professional bodies and international organisations such as Accreditation Board for Engineering and Technology (ABET), the International Engineering Alliance (IEA), and Engineering Council of South Africa (ECSA) emphasise the importance of producing engineers who are competent, capable of making informed, ethical judgements in diverse contexts (Kleine et al., 2023). Various modes of contextualisation—such as incorporating sociotechnical thinking, using professional case studies, and deploying "context tools"—help educators bridge the gap between technical training and professional practice (Kleine et al., 2023).

However, this emphasis on context must be balanced with a strong grounding in core scientific and technical disciplines. Some scholars caution that overly broad curricula may dilute essential competencies. To address this, integrating business, management, and leadership training can enhance engineers' ability to function effectively in interdisciplinary and professional environments (Hills & Tedford, 2003; Jørgensen, 2007).

The global context further reinforces the evolving role of engineering education. Different regions face distinct sustainability challenges, ranging from poverty and food insecurity in developing countries to infrastructure deficits and environmental degradation in more industrialised nations. These issues correspond with the United Nations' Sustainable Development Goals (SDGs), which offer a global framework for addressing interlinked social, economic, and environmental problems (United Nations, 2015).

Engineering solutions are central to meeting these goals. From innovations in agriculture and renewable energy to sustainable urban infrastructure and water management, engineers play a vital role in addressing sustainability challenges (Gupta & Gupta, 2020; Mills et al., 2021). These problems demand not only technical expertise but also interdisciplinary collaboration and an understanding of social justice, ethics, and long-term impact (Biswas, 2022).

To prepare graduates for this reality, engineering education must incorporate sustainability principles, encourage interdisciplinary knowledge, and foster innovation through experiential learning, research, and global engagement (Crawley et al., 2014; Mulder, 2017). Curricula should cultivate not only technical expertise but also critical thinking, ethical awareness, and a global perspective.

So, the purpose of engineering education is to produce competent professionals who can tackle complex problems, both engineering and non-engineering, socio-economic and environmental problems faced across local and global contexts. By equipping engineers to develop sustainable, equitable, and forward-thinking solutions, engineering education contributes to building a more habitable, just, and prosperous society (Ashford & Hall, 2011).

4.3 Understanding the Demands of the Contemporary and Future Society

Understanding the demands of contemporary and future society is essential for aligning engineering education with the realities of a rapidly evolving world. Modern societies are shaped by a range of global and local challenges—climate change, social inequality, food insecurity, public health crises, technological disruption, economic and financial instability—all of which require innovative, sustainable, and inclusive solutions (Godfray et al., 2010; Piketty, 2014; World Economic Forum, 2023). These demands call for a fundamental rethinking of how engineers are educated and what roles they are expected to play in both local communities and the broader global context (Ashford & Hall, 2011; United Nations, 2015).

One of the most pressing demands is the shift towards sustainability. Societies increasingly expect engineers to develop solutions that do not compromise environmental integrity, economic viability, or social equity. This is reflected in the United Nations' SDGs, which highlight global priorities such as clean energy, sustainable infrastructure, zero hunger, good health and well-being, and climate action (United Nations, 2015). Engineers must be capable of integrating these sustainability principles into their practice, addressing interconnected issues such as the impact of fossil fuels on climate change, the design of equitable food systems, or the management of public health infrastructure in times of crisis.

Closely linked to these sustainability concerns is the growing awareness of social inequality (Piketty, 2014). Many communities around the world, particularly in developing and underserved regions, face stark disparities in access to clean water, nutritious food, healthcare, energy, and employment. Addressing these inequalities requires engineers to consider not only technical feasibility but also the sociopolitical and economic contexts of their work. As Biswas (2022) argues, engineering education must instil a deep understanding of social equity and justice, enabling future engineers to create inclusive technologies and systems that serve all segments of society.

The industrial and employment landscape is also undergoing rapid transformation, driven by globalisation, digitalisation, and the rise of automation. Traditional engineering roles are being redefined by advances in artificial intelligence, robotics, and smart manufacturing systems, requiring engineers to possess adaptable skillsets and interdisciplinary knowledge (Brynjolfsson & McAfee, 2014; Mills et al., 2021; Mulder, 2017). At the same time, the nature of work itself is shifting, with emerging economies creating new markets for technological innovation, while mature economies focus on digital infrastructure, renewable energy, and advanced healthcare systems. In this context, engineering graduates must be prepared to operate in fluid employment environments, demonstrating agility, creativity, and a lifelong commitment to learning.

In tandem with technological advancements, economic and financial considerations shape the feasibility and sustainability of engineering solutions. Engineers must be able to work within budget constraints, evaluate cost–benefit trade-offs, and engage with stakeholders, including governments, businesses, and communities, to ensure the financial viability of projects (Hills & Tedford, 2003; Schwab, 2016; Stiglitz, 2010). In an increasingly resource-constrained world, this also means prioritising long-term resilience and risk management, especially in areas such as infrastructure development, climate adaptation, and public health preparedness.

Health-related challenges, including the global burden of disease, ageing populations, and pandemic preparedness, also demand engineering solutions. Biomedical engineering, health systems design, and digital health technologies are becoming central to addressing these issues. Engineers need to work closely with medical professionals and policymakers to create solutions that are both technologically sound and socially responsive (Gupta & Gupta, 2020; Schwab, 2016).

Incorporating all these dimensions—sustainability, social justice, health, climate, technological change, and economic resilience—engineering education must evolve to meet the multifaceted expectations of society. This involves reconfiguring curricula to emphasise interdisciplinary collaboration, systems thinking, ethical awareness, and real-world engagement through project-based and experiential learning models (Crawley et al., 2014; Kleine et al., 2023).

Ultimately, understanding the demands of contemporary and future society requires a holistic approach to engineering education. Future engineers must not only solve technical problems but also act as change agents who can navigate complex socio-technical systems, drive sustainable economic growth, and contribute to a more just and habitable world.

4.4 Characteristics of Engineering Problems

Engineering problems are inherently complex, diverse, and rooted in real-world contexts where practical, often-innovative solutions are essential. Unlike scientific problems, which seek to uncover theoretical truths, engineering problems are oriented towards utility and application, solving tangible issues that arise in nature, the environment, society, and industry (Aslaksen, 2013).

These problems typically require applying scientific and mathematical knowledge within the constraints of feasibility, resources, ethics, and societal expectations.

4.4.1 Purpose and Utility

The primary aim of engineering is to create solutions that meet societal needs through the development of products, systems, and services. Scientific knowledge is therefore valued in engineering not for its theoretical depth, but for its practical applicability. For instance, while the concept of an expanding universe is scientifically profound, it holds little direct relevance to engineering design. In contrast, principles such as the first law of thermodynamics are fundamental to numerous engineering applications (Aslaksen, 2013).

4.4.2 Problem Types and Structures

Engineering problems can vary widely in structure, from well-defined to ill-posed. Schuelke-Leech (2020) classifies them into four main categories:

- **Routine problems**: Well-structured problems with known solutions and established methodologies.
- **Originative problems**: Well-structured but lacking standard solutions, requiring original thinking.
- **Process-oriented problems**: Ill-structured problems where the path to a solution is known, though the problem itself may be ambiguous.
- **Wicked problems**: Ill-structured, lacking clear definitions or solutions, and often involving conflicting stakeholder goals and complex trade-offs.

However, these problems may be categorised into the three types of problems, such as well-defined, broadly defined and ill-defined problems.

Additionally, engineering problems may be categorised in terms of their specificity (Jonassen, 2014):

- **General problems**: Problems for which established solutions exist, e.g., designing a standard bicycle frame.
- **Apparent unique problems**: Initially perceived as unique but ultimately solvable using existing approaches, for example, creating a lightweight bicycle frame through known materials and design principles.
- **Specific problems**: Problems without existing solutions or methodologies, demanding novel or interdisciplinary approaches.
- **Complex or system-level problems**: Those involving multiple subsystems, stakeholders, or objectives, often addressed through systems engineering and iterative design.

4.4.3 *Complexity and Ambiguity*

A defining feature of many engineering problems is their complexity—multiple variables, interdependent systems, and uncertainty, and ambiguity, where information may be incomplete or contradictory. Practising engineers are generally accustomed to navigating ambiguity as a natural part of the design and problem-solving process, whereas students often struggle with it, having been primarily exposed to well-structured academic exercises (Douglas et al., 2022).

4.4.4 *Practical Constraints and Trade-Offs*

Engineering problems are rarely solved in ideal conditions. Engineers must address real-world constraints such as cost, time, regulatory standards, safety requirements, and material limitations. Design work often involves trade-offs—balancing performance, sustainability, and affordability—while aiming to maximise value for users or clients (Tranquillo, Goldberg, & Allen, 2023). Recognising and managing these constraints early in the design process is essential to avoid costly revisions and inefficiencies.

4.4.5 *Design, Innovation, and Systems Thinking*

Many engineering problems fall under the category of design challenges, which are inherently ill-structured and complex. These problems often lack a single correct solution or linear approach. Effective problem-solving in design requires not only technical proficiency but also creativity, contextual understanding, and the ability to synthesise systems, procedural, and strategic knowledge (Wilson et al., 2024; Zhao, 2015). Engineers must also be capable of identifying opportunities for innovation, especially in high-impact areas such as healthcare, energy systems, smart infrastructure, and climate adaptation.

4.4.6 *Optimisation and Advanced Techniques*

Modern engineering increasingly relies on optimisation and simulation tools to manage complex, nonlinear problems, such as improving automotive crashworthiness. These problems often diverge from textbook examples, requiring advanced computational methods like Exploratory Landscape Analysis (ELA) to characterise and solve them effectively (Long et al., 2022).

4.4.7 *Real-World Relevance and Transition from Education to Practice*

Although students often encounter structured problems in educational settings, real-world engineering demands the ability to handle ill-defined, open-ended challenges. The transition from academia to practice can thus be jarring, underscoring the importance of aligning engineering education with professional realities. Jonassen (2014) argues that curricula should better reflect the complexity, ambiguity, and contextual factors inherent in professional engineering work.

Engineering problems are defined not just by technical complexity but also by their embeddedness in social, environmental, and economic contexts. They are characterised by ambiguity, the need for practical and innovative solutions, and a constant balancing of constraints and trade-offs. Understanding and addressing these multifaceted challenges requires engineers to be technically skilled, adaptable, and contextually aware. Engineering education must, therefore, evolve to develop these capabilities, ensuring that graduates are prepared to solve the pressing problems of contemporary and future society.

4.5 Non-Engineering Problems

Non-engineering problems encompass a broad range of complex socio-economic and environmental challenges that directly impact human well-being and global sustainability. These issues often fall outside the technical boundaries of traditional engineering, yet they are intricately connected to the work engineers do. Many of the global concerns outlined in the UN-SDGs—such as poverty, inequality, health, food security, and climate change—can be categorised as non-engineering problems due to their complex, systemic, and value-laden nature.

These problems are typically ill-posed or even wicked, meaning they lack clear definitions, have no definitive solutions, and often involve conflicting stakeholder interests. The behaviour of such problems is difficult to predict, and any intervention may lead to unintended consequences.

While engineers may not be able to solve non-engineering problems directly, their work is instrumental in developing tools, technologies, and systems that can contribute to broader solutions. For instance, poverty, a deeply rooted socio-economic problem, may arise from a lack of employment opportunities or food insecurity. While engineers cannot eliminate poverty on their own, they can contribute by designing low-cost food production systems, affordable infrastructure, or technological platforms that enhance economic access and opportunity.

4.5.1 Key Characteristics of Non-Engineering Problems

The key characteristics of non-engineering problems are discussed below.

4.5.1.1 Pervasiveness and Systemic Nature

Non-engineering problems are pervasive and deeply embedded in societal structures. They cut across social, economic, and environmental domains and often persist over long periods. Their resolution requires more than technical fixes—it demands collective action, policy reform, and behavioural change.

4.5.1.2 Social Characteristics

- **Collective Action Challenges**: Many environmental issues are externalities that require collective solutions. Individuals may not directly experience the consequences of their actions, making it difficult to mobilise community-wide change without institutional support (DeSombre, 2018).
- **Community Engagement**: Social dynamics play a vital role in addressing non-engineering issues. Projects such as local renewable energy initiatives succeed when there is strong community involvement and buy-in, which often determines the scalability and longevity of such solutions (Courrier, 1980).
- **Behavioural and Organisational Factors**: In disciplines like software engineering, non-technical challenges—such as leadership, communication, and team dynamics—significantly influence project outcomes (Kellner et al., 1991).

4.5.1.3 Economic Characteristics

- **Resource Dependency**: Societies' reliance on non-renewable resources creates long-term vulnerabilities. The criticality and availability of such resources influence economic models and sustainable growth (Sinding-Larsen & Wellmer, 2012).
- **Affordability and Adoption**: For example, in renewable energy transitions, success depends not only on the technology but also on the cost of investment, financial incentives, and consumer willingness to adopt solutions like residential photovoltaics.
- **Legal and Market Structures**: Economic issues often intersect with legal and regulatory frameworks. In network infrastructure design, for instance, market dynamics and policy constraints are as influential as technical feasibility (Enslow, 1973).

4.5.1.4 Environmental Characteristics

- **Externalities and Sustainability**: Environmental degradation often results from unintended externalities of economic activity. Effective solutions require integrating sustainability into social and economic planning (DeSombre, 2018; Mata-Lima, 2024).
- **Ecosystem Impact**: Challenges such as pollution and resource depletion threaten long-term ecosystem viability. These problems demand interdisciplinary collaboration involving environmental science, policy, and engineering (Ni, 2022).

- **Climate Change and Innovation**: Global warming and environmental stressors demand technological innovation, such as carbon capture, sustainable energy systems, and adaptive resource management strategies (Ni, 2022).

4.6 Demands for Engineering Solutions for Non-Engineering Complex Problems

As mentioned previously, the modern world faces an array of complex socio-economic, environmental, and ecological challenges that are central to sustainable development. While non-engineering problems are shaped by social, political, economic, and environmental conditions, engineering and technological interventions are increasingly essential for achieving practical, scalable, and sustainable solutions. For example, supplying clean water to communities not only involves socio-political decisions—such as who builds and funds the infrastructure, who benefits from it, and what the environmental consequences might be—but also demands engineering expertise in feasibility analysis, system design, construction, and ongoing maintenance.

Even with sufficient funding, political will, and community participation, solutions, like dam construction or water treatment systems, cannot materialise without the contributions of engineers. The design, operation, and management of infrastructure such as reservoirs, waterworks, and distribution networks depend on applied engineering principles, technology, and innovation.

4.6.1 The Evolving Role of Engineering in Addressing Non-Engineering Problems

The UN-SDGs highlight a range of complex, non-engineering challenges—such as poverty, hunger, climate change, and inequality—that require support beyond traditional policy or social interventions. Although these problems are not inherently technical, engineering plays a critical enabling role in addressing them. By designing practical tools, systems, and infrastructure, engineers help translate high-level policy goals into effective on-the-ground solutions.

As DeSombre (2018) and Mata-Lima (2024) argue, many environmental problems arise as externalities of other human activities and cannot be resolved without integrated, systemic change. In this context, engineering serves as a facilitator, supporting broader socio-economic and environmental strategies through innovation, infrastructure, and technical expertise.

Key contributions of engineering to non-engineering problems include:

- **Technological interventions**: Development of water purification systems, renewable energy technologies, agricultural innovations, and sanitation solutions for underserved communities.

- **Affordable and scalable infrastructure**: Design and implementation of sustainable housing, resilient transportation networks, and decentralised energy systems.
- **Data and information systems**: Creation of tools to monitor and analyse socio-economic and environmental trends, supporting evidence-based decision-making and governance.
- **Sustainable design practices**: Integrating environmentally responsible engineering approaches that minimise ecological impact while supporting human well-being.

These contributions do not solve the root causes of non-engineering problems on their own. However, they are essential in enabling progress, supporting systemic reforms, and empowering stakeholders—from governments to grassroots communities—to respond more effectively to today's global challenges.

Furthermore, engineering increasingly plays a vital role in addressing non-engineering problems by integrating non-technical factors such as economic conditions, political structures, behavioural dynamics, and legal frameworks. For instance, energy efficiency initiatives often underperform when implemented without considering broader behavioural and urban contexts (Mahdavi et al., 2024), while success in software engineering depends heavily on organisational and managerial dynamics rather than purely technical execution (Kellner et al., 1991). Solving such complex challenges demands transdisciplinary collaboration, drawing on insights from life sciences, social sciences, computational tools, and engineering. This interdisciplinary convergence is essential for sustainable innovation, as seen in medical engineering, which combines neuroscience and biomedical technology to tackle intricate health issues (Cascini, 2016; Wu & Hallett, 2007).

To meet these demands, engineering education must evolve, fostering not just technical skills but also systems thinking, ethics, and social awareness. Educational strategies like capstone projects, innovation labs, and Problem-Based Learning aim to prepare students for the uncertainty and complexity of real-world problems, though full curricular integration remains a challenge (Akinci-Ceylan et al., 2018; Kamaruzaman et al., 2017; Wang et al., 2018).

Furthermore, modern engineering challenges increasingly involve ethical and behavioural dimensions—such as digital addiction and environmental degradation—that require a socio-technical perspective sensitive to human values and autonomy (Varshney, 2014). While the expanding role of engineers offers new opportunities for social impact and systemic innovation, it also presents limitations. Integrating non-technical considerations into design processes is inherently complex and demands reform in education, governance, and professional norms. There is also a risk of overreliance on technological solutions, which may neglect the deeper need for policy reform, cultural change, and community engagement. Therefore, engineering should be seen as an enabler within a broader, systemic response, not as a substitute for it.

4.7 Understanding Engineering Education and Students' Learning

Understanding engineering education and how students learn requires examining diverse pedagogical strategies, learning models, and the specific challenges faced by engineering students. Due to its inherent complexity, engineering education demands a multifaceted approach that addresses both cognitive and practical aspects of learning. This includes attention to student learning processes, the effectiveness of active learning, and the integration of engineering principles from early education through higher education.

The global labour market is undergoing rapid transformation, driven by technological advancements and increasing complexity in real-world problems. Today's challenges—ranging from climate change to socio-economic inequality—are multidisciplinary in nature and require not just technical solutions but also innovation, stakeholder engagement, and systemic thinking. As such, engineering knowledge alone is no longer sufficient. Engineers must be equipped to work across disciplines and collaborate with professionals from diverse fields to develop sustainable, inclusive, and impactful solutions.

To keep pace with these evolving demands, engineering education must broaden its scope. In addition to foundational knowledge in mathematics, science, and core engineering disciplines, curricula must include communication skills, teamwork, ethical reasoning, socio-economic awareness, entrepreneurship, intellectual property rights, environmental stewardship, and sustainable development. These areas are essential for preparing engineers to navigate the complex, interconnected challenges of the 21st century.

Recognising this need, organisations such as the IEA and the Engineering Council of South Africa (ECSA) have identified 11 critical competency areas, organised into five overarching groups (Table 4.1). These competencies reflect the skills and attributes engineering graduates must develop to remain effective and adaptable in a rapidly changing professional landscape.

In other words, in addition to a strong foundation in science, mathematics, and engineering principles, students must be equipped with:

- **Leadership**: Understanding team dynamics, resource constraints, and self-awareness.
- **Critical thinking**: Processing and evaluating information to make informed, unbiased decisions.
- **Effective speaking and writing**: Communicating ideas clearly across various media and audiences.
- **Creativity and design thinking**: Generating innovative solutions from diverse perspectives.
- **Decision-making**: Weighing alternatives and synthesising input to act decisively.
- **Problem-solving**: Identifying problems, analysing solutions, and executing efficiently.

Table 4.1 Outcomes Groups and Competencies

Group A: Engineering Problem-Solving

- Outcome 1 – Define, investigate, and analyse complex engineering problems
- Outcome 2 – Design or develop solutions to complex engineering problems
- Outcome 3 – Comprehend and apply advanced knowledge of the widely applied principles underpinning good engineering practice, specialist knowledge and knowledge specific to the jurisdiction and local conditions

Group B: Managing Engineering Activities

- Outcome 4 – Manage part or all of one or more complex engineering activities
- Outcome 5 – Communicate clearly with others in the course of a professional engineer's engineering activities

Group C: Risk and Impact Mitigation

- Outcome 6 – Recognise and address the reasonably foreseeable social, cultural, and environmental effects of the complex engineering activities
- Outcome 7 – Meet all legal and regulatory requirements and protect the health and safety of persons in the course of the complex engineering activities

Group D: Act Ethically, Exercising Judgement and Taking Responsibility

- Outcome 8 – Conduct engineering activities ethically
- Outcome 9 – Exercise sound judgement in the course of complex engineering activities
- Outcome 10 – Be responsible for making decisions on part or all of complex engineering activities

Group E: Professional Development

- Outcome 11 – Undertake sufficient professional development activities to maintain and extend the competence

---

- **Personal branding**: Building a unique professional identity and presence.
- **Excellence and productivity**: Managing time, prioritising tasks, and maintaining high performance.
- **Workplace navigation**: Understanding organisational culture, resolving conflicts, and fostering collaboration.

Furthermore, as engineering challenges grow more complex and interdisciplinary, how engineers are educated must evolve accordingly, prompting a shift towards more dynamic, student-centred pedagogies that align with real-world demands. To be effective, engineering education must critically consider and integrate several key pedagogical innovations and learning approaches as illustrated in Figure 4.1.

4.7.1 Pedagogical Shifts in Engineering Education

To meet the demands of the real world and competency needs, engineering education must embrace pedagogical innovation, drawing on both established and emerging learning theories. Traditional models such as those by Felder

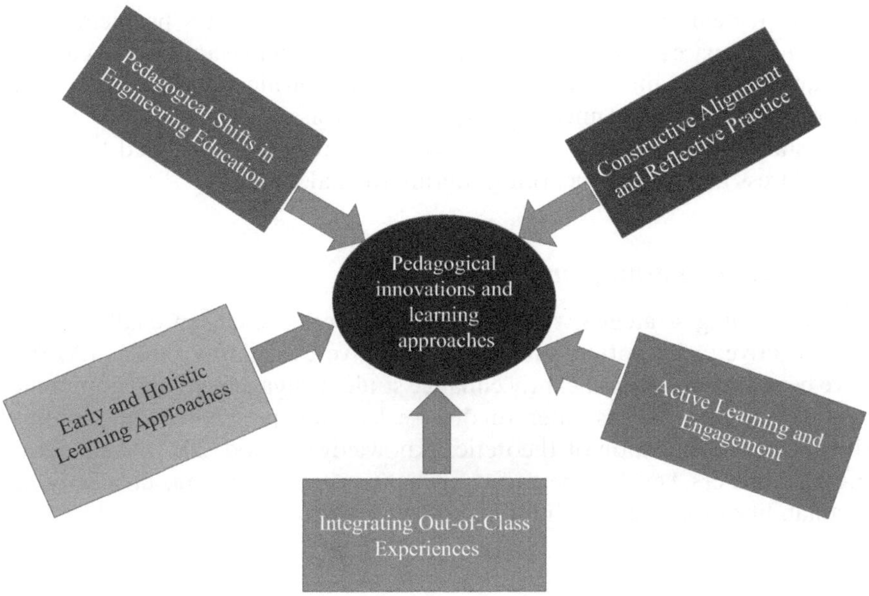

Figure 4.1 Key Pedagogical Innovations and Learning Approaches.

and Silverman, which classify students based on learning styles, offer limited flexibility and fail to reflect the dynamic, context-dependent nature of learning (Harvey, 2004).

A more effective alternative is the Interactive Learning Model, which emphasises the interplay between students and their learning environments. This model recognises that learning is shaped by social, institutional, and cognitive factors, making it better suited to the complexities of engineering education (Harvey, 2004).

4.7.2 Constructive Alignment and Reflective Practice

Constructive alignment is a critical component of effective pedagogy. It ensures that teaching methods and assessments are purposefully aligned with intended learning outcomes, enabling students to apply their knowledge in practical scenarios (Cropley & Sitnikova, 2005). Complementing this, reflective practice—both at individual and institutional levels—allows educators to continuously adapt and refine their strategies to meet the evolving needs of students and disciplinary expectations (Cropley & Sitnikova, 2005).

4.7.3 Early and Holistic Learning Approaches

Integrating engineering concepts at the K–12 level is essential for sparking early interest in STEM fields and building a strong foundation for technical literacy (Katehi et al., n.d.). This early exposure cultivates problem-solving skills and prepares students for more advanced academic and career paths.

Equally important is the adoption of holistic learning approaches at all levels of education. By connecting interrelated ideas, students develop a deeper understanding of engineering principles and are better equipped to identify and address gaps in their learning (Bernhard et al., 2010).

4.7.4 Active Learning and Engagement

Active learning strategies, such as hands-on projects, design challenges, and collaborative tasks—often referred to as Active Learning Products (ALPs)—have been shown to significantly enhance student engagement and comprehension. These approaches cater to diverse learning preferences and promote the practical application of theoretical knowledge (Wood, 2023). Active learning also fosters key competencies such as critical thinking, creativity, and adaptability—essential qualities for today's engineers.

4.7.5 Integrating Out-of-Class Experiences

Learning extends beyond the classroom. Co-curricular and extra-curricular experiences—including internships, design competitions, industry projects, and service-learning—play a vital role in developing essential skills like teamwork, leadership, and problem-solving (Strauss & Terenzini, 2005). These real-world engagements reinforce academic learning and contribute to broader professional readiness.

Thus, a forward-thinking engineering education must combine technical rigour with soft skills, ethical awareness, sustainability, and interdisciplinary collaboration. As the scope of engineering expands, so too must the educational systems that prepare future engineers. Tailored pedagogies, early STEM exposure, active learning, and competency-based frameworks are essential for preparing graduates to meet the demands of an increasingly complex and interconnected world.

4.8 Challenges of Engineering Education

Engineering education today faces a range of complex and evolving challenges, driven by rapid technological advancements, shifting societal expectations, and a dynamic global job market (Vieira et al., 2024). These challenges span curriculum design, student engagement, ethics, competency development, and structural reform. Addressing them requires a comprehensive understanding of educational practices, industry demands, and diverse learner needs. The key challenges identified in recent literature (Figure 4.2), along with insights into possible responses, are discussed below.

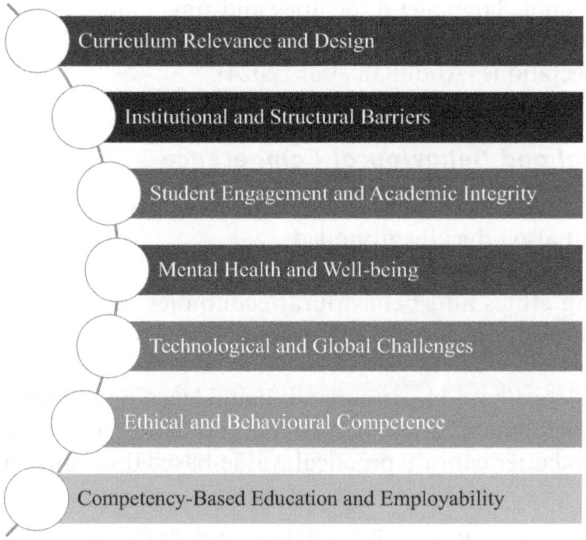

Figure 4.2 Challenges of Engineering Education.

4.8.1 Curriculum Relevance and Design

A major challenge in engineering education is ensuring curricula remain current and aligned with real-world technological advancements and industry expectations. As the pace of innovation accelerates, static or outdated curricula risk disengaging students and producing graduates with misaligned skill sets.

- The MEC3203 Materials Technology course at the University of Southern Queensland, for instance, was redesigned to integrate real-world engineering practices. This led to increased student engagement and academic integrity, demonstrating the value of curricular relevance (Islam, 2024).
- Modern engineering programmes must also strike a balance between technical knowledge and transferable skills, such as creativity, leadership, and ethical reasoning (Arciszewski & Rebolj, 2007; Garcia & Pinela, 2016).
- Embedding sustainability and ethics into the curriculum is increasingly vital, given the socio-environmental implications of engineering decisions (Garcia & Pinela, 2016; Veza et al., 2023).

4.8.2 Student Engagement and Academic Integrity

Maintaining high levels of student engagement and upholding academic integrity are ongoing challenges:

- Integrating real-world content and interactive learning methods has been shown to improve both engagement and ethical behaviour, as evidenced by curriculum updates in the MEC3203 course (Islam, 2024).

- Barriers such as language difficulties and time management issues further impede engagement, with many students struggling to keep pace with academic expectations (Abdullah et al., 2024).

4.8.3 Ethical and Behavioural Competence

In an age of accelerated innovation, engineers must not only be technically competent but also ethically grounded.

- Embedding ethics and behavioural economics into the curriculum supports improved ethical decision-making and fosters inclusive engineering practices.
- While challenge-based learning and digital learning environments promote ethical development, they necessitate rethinking traditional assessment methods to better capture practical, value-based decision-making.

4.8.4 Competency-Based Education and Employability

To enhance employability, engineering education must shift from content-heavy instruction to competency-based teaching focused on both professional and interpersonal skills.

- This shift faces resistance due to deeply entrenched educational structures and a reluctance to adopt unfamiliar pedagogies (Valero, 2022).
- Competency-based education is essential in preparing students for Industry 4.0, where adaptability, digital literacy, and innovation are crucial (Vega et al., 2023)

4.8.5 Technological and Global Challenges

Engineering curricula must be designed to adapt to the realities of digitalisation and globalisation:

- Flexible, interdisciplinary curricula that reflect technological evolution and international standards are key to preparing globally competent graduates (Gottlieb et al., 2019; Veza et al., 2023).
- Addressing geographic disparities in educational access—particularly in low-resource settings—is crucial for fostering diversity and innovation in engineering (Veza et al., 2023).

4.8.6 Institutional and Structural Barriers

Implementing transformative reforms often collides with institutional inertia and structural limitations:

- In regions like sub-Saharan Africa, barriers include curriculum irrelevance, lack of resources, and limited engineering capacity, all of which hinder responsiveness to local and global challenges (Mashiyane et al., 2023).
- Even in developed contexts, reforms—such as those initiated in China or the European Higher Education Area—highlight how change at the micro level requires a cultural shift and institutional support (Valero, 2022; Zhuang, 2023).

4.8.7 Mental Health and Well-being

The mental health of engineering students is an emerging concern. High academic pressure, long hours, and a culture of perfectionism often lead to stress and burnout.

- Recent studies show increasing reports of stress-related physical and psychological symptoms among students, indicating the need for enhanced mental health support systems in engineering programmes (Abdullah et al., 2024).

The challenges facing engineering education are multifaceted and deeply interconnected. From curriculum reform and ethics integration to competency development and student well-being, the path forward requires not only innovation but also institutional willingness to embrace change. Importantly, these challenges are context-specific; effective solutions must be tailored to local realities, resources, and stakeholder expectations. A collaborative effort involving academia, industry, and policymakers is essential to reshape engineering education and ensure that future graduates are equipped to navigate the complexities of a rapidly evolving world.

4.9 PBL as an Integral Part of Engineering Pedagogy

Given the evolving nature of the modern workplace, the increasing complexity of engineering problems, and the growing need to manage diverse stakeholders, engineers must possess more than just technical expertise. In addition to core competencies, they require a broad range of skills such as critical thinking, adaptability, and effective communication. To meet these demands, engineering education must become more flexible, interdisciplinary, and responsive to real-world challenges.

PBL has emerged as a transformative pedagogical approach in this context, offering a dynamic alternative to traditional lecture-based instruction. By engaging students in solving real-world engineering problems, PBL not only enhances conceptual understanding but also fosters essential skills such as creativity, teamwork, problem-solving, and self-directed learning. It equips students with the competencies needed to navigate the complexities of contemporary engineering practice.

Research has shown that PBL significantly improves knowledge retention and conceptual understanding by actively involving students in their learning. Students engaged in PBL consistently report higher satisfaction and deeper engagement, both of which are vital for long-term academic success. For example, electrical engineering students exposed to PBL demonstrated superior conceptual mastery compared to those taught through conventional methods (Riazifar & Khajehsaeid, 2024; Tanna et al., 2022; Yadav et al., 2010).

Beyond academic content, PBL supports the development of critical professional and cognitive skills. First-year engineering students participating in PBL have reported notable gains in critical thinking, leadership, creativity, and teamwork (Naik et al., 2024). It also promotes self-directed learning, empowering students to take ownership of their educational journey—a key competency for lifelong learning in an ever-changing field (Krishnan & Ruhizan, 2009).

A key strength of PBL lies in its real-world relevance. By simulating authentic engineering challenges, PBL bridges the gap between theoretical knowledge and practical application. This is particularly evident in disciplines such as civil and highway engineering, where students work on projects that closely mirror professional scenarios (Noor et al., 2019). Furthermore, integrating real-world contexts has been shown to increase student motivation and attendance, with PBL classes often achieving higher participation rates than traditional formats (Naik et al., 2024; Tanna et al., 2022).

PBL also yields benefits at the faculty and institutional levels. For educators, it encourages interdisciplinary collaboration and contributes to professional development, enriching their teaching practice (Naik et al., 2024). Institutions implementing PBL report improved student outcomes and satisfaction, which supports broader efforts to innovate and reform engineering curricula (Kolmos & Graaff, 2014).

However, the adoption of PBL is not without challenges. Successful implementation requires significant investment in faculty training and curriculum redesign. Moreover, some studies suggest that traditional methods may still be more effective for certain theoretical components and for developing individual learning skills (Marcinauskas et al., 2024). As such, a blended approach that combines the strengths of both PBL and traditional instruction may offer the most comprehensive educational experience, balancing conceptual depth with practical, skill-based learning.

4.10 Conclusion

Engineering education stands at a critical juncture, faced with the pressing need to evolve in response to the rapidly changing technological, socio-economic, and environmental landscape. As engineering problems grow more complex—often classified as "wicked" due to their ambiguity, systemic nature, and stakeholder entanglement—the role of engineers must expand beyond technical problem-solving to encompass ethical reasoning, interdisciplinary collaboration, and societal responsibility.

This chapter has explored the shifting context and purpose of engineering education, highlighting the necessity of integrating sustainability, ethics, and stakeholder engagement into engineering curricula. It has also examined the nature of engineering and non-engineering problems, emphasising the importance of equipping future engineers with the ability to address both through innovation, systems thinking, and contextual understanding.

Given these challenges, pedagogical transformation is essential. Active, student-centred approaches—particularly PBL—offer promising pathways for bridging the gap between theory and practice. PBL fosters deeper conceptual understanding, enhances critical and collaborative skills, and simulates the complexities of real-world engineering contexts. However, successful implementation requires institutional commitment, faculty training, and a willingness to rethink traditional structures.

To remain relevant and impactful, engineering education must become more adaptive, inclusive, and future-oriented. This involves not only updating curricula and teaching methods but also fostering a culture of continuous improvement and reflective practice. By doing so, engineering programmes can prepare graduates who are not only technically proficient but also socially conscious, ethically grounded, and capable of leading transformative change in a rapidly evolving world.

References

Abdullah, S. N., Kwad, A. M., Al-Obaidi, M. K., Albaker, B. M., Al Ghamdi, M. A., & Alsaedi, M. A. (2024). Challenges in engineering education: Addressing language barriers, time management issues, and mental health concerns. In *2024 International Conference on Vision, Vlsi, and Signal Processing (ICVVS)*, 25–30. https://doi.org/10.1109/icvee63912.2024.10823658

Ahrens, F., Cherubini, P., Boulton, J., Bartholomew, M., & Krumdieck, S. (2024). The first rule of transition engineering is to define the wicked problem. *Advances in Transdisciplinary Engineering*. https://doi.org/10.3233/atde240879

Akinci-Ceylan, S., Cetin, K. S., Fleming, R., Ahn, B., Surovek, A. E., Cetin, B., & Taylor, P. (2018). *Bridging the gap between academia and industry in approaches for solving ill-structured problems: Problem formulation and protocol development*. American Society for Engineering Education.

Arciszewski, T., & Rebolj, D. (2007). Challenges of civil engineering education. [Paper presentation]. In *International Conference on Engineering Education – ICEE 2007*, Coimbra, Portugal. https://itc.scix.net/data/works/att/w78-2007-121-999-Arciszewski.pdf

Ashford, N. A., & Hall, R. P. (2011). *Technology, globalisation, and sustainable development: Transforming the industrial state*. Yale University Press.

Aslaksen, E. W. (2013). Characteristics of engineering. In *The changing context of engineering* (pp. 79–106). Springer Berlin Heidelberg. https://doi.org/10.1007/978-3-642-32169-6_7

Barsoum, G. (2017). Engineering education and the development of professional competencies. *Engineering Education, 12*(1), 15–25.

Bernhard, J., Carstensen, A. K., & Holmberg, M. (2010). *Investigating engineering students' learning—learning as the 'learning of a complex concept'*. Linköping University, Department of Social and Welfare Studies. http://www.diva-portal.org/smash/record.jsf?pid=diva2:397178

Bhogal, R. (2022). *Increasing the societal impact of science, technology, engineering, and math with general collective intelligence*. PsyArXiv. https://doi.org/10.31730/osf.io/dwg73

Biggs, J. (1999). *Teaching for quality learning at university*. Society for Research into Higher Education & Open University Press.

Biggs, J., & Tang, C. (2007). *Teaching for quality learning at university* (3rd ed.). Open University Press.

Biswas, A. K. (2022). Water security: Challenges and opportunities. *Water International, 47*(1), 12–25.

Brynjolfsson, E., & McAfee, A. (2014). *The second machine age: Work, progress, and prosperity in a time of brilliant technologies*. W. W. Norton & Company.

Cascini, G. (2016). Engineering grand challenges demand for trans-disciplinary design science. *Journal of Integrated Design & Process Science, 19*(4), 1–2. https://doi.org/10.3233/JID-2015-0020

Courrier, K. (1980). *Non-technical characteristics of successful community renewable energy programs*. Office of Scientific and Technical Information (OSTI). http://www.osti.gov/scitech/biblio/6472699-non-technical-characteristics-successful-community-renewable-energy-program

Crawley, E. F., Cha, J., Malmqvist, J., & Brodeur, D. R. (2008). *The context of engineering education*. Chalmers University of Technology. https://research.chalmers.se/en/publication/71994

Crawley, E. F., Malmqvist, J., Östlund, S., & Brodeur, D. R. (2014). *Rethinking engineering education: The CDIO approach*. Springer.

Cropley, D. H., & Sitnikova, E. (2005). *Teaching and learning in engineering education: Constructive alignment*. [Paper presentation]. In *16th Annual Conference for the Australasian Association for Engineering Education*, Sydney, Australia.

DeSombre, E. R. (2018). Problem characteristics. In *Why good people do bad environmental things*. Oxford Academic. https://doi.org/10.1093/oso/9780190636272.003.0002

Douglas, E. P., Therriault, D. J., Berry, M. B., & Waisome, J. A. M. (2022). Comparing engineering students' and professionals' conceptions of ambiguity. In *2022 IEEE Frontiers in Education Conference (FIE)*, 1–4. https://doi.org/10.1109/FIE56618.2022.9962415

Enslow, P. H. (1973). Non-technical issues in network design—Economic, legal, social, and other considerations. *IEEE Computer, 6*(8), 21–30. https://doi.org/10.1109/MC.1973.6541671

Garcia, C., & Pinela, A. (2016). Engineering challenges in terms of academic and professional training. In *Advances and new trends in environmental informatics* (pp. 191–203). Springer, Cham. https://doi.org/10.1007/978-3-319-60937-9_15

Godfray, H. C. J., Beddington, J. R., Crute, I. R., Haddad, L., Lawrence, D., Muir, J. F., Pretty, J., Robinson, S., Thomas, S. M., & Toulmin, C. (2010). Food security: The challenge of feeding 9 billion people. *Science, 327*(5967), 812–818.

Gottlieb, M., Utesch, M., & Böhm, M. (2019). Beyond 2030 challenges of engineering education in an information systems-driven world—an extraction based on research topics. In *2019 IEEE Global Engineering Education Conference (EDUCON)*, 458–466. https://doi.org/10.1109/EDUCON.2019.8725245

Graves, H., & Garrett, R. K. (2014). A template for solving wicked problems. *Journal of Professional Communication*, *24*(1), 104–118. https://doi.org/10.1002/J.2334-5837.2014.00009.X

Gupta, A., & Gupta, S. K. (2020). Engineering solutions for food and water security in the 21st century. *Sustainability*, *12*(6), 2458.

Harrison, A. J. (1986). Roles of scientists, engineers and the public in the resolution of societal issues. In *The role of scientists and engineers in the legislative process* (pp. 37–41). Pergamon. https://doi.org/10.1016/B978-0-08-033911-5.50009-2

Harvey, R. (2004). *Beyond learning styles: Understanding the learning processes of engineering students through the interactive learning model.* [Paper presentation]. American Society for Engineering Education Annual Conference & Exposition, Salt Lake City, Utah. https://peer.asee.org/beyond-learning-styles-understanding-the-learning-processes-of-engineering-students-through-the-interactive-learning-model

Hills, G., & Tedford, D. (2003). The education of engineers: The uneasy relationship between engineering, science and technology. *Global Journal of Engineering Education*, *7*(1), 17–28.

Hipólito, I., & Khanduja, A. (2024). *The wicked and the complex: A new paradigm for societal problem-solving*. Preprints.org. https://doi.org/10.20944/preprints202407.1438.v2

Hmelo-Silver, C. E. (2004). Problem-based learning: What and how do students learn? *Educational Psychology Review*, 16, 235–266. https://doi.org/10.1023/B:EDPR.0000034022.16470.f3

Islam, M. M. (2024). Addressing challenges in engineering education: A case study of materials technology course. *Education Research and Applications*, *9*, 232. https://doi.org/10.29011/2575-7032.100232

Jonassen, D. H. (2014). Engineers as problem solvers. In *Cambridge handbook of engineering education research* (pp. 103–118). Cambridge University Press. https://doi.org/10.1017/CBO9781139013451.009

Jørgensen, U. (2007). Historical accounts of engineering education. In *International handbook of engineering education* (pp. 216–240). Springer. https://doi.org/10.1007/978-0-387-38290-6_10

Kamaruzaman, F., Hamid, R., & Mutalib, A. A. (2017). A review on issues and challenges in incorporating complex engineering problems in engineering curriculum and proposed solutions. In *2017 World Engineering Education Forum (WEEF)*, 697–701. https://doi.org/10.1109/WEEF.2017.8467167

Katehi, L., Pearson, G., & Feder, M. (Eds.). (n.d.). *Engineering in K-12 education: Understanding the status and improving the prospects*. The National Academies Press. https://doi.org/10.17226/12635

Kellner, M. I., Curtis, B., DeMarco, T., Kishida, K., Schlumberger, M., & Tully, C. (1991). Nontechnological issues in software engineering, In *Proceedings of 13th International Conference on Software Engineering*, Austin, TX, USA, pp. 144–146, doi: 10.1109/ICSE.1991.130632

Kleine, M. S., Zacharias, K., & Ozkan, D. (2023). Contextualization in engineering education: A scoping literature review. *Journal of Engineering Education*, *112*(4), 834–862. https://doi.org/10.1002/jee.20570

Kolmos, A., & de Graaff, E. (2014). Problem-based and project-based learning in engineering education: Merging models. In *Cambridge handbook of engineering education research* (pp. 141–161). Cambridge University Press. https://doi.org/10.1017/CBO9781139013451.012

Krishnan, M., & Ruhizan, R. M. Y. (2009). Problem-based learning in engineering education at Malaysian polytechnics: A proposal. In *2009 International Conference on Engineering Education (ICEED)*, 122–124. https://doi.org/10.1109/ICEED.2009.5490599

Krumdieck, S. (2024). Tackling 'wicked' problems calls for engineers with social responsibility. *Nature*, *629*(8014), 994–995. https://doi.org/10.1038/d41586-024-01519-1

Long, F. X., van Stein, B., Frenzel, M. A., Krause, P., Gitterle, M., & Bäck, T. (2022). Learning the characteristics of engineering optimisation problems with applications in automotive crash. In *Proceedings of the Genetic and Evolutionary Computation Conference*, 1373–1381. https://doi.org/10.1145/3512290.3528712

Mahdavi, A., Berg, C., & Ghiassi, N. (2024). Integrated energy solutions: A call for inclusion of non-technical factors. In *Proceedings of the 1st International Sustainable Energy Conference*, 1, 1093. https://doi.org/10.52825/isec.v1i.1093

Marcinauskas, L., Iljinas, A., Čyvienė, J., & Stankus, V. (2024). Problem-based learning versus traditional learning in physics education for engineering program students. *Education Sciences*, *14*(2), 154. https://doi.org/10.3390/educsci14020154

Mashiyane, T., Salifu, S., Ogunbiyi, O. F., & Oketola, A. (2023). Challenges, strategies and recommendations for reengineering engineering education in sub-Saharan Africa: A review. *Science, Engineering and Technology*, *4*(1). https://doi.org/10.54327/set2024/v4.i1.102

Mata-Lima, H. (2024). Barriers to sustainability: Reflections on social and environmental dimensions. *RGSA: Revista de Gestão Social e Ambiental*, *18*(12), e010300. https://doi.org/10.24857/rgsa.v18n12-070

Mills, E., Smith, J., & Taylor, P. (2021). Smart cities and sustainable urban development. *Urban Studies*, *58*(4), 823–838.

Moriarty, G. (1993). Contextualising engineering education. In *Proceedings of Frontiers in Education Conference*, 112–116. https://doi.org/10.1109/FIE.1993.405557

Mulder, K. F. (2017). Engineering education in sustainable development: Sustainability as a key competence. *Journal of Cleaner Production*, *168*, 1–8.

Naik, S. M., Bandi, S., & Reddy, C. (2024). Enhancing engineering education: Exploring the impact of problem-based learning on freshman students' skills and engagement – A case study. *Journal of Engineering Education Transformations*, *38*(Special Issue 1), 133–139. https://doi.org/10.16920/jeet/2024/v38is1/24222

Ni, B. J. (2022). Grand challenges in environmental engineering. *Frontiers in Environmental Engineering*, *1*, 1052154. https://doi.org/10.3389/fenve.2022.1052154

Noor, N. A., Alias, A., Ariffin, K., Bhkari, N. M., & Hashim, A. H. (2019). Employing the problem-based learning approach in civil engineering education: The highway engineering experience. In *Revolutionising K-12 blended learning through the i²Flex classroom model* (pp. 519–528). Springer. https://doi.org/10.1007/978-981-10-8730-1_52

OECD. (1996). *The knowledge-based economy*. Organisation for Economic Co-operation and Development

Piketty, T. (2014). *Capital in the twenty-first century*. Harvard University Press.

Riazifar, N., & Khajehsaeid, H. (2024). Enhancing engineering education through problem-based learning. In *Proceedings of UK and Ireland Engineering Education Research Network Conference 2023. University of Warwick*, https://journals.warwick.ac.uk/index.php/eern/article/view/1636

Rittel, H. W. J., & Webber, M. M. (1973). Dilemmas in a general theory of planning. *Policy Sciences, 4*(2), 155–169.

Rolstadås, A., & Paci, A. M. (2013). Editorial: Special issue on engineering education. *Journal of Intelligent Manufacturing, 24*(3), 421–422. https://doi.org/10.1007/S10845-012-0649-2

Schuelke-Leech, B. A. (2020). The place of wicked problems in engineering problem solving: A proposed taxonomy. In *2020 IEEE International Symposium on Technology and Society (ISTAS)*, 361–372. https://doi.org/10.1109/ISTAS50296.2020.9462174

Schwab, K. (2016). *The fourth industrial revolution*. Crown Business.

Sinding-Larsen, R., & Wellmer, F.-W. (2012). Non-renewable resource issues: Geoscientific and societal challenges: An introduction. In *Non-renewable resource issues* (pp. 1–19). Springer. https://doi.org/10.1007/978-90-481-8679-2_1

Stiglitz, J. E. (2010). *Freefall: America, free markets, and the sinking of the world economy*. W. W. Norton & Company.

Strauss, L. C., & Terenzini, P. T. (2005). *Engineering learning: Multiple influences on the development of analytical and group skills* (ED491040). ERIC. https://files.eric.ed.gov/fulltext/ED491040.pdf

Tanna, P., Lathigara, A., & Bhatt, N. (2022). Implementation of problem-based learning to solve real-life problems. *Journal of Engineering Education Transformations, 35*(Special Issue 1), 103–111. https://doi.org/10.16920/jeet/2022/v35is1/22015

Tranquillo, J., Goldberg, J., & Allen, R. (2023). Defining the engineering problem. In J. Tranquillo, J. Goldberg, & R. Allen (Eds.), *Biomedical engineering design* (pp. 103–121). Academic Press. https://doi.org/10.1016/B978-0-12-816444-0.00004-3

United Nations. (2015). *Transforming our world: The 2030 Agenda for Sustainable Development*. UN General Assembly.

Valero, M. (2022). Challenges, difficulties and barriers for engineering higher education. *Journal of Technology and Science Education, 12*(3), 551. https://doi.org/10.3926/jotse.1696

Varshney, L. R. (2014). Engineering for problems of excess. In *2014 IEEE International Symposium on Ethics in Science, Technology and Engineering*, 1–5. https://doi.org/10.1109/ETHICS.2014.6893467

Vega, Y. L. C., Lopez-Cuadra, Y., Zavaleta, R. M., Ibanez, J. C. A., & Alberto, M. M. (2023). Challenges of comprehensive training in engineering schools. *Universidad, Ciencia y Tecnología, 27*(118), 691. https://doi.org/10.47460/uct.v27i118.691

Veza, I., Yusrizal, Y., Idris, M., Putra, N. R., & Balogun, S. A. (2023). Current status of global engineering education: Mini review. *Radial: Jurnal Peradaban Sains, Rekayasa dan Teknologi, 1*(2), 8. https://doi.org/10.62024/radial.v1i2.8

Vieira, C. S., Gabriel, B., Robertt, V., & Figueiredo, E. (2024). Engineering education challenges and strengths: Reflecting on key stakeholders' perspectives. *Frontiers in Education, 9*, 1297267. https://doi.org/10.3389/feduc.2024.1297267

Wang, Q., Li, L.-Q., Chen, X.-H., Sun, H., Tang, Y., Wang, X.-Y., Zheng, S.-Y., Wang, D., & Yu, D. (2018). Refining complex engineering problems from students' innovative practices for engineering education in higher schools. In *DEStech Transactions on Social Science, Education and Human Science (ICSS)*. https://doi.org/10.12783/DTSSEHS/ICSS2017/19441

Wilson, E., Dunne, D., Bergen, T., & Mukhopadhyaya, P. (2024). Reframing engineering design: Integrating design thinking and systems thinking in engineering education and practice to address wicked problems. *Canadian Journal of Civil Engineering, 51*(11). https://doi.org/10.1139/cjce-2023-0433

Wood, A. K. (2023). Active learning in large classes. In *Effective teaching in large STEM classes* (pp. 2.1–2.18). Institute of Physics Publishing. https://doi.org/10.1088/978-0-7503-5231-4ch2

World Commission on Environment and Development. (1987). *Our common future.* Oxford University Press.

World Economic Forum. (2023). The Global Risks Report 2023, 18th Edition, Insight Report, World Economic Forum, Geneva, Switzerland.

Wu, J., & Hallett, M. (2007). A new research field and a new society, complex medical engineering. In *2007 IEEE/ICME International Conference on Complex Medical Engineering*, 1–4. https://doi.org/10.1109/ICCME.2007.4381680

Yadav, A., Lundeberg, M., Subedi, D., & Bunting, C. F. (2010). Problem-based learning in an undergraduate electrical engineering course. [Paper presentation]. *American Society for Engineering Education Annual Conference & Exposition*, Louisville, Kentucky. https://peer.asee.org/problem-based-learning-in-an-undergraduate-electrical-engineering-course

Yin, L., Yan, J., & Li, G. (2012). Application of eight engineering education management principles in the application-oriented talents training. In *Proceedings of the 2011 International Conference on Informatics, Management, and Industrial Engineering* (pp. 575–579). Springer, Berlin, Heidelberg. https://doi.org/10.1007/978-3-642-24823-8_91

Zhao, Y.Y. (2015). Towards innovative system development: A joint method of design thinking and systems thinking. *INCOSE International Symposium, Seattle, USA, 25*(1), 1427–1437. https://doi.org/10.1002/J.2334-5837.2015.00140.X

Zhuang, T. (2023). Impact, challenges and prospects of the engineering education reform. In: *Modernising China's undergraduate engineering education through systemic reforms. springerbriefs in education*, pp. 91–100. Springer. https://doi.org/10.1007/978-981-99-6388-1_8

Part II

Methodological Aspects of PBL

Chapter 5

Alignment of Problem-Based Learning with Outcomes-Based Learning

5.1 Introduction

The evolution of engineering education reflects a growing emphasis on student competence, industry readiness, and the ability to solve complex sociotechnical problems (Barsoum, 2017; Divjak, & Kukec, 2008; Moultrie, 2022). Outcomes-Based Education (OBE) has emerged as a dominant educational framework in this context, aiming to ensure that graduates achieve clearly defined learning outcomes aligned with societal and professional expectations (Aggarwal, 2024; Biggs & Tang, 2007). Its implementation across universities has sought to enhance the quality, accountability, and relevance of engineering programmes by making the teaching and learning process more transparent, outcomes-driven, and competency-focused (Hamidi et al., 2024; Kumar, 2023).

Despite its widespread adoption, OBE faces criticism for its rigid structure and limited ability to foster deep, critical, and integrative learning. Scholars argue that traditional OBE frameworks often overemphasise measurable outputs at the expense of student exploration and epistemic diversity (Brady, 1996; McCarthy, 1993). In engineering education—where students are expected to master both abstract theoretical concepts and hands-on problem-solving skills—these shortcomings become particularly evident (Donnelly, 2007).

To address these limitations, PBL has emerged as a significant alternative or complement to OBE. PBL places students in the role of active problem-solvers, working collaboratively to explore real-world, open-ended problems. The instructor acts as a facilitator rather than a transmitter of knowledge, guiding learners through discovery and critical reflection (Chen, 2008; Duch et al., 2001). This approach nurtures self-directed learning, communication, and practical competence—skills essential for professional practice in dynamic and interdisciplinary environments (Biley & Smith, 1998; Singh & Ramya, 2011; Williams, 2001).

The integration of OBE and PBL thus can offer a promising path forward: combining the structure and clarity of learning outcomes with the flexibility and depth of inquiry-based learning. Therefore, this chapter examines the potential for aligning OBE and PBL to enhance student achievement of

DOI: 10.1201/9781003529743-7

targeted competencies in engineering education. It focuses specifically on how PBL can be integrated within OBE frameworks to support deeper learning and skill development. Drawing on a case study from South Africa, the chapter offers critical insights into how the complementary strengths of these two pedagogical approaches can be harnessed to improve educational outcomes and better prepare engineering graduates to meet the complex demands of their profession and real-world complexity and lifelong learning.

5.2 Theoretical Basis and Implications of OBE

OBE is grounded in the principle of constructive alignment, a theory developed by Biggs (1999) and further elaborated by Biggs and Tang (2007). This pedagogical approach emphasises the alignment between intended learning outcomes (ILOs), teaching and learning activities (TLA), and assessment tasks (AT). The core idea is that when these three components are systematically aligned, students are more likely to engage in meaningful learning activities that directly support the achievement of the desired competencies (Biggs & Tang, 2007; Das, 2021). Constructive alignment draws on constructivist theories of learning, where students actively construct knowledge through engagement and reflection.

In the context of engineering education, constructive alignment is particularly important due to the field's dual emphasis on theoretical understanding and practical application. According to Tagg (2003) and Barsoum (2017), OBE provides a structured framework that helps ensure engineering graduates are equipped to address not only disciplinary knowledge but also broader socio-economic and environmental challenges. Frameworks such as Bloom's or Revised Bloom's Taxonomy and the Structure of Observed Learning Outcomes (SOLO) taxonomy guide the design of ILOs, enabling institutions to develop curricula that reflect varying levels of cognitive and functional complexity (Brabrand & Dahl, 2009).

Key features of OBE include its focus on learning outcomes, its student-centredness, and its alignment with industry and societal demands (Aggarwal, 2024; Hamidi et al., 2024;). OBE is characterised by a shift from input-focused education (what is taught) to output-focused learning (what students can do). Institutions employing OBE often design curricula around Competency Outcomes and Performance Outcomes (COPO), which are intended to reflect not just academic achievement but also readiness for professional practice (Cantika & Alhayat, 2023; Kumar, 2023).

Empirical studies have demonstrated the positive impact of OBE on student engagement, academic motivation, and skill development. For instance, examples from institutions like SMA Taruna Bakti Bandung and the Maharaja Agrasen Institute of Management Studies show improvements in learning outcomes and student performance when OBE is implemented with strong institutional support and teacher training (Aggarwal, 2024; Cantika & Alhayat, 2023).

Nevertheless, OBE is not without limitations. Scholars have critiqued OBE for reducing education to measurable competencies, potentially overlooking the holistic and transformative aims of higher learning (Brady, 1996; McCarthy, 1993). The emphasis on outcome assessment can lead to narrow curriculum design, with limited room for exploration and interdisciplinary learning. Moreover, assessing complex cognitive and practical competencies—particularly in theoretical courses—remains a persistent challenge (Donnelly, 2007).

These limitations suggest the need for pedagogical approaches that can better accommodate complexity, creativity, and collaborative problem-solving—qualities that are not easily captured by conventional outcome-based metrics. This recognition has led to increased interest in integrating OBE with more flexible, experiential learning models such as PBL.

PBL is a student-centred educational approach that emphasises active engagement with real-world, complex problems as the primary vehicle for learning. PBL has gained traction across a variety of disciplines, including law, engineering, science, and the social sciences (Özbıçakçı et al., 2012; Pepper, 2014). In contrast to traditional didactic methods, PBL shifts the focus from teacher-led instruction to student-led inquiry, where learners work collaboratively in small groups to explore open-ended problems. The role of the instructor in this model is not to provide direct answers but to act as a facilitator who supports critical thinking, guides reflection, and encourages the development of problem-solving strategies (Chen, 2008; Duch et al., 2001).

At its core, PBL fosters key cognitive and metacognitive skills such as identifying knowledge gaps, conducting independent research, integrating information from diverse sources, and refining understanding through discussion and application (Biley & Smith, 1998; Williams, 2001). This pedagogical method also promotes communication, collaboration, and self-directed learning, aligning closely with the intended outcomes of OBE, particularly in terms of competency development and real-world relevance (Solomon, 2011; Wolf, 2007). Through the exploration of authentic problems that mirror professional challenges, PBL encourages learners to apply theoretical knowledge in practical contexts, thereby enhancing their readiness for complex, interdisciplinary work environments (Hung & Amida, 2020; Hunter et al., 2025).

As discussed previously, several core characteristics define effective PBL implementation. First, authentic and ill-structured problems serve as the foundation of the learning process. These problems are designed to resemble real-life scenarios, requiring students to analyse information, make decisions, and justify their solutions (Pepper, 2014). Second, student-centred learning is paramount, as students take responsibility for identifying what they need to learn and seek out relevant resources independently (Hung & Amida, 2020; Ramadhani et al., 2019). Third, collaborative learning is central to PBL, with students working in small groups to share knowledge, test ideas, and develop solutions, thereby strengthening their communication and teamwork skills (Lalwani, et al., 2024; Zumbach & Prescher, 2023). Fourth, the teacher acts as a facilitator rather than a lecturer, guiding students through the learning process and encouraging reflective inquiry (Hunter et al., 2025; Lalwani, et al., 2024; Moore, 2009).

Effective implementation of PBL involves strategic design and assessment practices. Well-crafted problems that are contextually relevant and intellectually stimulating are essential for meaningful engagement (Pepper, 2014; Zumbach & Prescher, 2023). Assessment in PBL goes beyond evaluating the final product to include process-oriented measures such as peer evaluation, reflective journals, and group presentations, all of which help students develop deeper insights and transferable skills (Hunter et al., 2025). Additionally, PBL encourages interdisciplinary integration, allowing students to draw upon knowledge from multiple fields to address complex challenges (Pepper, 2014).

PBL offers several educational benefits. It has been shown to improve knowledge retention, foster critical thinking, and enhance the ability to apply learning in unfamiliar contexts (Hung & Amida, 2020; Lalwani, et al., 2024). Moreover, the relevance and immediacy of problem-based tasks tend to increase student motivation and engagement (Ramadhani et al., 2019). By promoting autonomy, accountability, and real-world thinking, PBL assists students cultivate essential soft skills such as leadership, adaptability, and professionalism—skills that are increasingly in demand in contemporary workplaces (Hung & Amida, 2020; Meifitri, 2020).

Despite its advantages, PBL also presents several challenges. It is often resource-intensive, requiring significant time, preparation, and facilitation skills from instructors, as well as sustained engagement from students (Adamyan, 2024; Lalwani, et al., 2024). Lectures or teachers (facilitators) must be well-trained in facilitation techniques and capable of guiding learning without overt control—an adjustment that can be difficult for those accustomed to lecture-based teaching (Ramadhani et al., 2019). Moreover, integrating PBL into traditional curricula may necessitate structural changes in course delivery, assessment formats, and institutional policies (Adamyan, 2024; Pepper, 2014).

Nonetheless, when thoughtfully implemented, PBL can serve as a transformative educational strategy. It not only enhances student learning outcomes but also aligns well with the broader aims of OBE, particularly in equipping learners with the critical competencies required for lifelong learning and professional adaptability. In the context of engineering education, PBL presents an opportunity to rethink conventional pedagogical approaches and better prepare graduates for complex, real-world challenges. Thus, it is argued that the alignment of OBE with PBL represents a strategic effort to harness the strengths of both paradigms: the clarity and accountability of OBE with the learner-centred, inquiry-driven nature of PBL.

5.3 Methodological Approach

A mixed-methods approach was adopted in this study, combining survey research, PBL-based workshops, and focus group discussions to gather both quantitative and qualitative data. The overarching aim was to explore the integration of PBL within an OBE framework in undergraduate engineering education. This methodological triangulation enabled a comprehensive understanding of the pedagogical alignment and its practical implementation.

5.3.1 Survey Design and Data Collection

A structured survey was administered to 120 participants who attended PBL workshops. These participants included both lecturers and students from three engineering disciplines: Civil Engineering, Mechanical Engineering, and Electrical, Electronics, and Computer Engineering. A convenience sampling technique was employed, and the questionnaire was pre-tested to ensure clarity and relevance. The design of the instrument was informed by findings from the literature review and lessons learned from PBL implementations at institutions such as University College London (UCL), UK.

The survey aimed to explore key dimensions associated with PBL, including the perceived acceptability and usefulness of the approach; the prerequisites for developing effective PBL assignments, such as required lecturer knowledge and competencies; the foundational knowledge and skills expected from students; the role of industry and challenge partners; and the importance of multidisciplinary collaboration. The survey employed a five-point Likert scale—ranging from 1 (not important) to 5 (most important)—to capture participants' perceptions regarding key aspects of the PBL experience.

5.3.2 Workshops on OBE-Aligned PBL

To provide experiential learning opportunities, four intensive workshops were conducted focusing on industry-oriented PBL tasks. Two workshops spanned two days and included 12 lecturers and 43 students. The other two workshops span a week each, with more than 100 students and 15 facilitators. Participants were grouped into interdisciplinary teams of three to six students, ensuring at least one member from each engineering discipline. Each group was supported by a designated facilitator acting as a mentor.

The workshops followed a structured, time-bound sequence of activities. These included: an introductory session on PBL and its alignment with OBE principles; the identification and selection of real-world problems by student groups; the preparation and presentation of a design brief; and consultations with challenge partners to refine and contextualise the chosen problems. The groups then moved on to conceptual development, evaluated alternative solutions, engaged in technical discussions with industry partners, and finalised their designs for presentation and peer evaluation. Challenge partners—including municipal officials, entrepreneurs, and community stakeholders—provided feedback on the social, economic, and environmental relevance of the proposed solutions. Technical input was offered by industry experts, while academic mentors provided theoretical and methodological guidance.

5.3.3 Focus Group Discussions

Following the workshops, focus group discussions were held with students, facilitators, industry partners, and academic experts. These discussions were conducted through informal, unstructured interviews and open-ended group dialogue, offering rich, qualitative insights. The sessions focused on participants' reflections on the workshops, the practicality of implementing OBE-linked PBL in their contexts, and recommendations for effective curriculum integration. Emerging themes included stakeholder collaboration, challenges in contextualising problems, and institutional readiness for pedagogical innovation.

5.3.4 Development of PBL-Based Course Assignments

As a continuation of the initiative, the participating facilitators were assigned to develop teaching and learning tasks grounded in PBL principles and aligned with OBE objectives. To ensure relevance, lecturers engaged in field visits and discussions with industry partners to frame realistic, context-specific problems. The resulting assignments were reviewed and evaluated for their alignment with OBE concepts such as constructive alignment, clearly defined outcomes, and performance-based assessments, alongside the core features of PBL, including problem authenticity, collaboration, and student autonomy.

5.3.5 Data Analysis and Evaluation

The data collected from surveys, workshops, focus groups, and assignment submissions were analysed using a combination of quantitative and qualitative methods. Quantitative data from the surveys were evaluated using the Perception Index (PI), calculated as the mean of the Likert scale responses. To assess the consistency and significance of the responses, additional statistical measures, such as the Coefficient of Variation (CoV) and the Interquartile Range (IQR), were used. A nonparametric significance test (Chi-square test) was performed to validate the responses and establish their statistical significance. These analyses helped identify patterns and validate the reliability of the responses. Qualitative data were analysed thematically using interpretive methods, allowing for the extraction of in-depth insights from narrative feedback and participant reflections.

The analysis focused on three central aspects: the perceptions of students and lecturers regarding the relevance and effectiveness of PBL in engineering education; the performance and engagement of students during the workshop-based PBL activities; and the quality, coherence, and curriculum integration of the OBE-linked PBL assignments developed by the lecturers.

This methodological framework enabled an in-depth exploration of how PBL can be effectively embedded within an OBE structure in engineering education. The multi-source data collection and triangulated analysis provided valuable insights into the opportunities, constraints, and practical strategies for achieving meaningful pedagogical integration. The results of the study are presented and discussed in the subsequent sections.

5.4 Results

5.4.1 *Perception of Students and Lecturers on Acceptability and Relevancy*

Table 5.1 presents the perceptions of lecturers and students regarding the various issues related to the acceptability and relevancy of PBL in engineering education. The evaluation was made under three aspects: (1) acceptability, relevancy and usefulness for engineering education, (2) prerequisite knowledge and skill required, and (3) challenges with regards to PBL.

5.4.1.1 Acceptability, Relevancy, and Usefulness of PBL

The findings indicate that PBL is widely perceived as relevant, with a mean score (PI) of 4.25 and a low coefficient of variation (CoV = 16.2%), suggesting strong consensus among respondents. A lower CoV implies that respondents had relatively similar views regarding PBL's relevance. Both students (PI = 3.98, CoV = 18.4%) and lecturers (PI = 4.16, CoV = 18.2%) found PBL acceptable, though the slightly lower rating by students may indicate challenges in adapting to this learning approach. The significance test results ($p < 0.001$ for students and $p = 0.023$ for lecturers) confirm that these perceptions are statistically significant, indicating that the acceptability of PBL is not due to random variations in responses.

In terms of usefulness, PBL was rated more effective for enhancing practical knowledge (PI = 4.32, CoV = 15.3%) than theoretical knowledge (PI = 3.55, CoV = 23.1%), suggesting that respondents find PBL more suited for applied learning than for conceptual understanding. The combined usefulness of both theoretical and practical knowledge was rated as important (PI = 3.74, CoV = 23.1%), highlighting the need for additional instructional support to balance theoretical depth. The interquartile range (IQR = 1) across most responses indicates moderate variability, meaning that opinions were fairly consistent within a central range. Additionally, PBL was found to be enjoyable (PI = 4.11, CoV = 16.9%, IQR = 1), reflecting its potential to increase student engagement and motivation in learning.

5.4.1.2 Prerequisite Knowledge and Skills Required

The study highlights the essential knowledge and skills necessary for effective engagement in PBL. A strong foundation in theories and principles (PI = 4.01,

Table 5.1 Perceptions Regarding PBL in Engineering Education

Aspects	*PI*	*SD*	*CoV (%)*	*IQR*	*Significance Test Value (p-values)*
Acceptability, relevancy, and usefulness					
PBL is relevant	4.25	0.689	16.2	1	<0.001
PBL is acceptable to students	3.98	0.733	18.4	0	<0.001
PBL is acceptable to lecturers	4.16	0.756	18.2	1	0.023
PBL is useful for enhancing theoretical knowledge	3.55	0.818	23.1	1	<0.001
PBL is useful for enhancing practical knowledge	4.32	0.661	15.3	1	<0.001
PBL is useful to enhance both theoretical and practical knowledge	3.74	0.865	23.1	1	<0.001
PBL is enjoyable	4.11	0.696	16.9	1	<0.001
Prerequisite knowledge and skills required					
Theories and principles	4.01	0.783	19.1	2	0.397
Analytical skills	3.82	0.850	22.3	2	<0.001
Design and creativity	4.42	0.657	14.8	1	<0.001
Multidisciplinary knowledge	3.83	0.771	20.1	1	0.042
Presentation skill	3.61	0.955	26.2	1	<0.001
Challenges with regard to PBL					
Time constraints	4.57	0.545	11.9	1	<0.001
Adaptability to heterogeneous group work	2.69	0.915	34.0	1	<0.001
Narrowing down to a relevant industry-based problem linked to broader socio-economic challenges	3.27	0.764	23.4	1	<0.001
Availability of appropriate challenge partners	3.48	0.648	18.6	1	<0.001
Availability of appropriate industry partners	3.43	0.729	21.3	1	<0.001
PBL is intensive and stressful	3.95	0.620	15.7	0	<0.001
Sample size	N	120			

CoV = 19.1%) was deemed important, though the non-significant p-value (p = 0.397) suggests some divergence in opinions, possibly indicating that some students perceive theoretical knowledge as less critical in PBL. Design and creativity (PI = 4.42, CoV = 14.8%, IQR = 1) emerged as the most critical skill, indicating a strong consensus on the importance of innovative thinking in problem-solving.

Analytical skills (PI = 3.82, CoV = 22.3%) and multidisciplinary knowledge (PI = 3.83, CoV = 20.1%) were also considered important, though their slightly higher CoV suggests some variation in perception, likely due to differences in students' academic backgrounds or exposure to interdisciplinary learning. Presentation skills received a lower rating (PI = 3.61, CoV = 26.2%), with a high CoV indicating considerable variability in responses, suggesting that some students may struggle with effective communication. The significant p-values ($p < 0.001$ for analytical skills and presentation skills, $p = 0.042$ for multidisciplinary knowledge) confirm that these skills are crucial for successful PBL engagement.

5.4.1.3 Challenges with Regard to PBL

Several challenges associated with the implementation of PBL were identified. Time constraints emerged as the most significant challenge (PI = 4.57, CoV = 11.9%, IQR = 1), with a low CoV indicating strong agreement that PBL requires substantial time investment. The high significance level ($p < 0.001$) confirms that time-related constraints are a widespread concern among respondents.

Adaptability to heterogeneous group work was rated the lowest (PI = 2.69, CoV = 34.0%), with a high CoV indicating diverse opinions—some students may adapt well, while others struggle with collaborative learning in mixed-ability teams. Narrowing down industry-relevant problems (PI = 3.27, CoV = 23.4%) and securing appropriate challenge partners (PI = 3.48, CoV = 18.6%) and industry partners (PI = 3.43, CoV = 21.3%) were noted as moderate challenges. The relatively higher CoV values suggest some variability in respondents' experiences, likely due to differences in their access to industry exposure or institutional support structures. The perception of PBL as intensive and stressful (PI = 3.95, CoV = 15.7%, IQR = 0) indicates a fairly consistent concern about cognitive load and workload demands.

The findings suggest that while PBL is highly relevant and engaging, its effectiveness could be enhanced through strategic pedagogical interventions. The relatively higher CoV in theoretical knowledge acquisition suggests variability in students' preparedness, indicating a need for structured reinforcement to ensure conceptual depth alongside practical application. The lower CoV values for design and creativity indicate strong consensus on the necessity of innovation-driven problem-solving in PBL, which should be further emphasised in the curriculum.

The significant challenges related to time constraints and group adaptability necessitate institutional interventions, such as phased PBL implementation or structured collaboration training to improve teamwork dynamics. Furthermore, the moderate CoV in industry engagement challenges suggests that establishing stronger university–industry linkages could help bridge the gap between theoretical learning and real-world application. Addressing presentation skill gaps through targeted training programmes could also enhance students' ability to communicate solutions effectively.

5.4.2 Student Performance on PBL Assignments during the Workshops

Students who participated in the workshops were tasked with addressing an industry-related problem aligned with the United Nations Sustainable Development Goals (SDGs). The deliverables and ILOs were clearly defined and communicated to the participants. As outlined in the methodology section, students were assigned to heterogeneous groups and guided by lecturers throughout the process. Additionally, industry and challenge partners were engaged to interact with the groups, providing insights and constructive feedback on their problem-solving approaches.

To assess student performance, a panel conducted evaluations based on a structured rubric that considered factors such as relevance, design brief quality, conceptualisation, exploration of alternatives, design approach, and final product. However, given the time constraints, developing a fully detailed design or prototype proved challenging for most groups. As a result, assessments primarily focused on conceptual design outputs rather than finalised products. The evaluation of student competencies and performance was based on the depth, quality, and innovativeness of their conceptual solutions. A snapshot of one of the problems tackled by a student group during the workshop is presented in Table 5.2.

5.4.2.1 The Final Design

Figure 5.1 presents the final conceptual design of the affordable wheelchair that can be used by mobility-disabled people in the rural areas of South Africa.

The case study presented exemplifies the framing of an industry-related problem that is socio-economically relevant and aligned with the SDGs. The teaching and learning activities, along with the assessment tasks, were constructively aligned with the principles of OBE. The assessment of the conceptual design was conducted using a structured rubric, evaluating key parameters such as relevance, design brief clarity, conceptualisation, exploration of alternatives, overall design, and final product feasibility.

According to the assessment panel, the proposed product and design were deemed highly relevant to the targeted rural context. The design brief was well-structured and clearly articulated, while the conceptual design was considered appropriate given the constraints of rural infrastructure. In addition, a comparative cost analysis was undertaken, revealing that the proposed product could be significantly more affordable than currently available alternatives, making it accessible to a broader segment of the population.

Despite these promising findings, the design team estimated that the full development of the wheelchair, including prototype fabrication and refinement, would require at least four months. Consequently, while the complete design could not be finalised within the workshop timeframe, and competency levels could not be measured in absolute terms, the PBL approach demonstrated that students were capable of developing viable engineering solutions

Table 5.2 Snapshot of the PBL Problem Tackled by the Students

Challenge: Affordable Manufacturing of Medical Implants and Accessories for All in Africa

Problem Statement: Manufacturing Mobility Equipment for Physically Challenged Individuals in Rural South Africa

Access to affordable assistive devices remains a significant challenge for individuals with mobility impairments in South Africa, particularly in rural areas. According to the Census of South Africa 2011, approximately 2,870,130 South Africans (7.5% of the population) live with various disabilities, with black Africans being the most affected demographic. Despite the integration of mobility devices into the national healthcare system, many individuals still face barriers in acquiring necessary assistive technologies. This lack of access exacerbates social and economic marginalisation, restricting their participation in community activities and limiting opportunities for advancement. Although legislative frameworks exist to promote accessibility, a substantial gap remains in the provision of assistive devices, particularly among black South Africans in rural areas. Addressing this challenge necessitates innovative approaches to the affordable manufacturing of mobility equipment tailored to the needs of physically challenged individuals in South Africa.

Objective

The primary objective of this initiative was to design a motor-driven wheelchair that conforms to South African National Standards (SANS) and can be manufactured using locally available materials. The proposed wheelchair was designed to be suitable for rural environments, with the capability to traverse unpaved roads and gravel pathways, while remaining affordable for end-users.

Methodology and Learning Tasks

The design process involved a systematic review of existing wheelchair products, their structural and functional attributes, and an assessment of locally available materials for cost-effective manufacturing. The team conducted material analysis to identify appropriate components and employed design software for product conceptualisation and optimisation. The approach integrated interdisciplinary collaboration and industry engagement to ensure practical feasibility and relevance.

Team Composition

The project team comprised five students from different engineering disciplines: two from electrical engineering, two from civil engineering, and one from mechanical engineering. The team was mentored by two lecturers—one from electrical engineering with expertise in software and modelling, the other from mechanical engineering specialising in product manufacturing.

Stakeholder Engagement

Key challenge partners included government representatives from the Department of Health, hospitals, research institutions such as the Product Development Technology Station (PDTS) at the Central University of Technology, Free State, South Africa, as well as civil society organisations and individuals with disabilities. Industry partners were identified from local manufacturing sectors, contributing insights into cost-effective production and scalability.

Intended Learning Outcome (ILO)

Although the ultimate goal was to manufacture an affordable motor-driven wheelchair, the specific Intended Learning Outcome (ILO) for the workshop was the development of a conceptual wheelchair design with desirable functional characteristics, as envisioned by the team. This approach emphasised the integration of theoretical knowledge with practical problem-solving, fostering an experiential learning environment that aligned with real-world engineering challenges.

Figure 5.1 Conceptual Design of the Affordable Wheelchair.

to real-world industry challenges within a constrained period. The outcomes of this exercise suggest that students exhibited acceptable competency levels in higher-order cognitive skills emphasised in OBE, including analytical reasoning, design thinking, and complex problem-solving.

5.4.3 Integration of OBE in PBL Assignments

As part of the broader initiative to enhance teaching and learning through PBL, lecturers were tasked with developing PBL-based assignments to be incorporated into their respective subject curricula. To ensure the industry relevance of these assignments, lecturers conducted site visits to various industries and workplaces to explore real-world challenges that organisations were actively addressing or planning to undertake.

Subsequently, a series of structured discussions were held between lecturers and industry partners to refine and prioritise key industry challenges. This collaborative effort facilitated the development of well-defined PBL problems, which were then formulated as assignments aligned with the OBE principles. These assignments were designed to foster critical thinking, problem-solving, and application of disciplinary knowledge in authentic industry contexts.

A representative example of such a PBL-based assignment is provided in Table 5.3.

Based on the snapshot of the assignment, Table 5.4 provides OBE-premised constructive alignment of the assignment with regard to ILOs, TLAs and ATs.

Table 5.3 Example of a PBL-Based Assignment

Assignment Title: Temperature Control of a Refrigerator System for a Craft Beer Manufacturing Process

Purpose

This assignment aims to develop a control system for the temperature regulation of a craft beer brewing system. The control system ensures that the brewing temperature is maintained between 18°C and 23°C, which is essential for achieving optimal brewing conditions. The project involves two primary phases: first, the development of a mathematical model using MATLAB's Simulink to simulate the control system; and second, the physical implementation of the model using an Arduino interface, a solid-state relay, and a temperature sensor. A light bulb is used in the experimental setup to simulate the heating process, allowing students to validate the feasibility of their model in a real-life scenario.

Intended Learning Outcomes (ILOs)

Upon completing the assignment, students will be able to:

1. Design a mathematical model using MATLAB Simulink to develop a control system for switched systems in a craft beer brewing process.
2. Construct a hardware setup using an Arduino interface with a solid-state relay and a temperature sensor.
3. Implement and test the designed control system in a real-life scenario.

Required Equipment and Software

- **Hardware**: Arduino Uno, light bulb, solid-state relay, temperature sensor.
- **Software**: MATLAB and Arduino IDE.

Cognitive Learning Activities

The project requires students to engage in a series of structured learning tasks, including:

1. Developing a system block diagram to conceptualise the control process.
2. Designing the mathematical model of the system.
3. Creating a Simulink model to control the relay, which involves:
   a. Programming the model to read the temperature of the light bulb.
   b. Storing the reference values of 18°C (lower limit) and 23°C (upper limit).
   c. Switching the heating element ON at 18°C and OFF at 23°C.
4. Constructing the hardware setup with Arduino as the interface between the software and hardware components.
5. Testing the functionality of the model to validate its effectiveness.
6. Enhancing the control system by implementing proportional (P) control and analysing its advantages over a simple ON/OFF control mechanism.

Assessment Tasks

The assessment of this project will be conducted using a structured rubric that evaluates students' ability to:

- Develop an accurate software model using MATLAB Simulink.
- Construct a functional hardware prototype using the required components.
- Successfully design and implement the hardware setup.
- Test and analyse the performance of the control system.
- Present findings through a structured report and/or presentation.

Industry Collaboration and Partnerships

This assignment is designed with the involvement of industry stakeholders to ensure practical relevance. The challenge partners include craft beer manufacturing industries that require precise temperature control for brewing. Additionally, industry partners such as software development firms and relay manufacturing companies contribute expertise in automation and control systems.

This PBL-based assignment integrates theoretical knowledge with practical application, equipping students with essential engineering competencies in system modelling, hardware implementation, and control system design.

Table 5.4 OBE Premised Constructive Alignment of the PBL Assignment

ILOs	*Teaching and Learning Activities*	*Assessment Tasks*
Develop a control system for the switched systems in a craft beer brewer	• Preparation of the system block diagram, • Use of MATLAB for developing a mathematical model of the system, • Use the Simulink model to control the relay, • Use of Arduino as the interface between the software and hardware to construct the hardware setup. • Testing of the working of the model, • Re-creation of the model with the control and theorise the work by stating the necessity of P control over an ON/OFF control.	• Software modelling, Hardware modelling, • Design and construction of the hardware setup, • Testing of the working of the model and • Presentation.

The assignment illustrates the integration of an industry-relevant problem within the discipline of electronics and electrical engineering, requiring a multidisciplinary knowledge base that spans control systems, computer science, software development, thermodynamics, and mathematics. Its direct relevance to the craft brewing industry underscores the practical applicability of engineering principles in solving real-world challenges.

The ILOs of the assignment are effectively aligned with the corresponding teaching and learning activities, as well as assessment tasks, demonstrating a constructive alignment that is a fundamental characteristic of OBE. This alignment ensures that students engage in problem-solving tasks that develop critical technical and analytical competencies while addressing industry-specific needs.

Furthermore, the structured development of the assignment highlights the capacity of lecturers to formulate industry-based problems within the OBE framework. The process of assignment framing, undertaken in collaboration with industry partners, reinforces the importance of academia–industry engagement in enhancing the relevance of engineering education. This approach not only bridges the gap between theoretical knowledge and practical application but also fosters a learning environment that promotes innovation and industry readiness among students.

5.5 Discussion

The study's findings underscore the relevance and effectiveness of PBL as perceived by both students and lecturers in the context of engineering education.

The perception of students and lecturers indicates that PBL is not only acceptable but also highly relevant and enjoyable. PBL significantly contributes to the acquisition of practical knowledge and, to a slightly lesser extent, theoretical understanding. This perception aligns with broader pedagogical shifts towards learner-centred, competency-driven approaches, particularly within OBE frameworks (Aggarwal, 2024; Kumar, 2023). PBL was found to be especially effective in enhancing practical knowledge acquisition (PI = 4.32), while still contributing meaningfully to theoretical understanding (PI = 3.55), indicating its dual role in bridging conceptual and applied learning (Chen, 2008; Hunter et al., 2025).

However, the successful implementation of PBL hinges on the presence of certain critical prerequisites. These include foundational theoretical and practical knowledge, analytical thinking, creativity, and design skills—competencies essential for problem-solving in multidisciplinary settings (Hung & Amida, 2020; Lalwani et al., 2024). Among these, "design and creativity" was rated the most essential (PI = 4.42), signalling that engineering curricula should increasingly prioritise innovation and ideation as core learning goals.

Despite its pedagogical strengths, PBL faces significant implementation challenges. The important among these are time constraints (PI = 4.57) and the intensity and stressfulness (PI = 3.95). However, the difficulties students face when working in heterogeneous groups (PI = 2.69) are observed to be less challenging. These findings reflect broader concerns raised in the literature, where PBL is described as resource-intensive, requiring well-prepared facilitators and substantial student engagement (Adamyan, 2024; Ramadhani et al., 2019). Institutions seeking to integrate PBL must therefore address these structural demands through targeted interventions, such as facilitator training, phased project implementation, and support for group collaboration dynamics.

Another recurring challenge is the engagement of industry and challenge partners. While their involvement is crucial to ensuring the authenticity and relevance of PBL assignments, their availability and alignment with academic structures remain challenging (Hunter et al., 2025; Lalwani et al., 2024). Nonetheless, when industry collaboration is achieved, it significantly enriches student learning and promotes professional readiness. Focus group feedback and case evidence from this study reaffirm that PBL is highly applicable to real-world industry contexts and fosters deeper collaboration between academia and external stakeholders.

Further, scholarships emphasise that PBL's integration within an OBE framework supports higher-order cognitive learning through constructively aligned ILOs, TLAs, and ATs (Biggs & Tang, 2007; Das, 2021). The student projects and lecturer-designed assignments in this study reflect these principles. For instance, the temperature control system assignment and the affordable wheelchair project both required application, analysis, and design—key cognitive activities underpinning effective OBE delivery.

In this regard, the synergy between PBL and OBE is evident. PBL enhances OBE by providing contextualised, authentic problems that foster the development of transferable skills such as critical thinking, communication, and adaptability—competencies increasingly demanded by employers in the Fourth Industrial Revolution (Hamidi et al., 2024; Hung & Amida, 2020).

5.6 Conclusion

This study demonstrates that integrating PBL within an OBE framework can significantly enhance student engagement, competency development, and industry relevance in engineering education. The findings confirm that PBL is widely accepted by students and lecturers, particularly for its ability to support practical knowledge acquisition and foster higher-order thinking skills.

However, its effective implementation is conditioned by the presence of certain foundational skills and knowledge, as well as the availability of industry and challenge partners. Time constraints and adaptation to collaborative work also present ongoing challenges that require institutional strategies for mitigation.

Despite these limitations, the study provides a compelling argument for the broader adoption of OBE-aligned PBL approaches. It highlights the importance of designing assignments around real-world problems in consultation with industry stakeholders to ensure relevance and impact. To validate and expand on these findings, further research should explore the application of this model across multiple institutions and disciplines.

By fostering an ecosystem of learning that is both rigorous and responsive to societal needs, the integration of PBL and OBE holds significant promise for transforming engineering education and producing graduates equipped for complexity, collaboration, and innovation.

References

Adamyan, N. (2024). Solutions to problematic situations as an interactive learning tool and for the development of students' thinking. *SHPH Gitakan Teghekagir*, 380–391. https://doi.org/10.54151/27382559-24.1pb-380

Aggarwal, T. (2024). An empirical analysis of outcome-based education research. *International Journal for Multidisciplinary Research*. https://doi.org/10.36948/ijfmr.2024.v06i05.28001

Barsoum, G. (2017). Quality, pedagogy and governance in private higher education institutions in Egypt. *Africa Education Review*, *14*(1), 193–211.

Biggs, J. (1999). What the student does: Teaching for enhanced learning. *Higher Education Research & Development*, *18*(1), 57–75. https://doi.org/10.1080/0729436990180105

Biggs, J., & Tang, C. (Eds.). (2007). *Teaching for quality learning at university* (3rd ed.), pp. 22–26). SHRE & Open University Press.

Biley, F., & Smith, K. (1998). Exploring the potential of problem-based learning in nurse education. *Nurse Education Today*, *18*, 353–361.

Brabrand, C., & Dahl, B. (2009). Using SOLO taxonomy to analyse competence progression of university science curricula. *Higher Education, 58*, 531–549. https://doi.org/10.1007/s10734-009-9210-4

Brady, L. (1996). Outcome-based education: A critique. *The Curriculum Journal, 7*(1), 5–16. https://doi.org/10.1080/0958517960070102

Cantika, V. M., & Alhayat, A. (2023). Implementation of outcome-based education in a leading high school (a case study at SMA Taruna Bakti Bandung). *Proceedings Series on Social Sciences & Humanities, 12*, 52–59. https://doi.org/10.30595/pssh.v12i.773

Chen, N.C. (2008). An educational approach to problem-based learning. *Kaohsiung Journal of Medical Sciences, 24*(3 Suppl), S23–S30.

Das, D. K. (2021). Constructive alignment for deep learning in undergraduate civil engineering education. *African Journal of Research in Mathematics, Science and Technology Education*, 25(1), 77–90. https://doi.org/10.1080/18117295.2021.1924493

Divjak, B., & Kukec, S. K. (2008). Teaching methods for international R&D project management. *International Journal of Project Management, 26*, 251–257.

Donnelly, K. (2007). Australia's adoption of outcomes-based education: A critique. *Issues in Educational Research, 17*(2), 183–206.

Duch, B. J., Groh, S. E., & Allen, D. E. (2001). Why problem-based learning? A case study of institutional change in undergraduate education. In B. J. Duch, S. E. Groh, & D. E. Allen (Eds.), *The power of problem-based learning* (pp. 3–13). Stylus Publishing.

Hamidi, H., Hejran, A. B., Sarwari, A., & Edigeevna, S. G. (2024). The effect of outcome-based education on the behaviour of students. *European Journal of Theoretical and Applied Sciences, 2*(2), 764–773. https://doi.org/10.59324/ejtas.2024.2(2).68

Hung, W., & Amida, A. (2020). Problem-based learning in college science. In *Teaching and learning of science in higher education* (pp. 325–339). Springer. https://doi.org/10.1007/978-3-030-33600-4_21

Hunter, C., Archer, S., & Illingworth, C. (2025). Problem-based learning. In *Elgar encyclopedia of education* (pp. 302–305). Elgar Online. https://doi.org/10.4337/9781035302932.ch83

Kumar, A. B. R. (2023). Outcome-based education model (pp. 2–14). https://doi.org/10.9734/bpi/mono/978-81-19315-93-2/ch1

Lalwani, K., et al. (2024). How to design and moderate an effective problem-based learning discussion. In K. Lalwani et al. (Eds.), *Professional, ethical, legal, and educational lessons in medicine: A problem-based learning approach.* Oxford Academic. https://doi.org/10.1093/med/9780197655979.003.0052

McCarthy, M. M. (1993). Challenges to the public school curriculum: New targets and strategies. *The Phi Delta Kappan, 75*(1), 58–60.

Meifitri. (2020). Problem solving ability and communication skill through problem-based learning approach. In *Proceedings of the Eighth International Conference on Languages and Arts (ICLA-2019)* (pp. 146–149). https://doi.org/10.2991/ASSEHR.K.200819.028

Moore, J. (2009). An exploration of lecturer as facilitator within the context of problem-based learning. *Nurse Education Today, 29*, 150–156.

Moultrie, J. (2022). Educating engineering systems designers: A systems design competences and skills matrix. In: A. Maier, J. Oehmen, P.E. Vermaas (eds) *Handbook of engineering systems design*, pp. 1–35. Springer, Cham. https://doi.org/10.1007/978-3-030-46054-9_28-1

Özbıçakçı, Ş., Bilik, Ö., & İntepeler, Ş. S. (2012). Assessment of goals in problem-based learning. *Nurse Education Today, 32*, e79–e82.

Pepper, C. (2014). Problem-based learning (PBL). In M. Davies & M. Barnett (Eds.), *The Palgrave Handbook of critical thinking in higher education* (pp. 1–3). Springer. https://doi.org/10.1007/978-94-007-6165-0_128-2

Ramadhani, R., Huda, S., & Umam, R. (2019). Problem-based learning, its usability and critical view as educational learning tools. *Journal of Gifted Education and Creativity*, *6*(3), 219–231.

Singh, M., & Ramya, K. R. (2011). Outcome-based education. *International Journal of Nursing Education*, *3*(2), 87–91. http://www.indianjournals.com/ijor.aspx?target=ijor:ijone&volume=3&issue=2&article=025

Solomon, P. (2011). Problem-based learning. In M. J. Bradshaw & A. J. Lowenstein (Eds.), *Innovative teaching strategies in nursing and related health professions* (pp. 131–147). Jones and Bartlett Publishers.

Tagg, J. (2003). *The learning paradigm college* (pp. 63–85). Anker Publishing.

Williams, B. (2001). Developing critical reflection for professional practice through problem-based learning. *Journal of Advanced Nursing*, *34*(1), 27–34.

Wolf, A. (2007). Tutoring PBL: A model for student-centred teaching. In L. E. Young & B. L. Paterson (Eds.), *Teaching nursing* (pp. 231–243). Lippincott.

Zumbach, J., & Prescher, C. (2023). Problem-based learning and case-based learning. In J. Zumbach, D. A. Bernstein, S. Narciss, & G. Marsico (Eds.), *International handbook of psychology learning and teaching*. Springer. https://doi.org/10.1007/978-3-030-28745-0_58

Chapter 6

A Methodological Approach for Problem-Based Learning in Engineering Education

6.1 Introduction

Competence and skill are essential requirements for success in the modern industrial world, particularly within the engineering field. The responsibility for developing competent and skilled engineering graduates lies with universities. Problem-Based Learning (PBL) has emerged as an innovative pedagogical method that centres on real-world, complex problems to stimulate active engagement and critical thinking among students. In engineering education, PBL bridges the gap between theoretical knowledge and practical application, preparing students for the challenges they will face in the professional world. By presenting students with ill-structured problems, PBL encourages collaboration, research, and the application of engineering principles in a real-world context.

PBL shifts the focus from traditional teacher-led instruction to a student-centred approach, where students take responsibility for their learning by engaging with real-world challenges (Dias et al., 2022). This method fosters a deeper understanding of engineering concepts and enhances skills such as critical thinking, teamwork, and self-directed learning. PBL allows students to apply theoretical knowledge to practical situations, enhancing their problem-solving skills and understanding of engineering principles (Buentello-Montoya, 2022; Díaz et al., 2023). Through interdisciplinary collaboration, students work on projects requiring the integration of various engineering concepts, such as designing piezoresistive pressure sensors or simulating wind turbine voltage stability (Stoyanov et al., 2023; Wachholz & Rasia, 2023).

This collaborative environment reflects the nature of engineering work environments and prepares students for the challenges they will face in their professional careers (Riazifar & Khajehsaeid, 2024). Group activities such as case study analyses, ideathons, and hackathons foster teamwork and communication skills (Kabilan, 2024). Assessment in PBL is multifaceted, incorporating project presentations, peer reviews, and reflective journals, which provide a comprehensive view of student learning and development (Kabilan, 2024; d'Escoffier et al., 2024). Feedback plays an integral role in refining students'

DOI: 10.1201/9781003529743-8

problem-solving approaches and deepening their understanding (Utama et al., 2024). The use of technology, such as virtual laboratories and simulation tools, enhances student engagement by allowing them to experiment and visualise complex engineering processes (Stoyanov et al., 2023).

While PBL offers significant benefits, its implementation comes with challenges. It requires a shift in teaching philosophy, new assessment strategies, and substantial resource investment. Additionally, its success depends on instructors' ability to facilitate learning effectively and students' willingness to engage actively with the material. Despite these challenges, PBL's potential to transform engineering education by fostering critical skills and preparing students for real-world challenges makes it a valuable approach in modern educational settings.

Engineering education traditionally follows an Outcomes-Based Education (OBE) model. Work Integrated Learning (WIL) is a key component of the curriculum, though it is no longer seen as indispensable by some universities to be integrated into the curriculum. While work-based learning is encouraged, universities focus on familiarising students with real-life projects and practical problem-solving they will encounter after graduation. This form of learning, which involves tackling real-world industrial problems, often within the university context, does not necessarily require on-site work experience. However, the lack of mandatory WIL, coupled with the absence of internships or post-graduation workplace training, creates gaps that put students at risk when transitioning into the workforce.

Professional bodies and universities mandate that students must demonstrate competency in exit-level outcomes (Graduate Attributes) before they can graduate and qualify for professional registration. Consequently, it has been argued that PBL could offer an effective method for developing critical competencies, such as problem-solving, design, teamwork, and the ability to work in multidisciplinary environments—skills necessary for industry relevance (Buentello-Montoya, 2022; Díaz et al., 2023). Several universities have begun introducing PBL into engineering education, organising workshops and learning programmes to incorporate this approach. Despite these efforts, PBL has not been widely adopted, particularly in engineering, and lacks a well-defined methodology for effective implementation. There have been debates regarding the modalities of its application, highlighting the need for a more structured approach.

This study explores how PBL can be effectively integrated into undergraduate engineering education at universities. It examines a methodology that can engage students in deep, active learning, helping them develop the competencies required for the profession. Situated within the undergraduate engineering programmes of a University of Technology in South Africa, which follows the OBE framework, the study proposes a five-stage process for implementing PBL. Drawing on the action learning approach, insights from workshops conducted for engineering students, and a subsequent student perception survey, the study aims to enable the effective and successful implementation of PBL, helping students achieve the desired learning outcomes and competencies.

6.2 Implementation of PBL: Lessons Learned from Literature on Methodological Approaches

6.2.1 *PBL as a Pedagogical Approach*

As discussed previously, PBL in engineering education is a pedagogical approach emphasising student-centred learning through real-world problem resolution. This method enhances engagement, critical thinking, and the practical application of theoretical knowledge (Buentello-Montoya, 2022; Díaz et al., 2023; Stoyanov et al., 2023; Wachholz & Rasia, 2023). PBL is particularly effective due to its focus on real-world applications and interdisciplinary learning, which are essential for developing the skills needed in the rapidly evolving technological landscape. By immersing students in problem-solving activities, PBL fosters an active learning environment where they engage with meaningful challenges, leading to higher levels of satisfaction and retention of engineering concepts (Riazifar & Khajehsaeid, 2024). In laboratory settings, students apply theoretical knowledge to practical problems, strengthening their understanding and motivation (Díaz et al., 2023). The approach has proven especially effective in courses such as electricity and electromagnetism, where students design and simulate real-world applications, resulting in high performance and engagement (Buentello-Montoya, 2022).

Beyond technical knowledge, PBL fosters entrepreneurial skills by simulating business challenges through activities like ideathons, hackathons, and business model development (Kabilan, 2024). The collaborative nature of PBL enhances students' ability to innovate and adapt to complex problems, which are crucial skills in modern engineering (Kabilan, 2024). Additionally, PBL promotes higher-order thinking skills (HOTs) such as creativity, critical thinking, and problem-solving, essential for navigating contemporary technological challenges (Bhaumik et al., 2024). The integration of thematic-based case studies and group-based learning opportunities within PBL frameworks further strengthens students' cognitive and interpersonal abilities (Bhaumik et al., 2024).

Despite its advantages, PBL presents challenges, for example, in subjects like physics, where traditional methods may be more effective for theoretical knowledge acquisition (Marcinauskas et al., 2024). The integration of PBL with sustainability education is still evolving, requiring a systemic approach to develop sustainability skills (d'Escoffier et al., 2024). Institutions must assess their readiness for PBL adoption, considering faculty training and curriculum design for effective implementation (Dos Santos et al., 2022). While PBL enhances active learning, traditional methods remain valuable in providing a strong theoretical foundation for subjects requiring detailed conceptual understanding. A balanced integration of PBL and traditional methods may offer the most comprehensive educational experience for engineering students.

PBL, as a pedagogical strategy, supports active learning through engagement with real-life problems (Elaine & Karen, 2016). It encourages students to learn collaboratively, develop self-directed learning habits, and construct mental models for knowledge acquisition (Dochy et al., 2003; Schmidt & Moust, 2000). The underpinning philosophy of PBL aligns with constructivist learning principles, where students are active knowledge seekers and co-creators of knowledge through relevant experiences built on prior understanding (Derry, 1996; Mayer, 1996). Teachers act as facilitators, guiding students in problem analysis and solution development. PBL is particularly appealing to educators as it fosters both individual and group learning, aligning with social theories of learning that emphasise the importance of interaction in cognitive development (Glaser & Bassok, 1989; Palincsar, 1998; Vygotsky, 1978). Moreover, PBL is flexible and adaptable to various contexts, promoting competency-based learning (Mubuuke et al., 2017; Schmidt, 1998; Ward & Lee, 2002; Yeung et al., 2003).

The implementation of PBL varies across programmes and institutions. It is well-established in fields such as medicine and nursing and is being increasingly incorporated into the physical sciences, including physics and engineering (Servant-Miklos, 2019; Servant-Miklos & Kolmos, 2022). While the methodology may differ, PBL typically follows an iterative process involving problem analysis, self-directed learning, and a reporting phase, with facilitators scaffolding students' inquiry and sense-making (Barrows, 1988; Elaine & Karen, 2016; Schmidt, 1993; Schmidt & Moust, 2000).

6.2.2 *Models of PBL*

Various models of PBL have been developed to enhance their effectiveness across diverse educational contexts. These models differ in complexity, gradually incorporating deeper levels of student engagement and cognitive processing. The following sections present both foundational and advanced PBL models, outlining their key components and applications in educational settings.

6.2.2.1 Established PBL Frameworks

Two prominent models widely adopted in educational practice are the Barrows and Myers Model and the Fogarty Model:

- **Barrows and Myers Model**: This five-step framework includes: (1) introduction to the PBL concept and learning environment, (2) problem configuration and task allocation, (3) critical analysis of gathered information, (4) generation of possible solutions, and (5) abstraction of knowledge through self-evaluation and reflection (Kazemi & Ghoraishi, 2012; Savery & Duffy, 1995).

- **Fogarty Model**: This model proposes a seven-step process: (1) problem presentation, (2) definition of the problem, (3) establishment of assumptions, (4) information search, (5) revision of the problem statement, (6) generation of alternative solutions, and (7) evaluation of proposed solutions (Fogarty, 1997; Kazemi & Ghoraishi, 2012).

While these models have been effectively implemented across various disciplines, PBL was historically underutilised in engineering education. Until recent years, its application in engineering remained limited, and a standardised methodology for its integration in this field is still lacking.

Over time, educators and researchers have recognised the need to adapt and refine the traditional PBL framework to better suit varying educational objectives, learner profiles, and disciplinary demands. While the core philosophy of PBL—centred on inquiry, collaboration, and real-world problem-solving—remains consistent, its implementation has evolved into multiple models that offer varying levels of structure and complexity. These models provide flexible frameworks for integrating PBL into curricula, ranging from basic step-based approaches to more advanced, interdisciplinary formats. The following section thoroughly explores these models, highlighting their unique features, pedagogical intentions, and applicability across different educational contexts.

6.2.2.2 Basic Models of PBL

The basic models are typically used to introduce students to the PBL process and gradually build their problem-solving and collaborative skills:

- **Three-Step Model**: This basic model involves three main stages: problem definition, research and solution implementation, and evaluation of the outcome. It offers a clear and accessible entry point into PBL for beginners (Epler & Jacobs, 2022).
- **Four-Step Model**: Expanding on the previous model, this version includes a separate research phase before implementing solutions. This emphasises the role of evidence-based decision-making in problem-solving (Epler & Jacobs, 2022).
- **Five-Step Model**: A brainstorming phase is added to encourage the generation of multiple solutions. This fosters creative thinking and helps students critically assess the viability of various options before selecting one (Epler & Jacobs, 2022).
- **Six-Step Model**: This model incorporates a presentation stage, allowing students to communicate their findings to peers. It promotes the development of communication skills and peer-based learning (Epler & Jacobs, 2022).

6.2.2.3 Advanced Models of PBL

The advanced models are more comprehensive and are designed to support deeper learning, greater autonomy, and the integration of interdisciplinary skills. There are:

- **Seven- and Eight-Step Models**: These models emphasise student agency, self-directed learning, and iterative problem-solving. By involving more stages—such as revisiting and refining problem definitions—they foster deeper engagement and the development of metacognitive skills (Epler & Jacobs, 2022).
- **Problem-Based Project Organised Learning (PBPOL)**: This model is notably implemented at Aalborg University. It merges principles from both problem-based and project-based learning. Students work collaboratively on complex problems over extended periods, culminating in a comprehensive project report. This approach enhances both critical thinking and project management skills (Kolmos & de Graaff, 2014).

Each of these PBL models can be adapted to fit different educational levels, disciplines, and institutional contexts. Whether used for introductory exposure or advanced problem-solving, they provide structured pathways for scaffolding student learning and promoting lifelong skills such as collaboration, inquiry, and critical thinking.

6.3 Research Methods

To effectively address a wide range of complex problems, a more rigorous model or methodological approach is required—one that has been developed through action learning and research, and is presented in this chapter. This chapter develops a methodological approach grounded in an action learning research framework, structured around key phases—planning, implementation, and reflection. The objective is to explore a method or process suitable for integrating PBL into engineering education.

Action learning is defined as both a process and a dynamic programme, in which small groups collaborate to address real-world problems while simultaneously reflecting on their learning and its broader impact on individuals and the organisation (Marquardt, 1999, p. 4). Given this premise, the action learning framework emerged as highly relevant to this study, particularly because PBL was implemented through a series of workshops, with iterative reflection and refinements made at each stage to enhance the process. Furthermore, a survey was conducted among undergraduate students, along with an unstructured discussion among facilitators who participated in the workshops.

6.3.1 Action Learning

A five-step action learning process—pre-planning, planning, preparation, implementation, and reflection—was adopted and applied as follows (Figure 6.1).

6.3.1.1 Pre-Planning

During this stage, key discussions occurred with relevant stakeholders, including academic leaders, coordinators, instructors, industry partners, and PBL experts. Decisions were made regarding several critical aspects, such as the selection of student groups, the number of participants, identification of facilitators, involvement of technical and industry experts, scheduling, timeframes, delivery modes, types of problems to be addressed in the PBL activities, and resource requirements (both intellectual and financial).

6.3.1.2 Planning

A detailed roadmap for conducting the workshops was developed, outlining the sequential stages to follow, estimating the necessary human, instructional, and intellectual resources, and defining the roles of various stakeholders. Additional planning elements included budgeting, venue selection, and scheduling.

6.3.1.3 Preparation

In this phase, the final student groups and facilitators were selected, and the venue, schedule, and duration of the workshops were confirmed. Technical and industry experts were contacted, and their participation was secured. The broad problem types to be addressed by students were defined, and the delivery modalities for PBL were finalised. Student groups were formed, and participants were notified about their group membership and the assigned problems, allowing them to prepare by reviewing relevant prior knowledge.

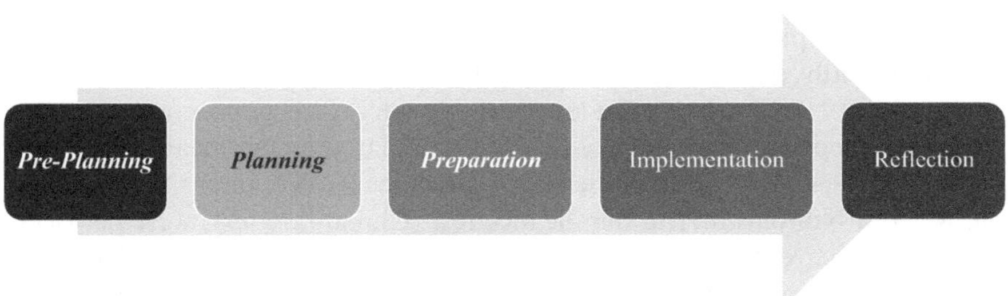

Figure 6.1 Steps in the Action Learning Process.

6.3.1.4 Implementation

On the scheduled dates, the PBL workshops were conducted. Prior to the workshops, an induction programme was organised to orient students and facilitators (instructors) regarding the programme's objectives, expected outcomes, and the roles and responsibilities of various stakeholders, including facilitators, experts, and students. The assessment methods were also explained.

Following the induction, students met in their respective groups to brainstorm potential solutions to the assigned problems. The workshops were conducted in a sequential manner, with continuous adjustments and improvements incorporated based on feedback and formative assessments conducted during and after each session.

6.3.1.5 Reflection

Throughout and after the workshops, facilitators maintained daily journals to document observations, key discussions, and reflections. Informal discussions were held with students during and after the workshops to gather their perceptions and opinions. Feedback was also collected from industry experts, participating faculty, and academic leaders through post-workshop interviews. These inputs were critically analysed to identify strengths, weaknesses, challenges, and areas for further improvement.

The insights from each workshop, along with reflections from all stages—from pre-preparation to implementation—were compiled into comprehensive journals. Lessons learned were reviewed, validated, and used to enhance the process for subsequent workshops.

6.3.2 Perception Survey and Data Analysis

A perception survey was conducted among undergraduate engineering students who participated in the PBL activities. Of the more than 150 students who participated in various workshops, 120 responded to the survey. The survey used a five-point Likert scale (1 = not important, 2 = marginally important, 3 = fairly important, 4 = significantly important, and 5 = most important) to gather perceptions on key aspects of the PBL experience.

Additionally, qualitative data were collected through non-structured discussions with 12 facilitators. The quantitative survey responses were analysed using a perception index (PI), which was calculated as the mean of all Likert scale responses. Also, the Coefficient of Variance (CoV), Inter Quartile Range (IQR), and Nonparametric significance test (Chi-square test) were performed to validate the responses and establish their statistical significance. The qualitative information gathered from discussions, journals, and stakeholder opinions was analysed using interpretative and narrative analysis techniques.

The combined quantitative and qualitative findings, along with the lessons learned and feedback from instructors, technical experts, and students, provided valuable insights for refining the PBL process and enhancing its implementation in future iterations.

6.4 The Proposed PBL Model: Stages and Steps

The development of an effective methodology for implementing PBL in undergraduate engineering education was accomplished through an iterative, workshop-based learning process. This approach enabled continuous refinement, as insights and lessons learned from each workshop were systematically incorporated to enhance the emerging model. The outcome was a structured, five-stage PBL framework consisting of 20 clearly defined steps, designed to support the efficient and effective implementation of PBL, achieve targeted learning outcomes, and cultivate essential student competencies.

The broader 20-step framework served as the foundational structure for this methodology. It was piloted and tested across multiple workshops, during which iterative feedback and practical experiences were used to further refine and streamline the approach (as illustrated in Table 6.1). This adaptive process not only improved the design and delivery of the workshops but also ensured that the evolving framework remained aligned with both pedagogical goals and the practical realities of implementing PBL in engineering education. However, in the final stages, the various steps within each stage were established, with their significance presented in Table 6.2 and discussed below. The importance and significance of each step were decided based on the concurrence evaluation of PI, SD, CoV, IQR and p-values.

Table 6.1 Plausible Initial Plan of a 20-Step Process

Steps	*Activity*
Step1	Understanding of philosophy
Step 2	Selection of challenges
Step 3	Selection of problems
Step 4	Formation of groups/teams
Step 5	Introduction to the groups and engagement with the groups
Step 6	Identification of the problem
Step 7	Preparation of draft design brief
Step 8	Swapping among teams
Step 9	Refinement of the draft design brief
Step 10	Presentation of design brief and feedback
Step 11	Discussion with stakeholders—challenge partners
Step 12	Preparation of the final design brief and submission
Step 13	Generating engineering ideas and discussing among team members
Step 14	Selection of ideas and generation of engineering concepts
Step 15	Discussion with technical partners
Step 16	Transformation of concepts to alternative engineering solutions
Step 17	Presentation of ideas and alternative solutions
Step 18	Evaluation of alternate solutions
Step 19	Design/develop a model/prototype of a sustainable engineering solution
Step 20	Presentation and showcasing—posters, Dragons' Den, videos

Table 6.2 Perception Index and the Level of Importance of Various Steps in the PBL Process

Steps in PBL	*PI*	*SD*	*CoV (%)*	*IQR*	*p-Value*	*Level of Importance*
Stage 1: Pre-PBL-Offering Preparation						
Contextualising PBL within the OBE	3.925	0.7237	18.4	1.0	<0.001	Fairly important
Reflection of the prerequisite knowledge and skills of the students	3.750	0.7247	19.3	1.0	<0.001	Fairly important
Selection of stakeholders (challenge (problem owners) and technical partners)	3.583	0.8460	23.6	1.0	<0.001	Fairly important
Selection of problems in alignment with learning outcomes	4.133	0.7771	18.8	1.0	<0.001	Significantly important
Stage 2: Team Building, Problem Formulation, and Design Brief						
Formation and preparation of heterogeneous teams to work on a complex problem	4.058	0.7702	19.0	2.0	<0.001	Significantly important
Engagement among the team members	3.983	0.8499	21.3	2.0	<0.001	Fairly important
Identification sub-problems	4.375	0.6868	15.7	1.0	<0.001	Significantly important
Preparation of design brief	4.258	0.7041	16.5	1.0	<0.001	Significantly important
Stage 3: Stakeholders' Engagement and Finalisation of The Design Brief						
Engagement among the members across the teams	3.483	0.8299	23.8	1.0	<0.001	Fairly important
Refinement of the design brief and presentation of the design brief	4.175	0.7956	19.1	1.0	<0.001	Significantly important
Engagement with stakeholders	3.608	0.7810	21.6	1.0	<0.001	Fairly important
Finalisation of the design brief	4.308	0.6711	15.6	1.0	<0.001	Significantly important
Stage 4: Idea Generation and Solutions						
Generation of ideas and engineering concepts	4.142	0.7810	18.9	1.0	<0.001	Significantly important
Engagement with technical experts	3.708	0.6533	17.6	1.0	<0.001	Fairly important
Transformation of concepts to alternative solutions (design or products)	4.275	0.7096	16.6	1.0	<0.001	Significantly important
Evaluation of the alternate solutions	4.050	0.7653	18.9	2.0	<0.001	Significantly important
Stage 5: Final Design/Development of Product and Showcasing						
Design of sustainable engineering solutions or the creation of sustainable products	4.258	0.6798	16.0	1.0	<0.001	Significantly important
Showcasing the design or product for evaluation	4.208	0.6723	16.0	1.0	<0.001	Significantly important

6.4.1 *Stage 1: The Pre-PBL Offering Preparation*

The first stage is the pre-PBL offering preparation, which includes contextualizing PBL within the OBE, a reflection of the prerequisite knowledge and skills of the students, selection of stakeholders (challenge (problem owners) and technical partners) and selection of problems in alignment with learning outcomes. As evidenced by the perception survey (Table 6.2), the respondents perceive that this stage is fairly important. For example, respondents perceive contextualizing PBL within the OBE (PI = 3.925, CoV = 18.4%, IQR = 1.0, & $p < 0.001$), the reflection of prerequisite knowledge and skill of the students (PI = 3.75, CoV = 19.3%, IQR = 1.0, & $p < 0.001$), and the selection of stakeholders (problem owners and technical partners) (PI = 3.58, CoV = 23.6, IQR = 1.0, & $p < 0.001$) as fairly important. However, the selection of problems in alignment with learning outcomes (PI = 4.133, CoV = 18.8%, IQR = 1.0, & $p < 0.001$) is perceived to be significantly important for PBL. In this context, two of the instructors express the view that:

> ...we think that prior knowledge and contextualisation of the problem are definitely important, but the alignment of the problem to the learning outcomes of the module or learning programme will help students to learn appropriate cognitive attributes such as analysis and design and achieve competency in exit level outcomes such as problem-solving.

6.4.2 *Stage 2: Team Building, Problem Formulation, and Design Brief*

The second stage consists of the formation and preparation of heterogeneous teams to work on a complex problem, engagement among the team members, identification of sub-problems, and preparation of a design brief. PBL occurs in teams, and, therefore, the creation of appropriate teams and team building is highly essential, as evidenced from the perception survey (PI = 4.058, CoV = 19%, IQR = 2.0, & $p < 0.001$). In order to build the team and understand the strengths, weaknesses, and challenges, engagement among the team members is significantly important (PI = 3.983, CoV = 19%, IQR = 2.0, & $p < 0.001$). However, as perceived by the respondents, the highly important activities are the identification of sub-problems (PI = 4.375, CoV = 15.7%, IQR = 1.0, & $p < 0.001$) and the preparation of the design brief (PI = 4.258, CoV = 16.5%, IQR = 2.0, & $p < 0.001$). Identification of the sub-problem from the broad and complex problem helps the students narrow down, focus, and work on a particular issue. Similarly, the preparation of a design brief provides the students with an outline or framework to work on the problem. It also helps them realize the intellectual resources, time, and limitations they have. These notions are corroborated by three of the instructors and one industry expert. According to them:

> ...the subproblem delimits the scope and focus of the work at hand. The students can know what they have to do.

...The preparation of draft design briefs provides a kind of rough outline, objectives to be achieved, and methodology to be followed. It also helps to understand the resources, limitations and constraints.

6.4.3 Stage 3: Stakeholders' Engagement and Finalization of the Design Brief

Engagement among the members across the teams, refinement of the design brief, presentation of the design brief, engagement with stakeholders and finalisation of the design brief are the steps that constitute the third stage of the learning. According to the perceptions of the respondents, engagement among the members across the teams (PI = 3.483, CoV = 23.8%, IQR = 1.0, & $p < 0.001$) and engagement with stakeholders (PI = 3.608, CoV = 21.6%, IQR = 1.0, & $p < 0.001$) are fairly important. Further, refinement of the design brief and presentation of the design brief (PI = 4.175, CoV = 19.1%, IQR = 1.0, & $p < 0.001$) and finalisation of the design brief (PI = 4.308, CoV = 15.6%, IQR = 1.0, & $p < 0.001$) are significantly important. Engagement among the members across the teams offers various teams the opportunity to learn about the intellectual and human resources that other teams have, which different teams can use. In other words, they can collaborate to solve any aspect by sharing knowledge and resources. As one of the industry experts and an academic leader who participated in the programme observed:

> ...Collaboration is vital to solving problems because every team may not have everything they need. For example, a team may have a student who is very good at solving mathematical equations, and another team may have someone who is good at using the software. So, they can collaborate and use each other's knowledge and skills.

Engagement with stakeholders offers insights into the complexity and intricacies of the problems. Students can also benefit and learn from the experiences and expertise of the stakeholders, particularly industry experts.

However, refinement of the design brief and presentation of the design brief (PI = 4.175, CoV = 19.1%, IQR = 1.0, & $p < 0.001$) and finalisation of the design brief (PI = 4.308, CoV = 15.6%, IQR = 1.0, & $p < 0.001$) are found to be significant steps in this stage. According to two of the instructors and two industry experts:

> ... presentation of the draft design brief helps improvement and refinement from the feedback and comments of their peers, instructors and industry experts. Also, the final design brief delineates the objectives, scope and methodology of the work. The students now know what exactly to do and how they should do it. So, these are very crucial steps in the learning process.

6.4.4 *Stage 4: Idea Generation and Solutions*

The fourth stage consists of the generation of ideas and engineering concepts, engagement with technical experts, the transformation of concepts to alternate solutions (design or products), and evaluation of the alternate solutions. According to the respondents' perception, generation of ideas and engineering concepts (PI = 4.142, CoV = 18.9%, IQR = 1.0, & $p < 0.001$), the transformation of concepts to alternate solutions (design or products) (PI = 4.275, CoV = 16.6%, IQR < 1.0 & $p < 0.001$), and evaluation of alternate solutions (PI = 4.050, CoV = 18.9%, IQR = 2.0, & $p < 0.001$) are significantly important. These perceptions are also corroborated by the instructors. As one instructor pointed out:

> ...I find students really get engaged seriously and deliberate a lot during this idea generation stage. This step provides them with the opportunity to express their ideas and opinions and also to appreciate each other's ideas.

Also, two of the instructors observed:

> ...During the development of alternate solutions, we find the students show their innovativeness and ingenuity...they get engaged with literature, different sources of knowledge, different existing products, etc....they really enjoy this step and learn.

Similarly, one of the instructors said:

> ...I found during the evaluation of alternate solutions step, they [students] argue a lot...[but] also try to be rational and logical to find different merits and demerits of the solutions. Definitely, they learn significantly...at least how to evaluate the different solutions.

However, engagement with technical experts, although fairly important (PI = 3.708, CoV = 17.6%, IQR = 1.0 & $p < 0.0013.72$), according to the industry experts, the impact on student learning seems to be limited.

6.4.5 *Stage 5: Final Design/Development of Product and Showcasing*

Design of sustainable engineering solutions or creation of sustainable products (PI = 4.258, CoV = 16.0%, IQR = 1.0 & $p < 0.001$) and showcasing the design or product for evaluation (PI = 4.208, CoV = 16.0%, IQR = 1.0, & $p < 0.001$) constitute the fifth and final stage of the learning process before the final assessment. These two steps in the final stages of the PBL are perhaps the most important, as this culminates the students' endeavour to a particular outcome and also makes their effort to be visualised by other stakeholders or evaluators. According to some students:

> ...Coming all the way from understanding a problem to developing a solution really makes us learn how to learn and how to use our knowledge. It broadened our outlook and made us work together and learn from each other.

Another student observed:

> ...Initially I was not sure and a bit sceptical. I thought about what I will learn and whether I can really bring out something as a solution with the help of...these [my team members] guys...but when I see people appreciate our work after our presentation, I feel we are good. More than that, we learned how to develop the solution and how to sell to the stakeholders.

6.5 Discussion

The methodological approach—or proposed model—for implementing PBL in this study (Figure 6.2) aligns with and extends the existing literature, particularly within the context of engineering education. Developed through a structured, iterative process, the model incorporates stages such as preparation, problem formulation, stakeholder engagement, idea generation, solution development, and solution showcasing. It emphasises active, student-centred learning and responds to challenges specific to engineering education, notably the limited availability of WIL opportunities, which has created gaps in practical, real-world exposure.

The **preparation phase** lays a crucial foundation by aligning PBL activities with OBE, considering students' prerequisite knowledge, selecting appropriate stakeholders, and defining meaningful problems. This aligns with recommendations from PBL scholars who stress the importance of connecting learning activities to curriculum goals and learners' prior knowledge to foster deeper engagement and understanding (Dochy et al., 2003; Schmidt & Moust, 2000). Survey results from this study indicated that selecting problems aligned with learning outcomes was considered the most critical step in this phase, reinforcing the importance of problem relevance—an emphasis also noted by Savery and Duffy (1995). Furthermore, stakeholder selection and preparation align with Bhaumik et al. (2024), who highlight the importance of industry engagement in contextualising real-world challenges. For engineering educators, this underscores the need to design problems that are not only complex and authentic but also directly linked to desired learning outcomes.

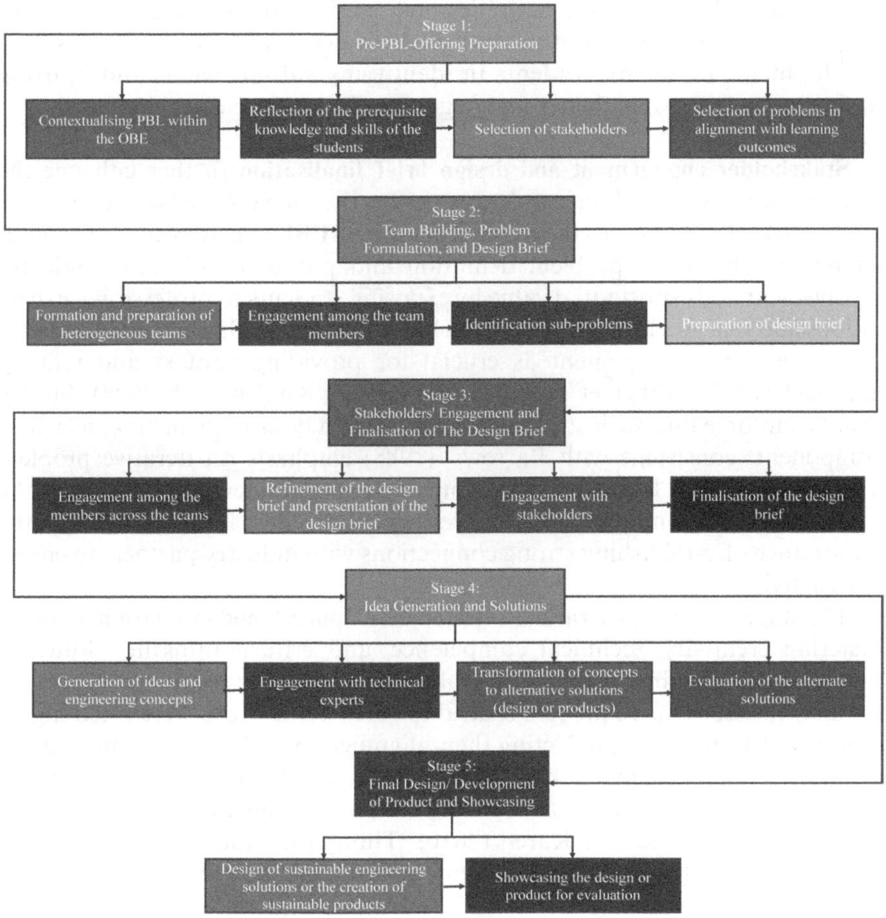

Figure 6.2 The Proposed Methodological Approach for Implementing PBL in Engineering Education.

Team building and problem formulation enhance collaboration and critical thinking by encouraging diverse teams to deconstruct complex problems, identify sub-problems, and prepare structured design briefs. This approach mirrors findings in the PBL literature regarding the cognitive benefits of collaborative problem-solving and guided structure. The importance of team diversity and sub-problem identification aligns with Vygotsky's (1978) theories on collaborative learning. It is supported by Schmidt (1998) and Bhaumik et al. (2024), who emphasise teamwork as a driver of cognitive engagement and higher-order thinking skills (HOTs). Preparing a design brief—a step rated highly in the survey—guides the problem-solving process and reflects the emphasis on structured formulation found in Fogarty's (1997) seven-step model.

In practical terms, these activities help engineering students build communication, teamwork, and analytical skills essential to their professional development. Engaging students in identifying sub-problems and drafting design briefs empowers them to manage complex tasks and set clear goals and methodologies.

Stakeholder engagement and design brief finalisation further enhance the learning process by enabling students to refine their ideas based on expert feedback. This stage embodies the iterative nature of PBL, where continuous input strengthens both the problem definition and proposed solutions. Industry involvement adds substantial value by exposing students to professional expectations and real-world constraints. The literature affirms this benefit, highlighting stakeholder engagement as crucial for providing context and refining approaches (d'Escoffier et al., 2024; Riazifar & Khajehsaeid, 2024). Survey results reinforce this, with students ranking design brief refinement as a critical component, consistent with Barrows' (1988) emphasis on iterative problem analysis. Similarly, Kabilan (2024) notes that stakeholder input enhances the authenticity of learning experiences. For educators, these findings highlight the importance of establishing strong connections with industry partners to enrich PBL activities.

The stages of **idea generation, solution development, and evaluation** focus on fostering creativity, technical competence, and critical thinking. Students' involvement in brainstorming and evaluating alternative solutions reflects the inquiry-based, iterative process central to PBL. These steps were rated highly in terms of importance, reflecting their alignment with literature emphasising innovation and problem-solving (Díaz et al., 2023). The evaluation phase underscores the need for critical reflection and refinement—an essential feature noted by Elaine and Karen (2016). Though interaction with technical experts was rated slightly lower, it remains valuable for providing specialised insights, supporting Stoyanov et al.'s (2023) findings on expert guidance in PBL. This phase reinforces the need for hands-on, reflective activities that promote solution-focused thinking and the development of higher-order competencies.

The **final stage**, which involves the development and showcasing of sustainable engineering solutions, serves as a capstone where students apply their learning, demonstrate competencies, and receive feedback. This phase is aligned with the reflective components of existing PBL models and emphasises experiential learning through real-world application. Designing and presenting sustainable products were identified as critical steps in this study, echoing the literature's focus on authentic assessment and application (Buentello-Montoya, 2022; Riazifar & Khajehsaeid, 2024). The presentation of solutions also aligns with the final phase of the Barrows and Myers Model, which stresses consolidation of learning through reflection and evaluation (Savery & Duffy, 1995). Pedagogically, this stage suggests that educators should adopt comprehensive assessment strategies—including peer reviews, project showcases, and reflective journals—that go beyond traditional examinations to holistically evaluate student learning and development.

Overall, the findings of this study affirm and extend existing PBL literature by addressing challenges and opportunities specific to engineering education. The proposed methodology enhances traditional PBL by introducing structured problem formulation, iterative refinement, and stakeholder engagement. It also aligns with competency-based learning and OBE, offering a scalable, flexible approach suited for contexts with limited WIL opportunities.

From a practical standpoint, the proposed model offers valuable guidance for engineering educators. It underscores the importance of clear problem definition, iterative design, real-world engagement, and hands-on learning in building students' cognitive, technical, and professional skills. By bridging gaps in conventional teaching methods and integrating PBL more effectively into the curriculum, this approach better prepares students for the complexities of modern engineering practice. Ultimately, it contributes to the growing body of evidence that well-implemented PBL can transform engineering education by fostering creativity, collaboration, critical thinking, and practical problem-solving skills.

6.6 Conclusions

The development of a methodology/model for implementing PBL in engineering education presents a significant challenge, necessitating a structured and iterative approach. A central concern is ensuring that each step in the process effectively supports student learning and skill development. Given that PBL is still not widely practised at the undergraduate level in engineering, there is a significant absence of established benchmarks to guide its implementation. Moreover, aligning PBL with an OBE framework adds another layer of complexity, as it requires careful planning to ensure that intended learning outcomes are achieved and that students demonstrate the necessary competencies aligned with exit-level outcomes. Without a clearly defined methodology/process including steps, there is a risk that PBL activities may lose focus, leading to misalignment between learning tasks and desired educational goals.

This study aimed to address these challenges by developing a comprehensive methodology for delivering PBL in undergraduate engineering education. Drawing on an action learning approach and insights gained from a series of workshops, the study gathered and analysed the perceptions of students, instructors, and key stakeholders, including industry partners. The findings led to the identification of five sequential stages essential for the effective implementation of PBL and the attainment of targeted learning outcomes.

The first stage, *pre-PBL preparation*, involves contextualising PBL within the OBE framework, reflecting on students' prior knowledge and skills, selecting appropriate stakeholders, and identifying complex problems that align with learning outcomes. The second stage focuses on *team formation and problem formulation*, emphasising the creation of diverse teams, fostering team engagement, breaking down problems into sub-problems, and drafting a structured design brief. The third stage, *team engagement and stakeholder interaction*, includes refining and finalising the design brief, presenting it, and incorporating feedback from stakeholders to ensure real-world relevance. The fourth stage, *solution development*, highlights idea generation, collaboration with technical experts, transformation of ideas into alternative solutions (designs or products), and critical evaluation of these alternatives. The final stage centres on the *creation and showcasing of sustainable engineering solutions*, where students design, build, and present their solutions for evaluation and assessment.

Formative assessment is integrated throughout the process, particularly at key transition points such as the preparation of the draft design brief, development of alternative solutions, and final product creation. While all steps are important, certain milestones were identified as particularly significant for successful PBL delivery. These include the selection of problems aligned with learning outcomes, formation of diverse teams, identification of sub-problems, finalisation of the design brief, idea generation, and the development of sustainable engineering solutions.

The study further indicates that students engaged in this PBL framework can develop a wide range of cognitive learning attributes, including the ability to understand and analyse complex problems, apply knowledge, evaluate alternatives, and design or create innovative solutions. In addition, students can achieve critical exit-level outcomes, such as problem-solving abilities, technical competence, conceptual understanding, commitment to sustainability, teamwork, and effective communication. These attributes are cultivated at various stages of the PBL process, with assessment carried out using a structured rubric that supports the formative evaluation of both cognitive skills and outcome-based competencies. This approach embeds assessment as an integral component of the learning journey, promoting active engagement and deeper learning.

Thus, when offered through a structured and systematic methodology, PBL has the potential to significantly enhance student learning and the achievement of core competencies in undergraduate engineering education. Through a focus on systematic problem formulation, stakeholder engagement, iterative refinement, and real-world application, the proposed methodology/model offers a robust framework for effectively implementing PBL, ultimately preparing engineering students to meet the complex demands of professional practice.

References

Barrows, H. S. (1988). *The tutorial process*. Southern Illinois University School of Medicine.

Bhaumik, K., Datta, A., Ismayel, G., & Sarif, B. M. (2024). Fostering higher-order thinking: Pedagogical strategies in engineering education. *Journal of Engineering Education Transformations*. https://doi.org/10.16920/jeet/2024/v38i1/24177

Buentello-Montoya, D. (2022). Problem-based learning as a tool to foster electricity and electromagnetism-related concepts in an engineering course. In *International Conference on Teaching, Assessment, and Learning for Engineering* (pp. 174–178). https://doi.org/10.1109/TALE54877.2022.00035

d'Escoffier, L. N., Guerra, A., & Braga, M. (2024). Problem-based learning and engineering education for sustainability. *Journal of Problem-Based Learning in Higher Education*. https://doi.org/10.54337/ojs.jpblhe.v12i1.7799

Derry, S. J. (1996). Cognitive schema theory in the constructivist debate. *Educational Psychologist*, *31*(3–4), 163–174.

Dias, S., Queiroz, K., & Dias, A. (2022). Proposta de metodologia para a aplicação do aprendizado baseado em problemas no ensino dos conteúdos de eletrônica digital. https://doi.org/10.20906/cba2022/3321

Díaz, O., Segredo-Morales, E., & Gonzalez, E. (2023). Problem-based learning: Application to a laboratory practice in the degree of industrial chemical engineering. *International Journal of Engineering Pedagogy (IJEP)*. https://doi.org/10.3991/ijep.v13i6.39737

Dochy, F., Segers, M., Van den Bossche, P., & Gijbels, D. (2003). Effects of problem-based learning: A meta-analysis. *Learning and Instruction*, *13*(5), 533–568.

Dos Santos, S. C., Bispo, E. L., de Carvalho Santos Filho, O., & de Oliveira R. P. A. (2022). Problem-based learning diagnosis in computing higher education: An overview from Brazilian public institutions. *SN Computer Science*, *3*(3). https://doi.org/10.1007/s42979-022-01106-z

Elaine, H. J., & Karen, G. Y. (2016). Problem-based learning: An overview of its process and impact on learning. *Health Professions Education*, *2*, 275–279.

Epler, P. & Jacobs, J. (2022). Problem-based learning classroom models in higher education. In: *Guide to integrating problem-based learning programs in higher education classrooms: Design, implementation, and evaluation*, pp. 109–130. IGI Global Scientific Publishing. https://doi.org/10.4018/978-1-7998-8177-3.ch007

Fogarty, R. (1997). *Problem-based learning and other curriculum models for the multiple intelligences classroom*. IRI/SkyLight Training and Publishing.

Glaser, R., & Bassok, M. (1989). Learning theory and the study of instruction. *Annual Review of Psychology*, *40*, 631–666.

Kabilan, S. J. (2024). A problem-based learning (PBL) approach in effective implementation of innovation and entrepreneurship education for engineering undergraduates. *Journal of Engineering Education Transformations*, *38*(is1), 120–126. https://doi.org/10.16920/jeet/2024/v38is1/24220

Kazemi, F., & Ghoraishi, M. (2012). Comparison of problem-based learning approach and traditional teaching on attitude, misconceptions and mathematics performance of university students. *Procedia – Social and Behavioural Sciences*, *46*, 3852–3856.

Kolmos, A., & de Graaff, E. (2014). Problem-based and project-based learning in engineering education: Merging models. In A. Johri & B. M. Olds (Eds.), *Cambridge handbook of engineering education research* (pp. 141–161). Cambridge University Press. https://doi.org/10.1017/CBO9781139013451.012

Marcinauskas, L., Iljinas, A., Čyvienė, J., & Stankus, V. (2024). Problem-based learning versus traditional learning in physics education for engineering program students. *Neveléstudomány*. https://doi.org/10.3390/educsci14020154

Marquardt, M. J. (1999). *Action learning in action: Transforming problems and people for world-class organisational learning*. Davies-Black Publishing.

Mayer, R. E. (1996). Learners as information processors: Legacies and limitations of educational psychology's second metaphor. *Educational Psychologist*, *31*(3–4), 151–161.

Mubuuke, A. G., Louw, A. J. N., & Van Schalkwyk, S. (2017). Cognitive and social factors influencing students' response and utilisation of facilitator feedback in a problem-based learning context. *Health Professions Education*, *3*, 85–98.

Palincsar, A. S. (1998). Social constructivist perspectives on teaching and learning. *Annual Review of Psychology*, *49*, 345–375.

Riazifar, N., & Khajehsaeid, H. (2024). Enhancing engineering education through problem-based learning. In *Proceedings of UK and Ireland Engineering Education Research Network Conference 2023*. University of Warwick. https://journals.warwick.ac.uk/index.php/eern/article/view/1636

Savery, J. R., & Duffy, T. M. (1995). Problem-based learning: An instructional model and its constructivist framework. *Educational Technology*, *35*(5), 31–38.

Schmidt, H. G. (1993). Foundations of problem-based learning: Some explanatory notes. *Medical Education*, *27*(5), 422–432.

Schmidt, H. G. (1998). Problem-based learning: Does it prepare medical students to become better doctors? *Medical Journal of Australia*, *168*, 429–430.

Schmidt, H. G., & Moust, J. H. C. (2000). Factors affecting small-group tutorial learning: A review of research. In D. H. Evensen & C. E. Hmelo-Silver (Eds.), *Problem-based learning: A research perspective on learning interactions* (pp. 19–52). Lawrence Erlbaum.

Servant-Miklos, V. F.C. (2019). Fifty years on: A retrospective on the world's first problem-based learning programme at McMaster University Medical School. *Health Professions Education*. 5(1). https://doi.org/10.1016/j.hpe.2018.04.002

Servant-Miklos, V. F. C., & Kolmos, A. (2022). Student conceptions of problem and project-based learning in engineering education: A phenomenographic investigation. *Journal of Engineering Education*, *111*(4), 792–812. https://doi.org/10.1002/jee.20478

Stoyanov, I., Ivanov, V., & Koşunalp, S. (2023). Applying a problem-based learning approach to the topic of modelling the power and improvement of voltage stability of wind turbines. https://doi.org/10.1109/elma58392.2023.10202556

Utama, N. Y., Mindarta, E. K., Komarudin, K., & Suhartadi, S. (2024). Implementasi problem-based learning untuk meningkatkan kompetensi perbaikan sistem bahan bakar siswa pusdiklat. *Jurnal Pendidikan Vokasi Otomotif*, *7*(1), 168–183. https://doi.org/10.21831/jpvo.v7i1.79356

Vygotsky, L. S. (1978). *Mind in society: The development of higher psychological processes*. Harvard University Press.

Wachholz, L. C., & Rasia, L. A. (2023). A proposition to teach engineering projects using problem-based learning (PBL) concepts. *Caderno Pedagógico*, *20*(2), 1115–1126. https://doi.org/10.54033/cadpedv20n2-023

Ward, J., & Lee, C. (2002). A review of problem-based Learning. *Journal of Family and Consumer Sciences Education*, *20*(1), 16–26.

Yeung, E., Au-Yeung, S., Chiu, T. et al. (2003). Problem design in problem-based learning: Evaluating students' learning and self-directed learning practice. *Innovations in Education and Teaching International*, *40*(3), 237–244.

Part III

Implications of PBL Engineering Education

Chapter 7

Problem-Based Learning for Reinforcing Engineering Education

7.1 Introduction

Problem-Based Learning (PBL) is a transformative, student-centred educational approach that has significantly reshaped engineering education. As discussed in previous chapters, it emphasizes active learning, critical thinking, and the practical application of theoretical knowledge (Duch et al., 2001; Hmelo-Silver, 2004; Hmelo-Silver & Barrows, 2006). Unlike traditional lecture-based methods, which often encourage passive absorption of information, PBL actively engages students in tackling real-world, open-ended problems. Through this process, students are required to investigate, analyze, and develop solutions to complex challenges that closely mirror those found in professional engineering contexts.

By placing students at the centre of their learning journey, PBL fosters not only technical proficiency but also essential professional skills such as teamwork, effective communication, leadership, and self-directed learning. These attributes are critical for success in the engineering field (Wolf, 2007).

In the context of engineering education—where bridging theory and practice is vital—PBL serves as a powerful tool to enhance students' problem-solving abilities and their capacity to apply core engineering concepts in practical scenarios. This method encourages multidisciplinary thinking, collaborative engagement, and innovation. Typically, the PBL cycle involves identifying and understanding a problem, conducting research, generating and evaluating potential solutions, and reflecting on outcomes. This iterative process enhances cognitive engagement and supports deeper knowledge retention.

Research underscores the effectiveness of PBL in improving both student competence and motivation. For example, a study at Pusdiklat Suzuki and the State University of Malang demonstrated that implementing PBL in automotive engineering, particularly in vehicle fuel system repair, significantly boosted both knowledge and skills mastery—from 56% to 100% and from 48% to 100%, respectively (Utama et al., 2024). Similarly, studies conducted in Malaysia found that PBL increased student engagement by presenting complex, real-world challenges that stimulated critical thinking, creativity, and collaborative learning (Hidayat & Osei Kwabena, 2023).

DOI: 10.1201/9781003529743-10

PBL has been successfully applied across various engineering disciplines, including electromagnetism, circuit design, and wind turbine engineering. For instance, students working on PBL-based projects related to supercapacitor circuit design or electromagnetic engine simulations reported high levels of satisfaction and improved performance, with average grades showing notable increases (Buentello-Montoya, 2022). In wind turbine engineering, the integration of PBL with virtual laboratories enabled students to address key issues such as reactive power compensation and voltage stability, thereby enhancing their understanding of energy systems and power quality (Stoyanov et al., 2023).

Despite its many benefits, PBL presents several challenges. Effective implementation requires careful planning, especially in large classes, and continuous evaluation to ensure the relevance and authenticity of problem scenarios. Moreover, while PBL excels at cultivating practical problem-solving and soft skills, traditional methods may remain more effective for delivering theoretical content, particularly when assessments involve closed-ended questions (Boud & Feletti, 1991; Marcinauskas et al., 2024). Thus, a balanced integration of PBL and traditional approaches may yield the most comprehensive educational outcomes in engineering curricula.

One ongoing challenge in engineering education is the limited integration of Work-Integrated Learning (WIL), a core element of Outcomes-Based Education (OBE). Although WIL is essential for providing students with workplace experience, it is often optional, leaving gaps in students' preparedness for industry. This lack of practical exposure can impede the development of key graduate attributes such as problem-solving, design competence, and teamwork. Furthermore, critics of OBE argue that it does not always effectively support cognitive development (Brady, 1996; McCarthy, 1993).

PBL presents a promising solution to these limitations by offering hands-on, practice-based learning experiences within academic settings. When aligned with real-life socio-economic and environmental challenges—particularly those linked to the United Nations Sustainable Development Goals (SDGs)—PBL enables students to develop both industry-relevant skills and higher-order cognitive abilities.

To explore the practical relevance, implications, and challenges of PBL, a study was conducted involving industry-based PBL workshops for undergraduate engineering students. This investigation addressed the following research questions:

- RQ1 What is the relevance of PBL for engineering education?
- RQ2 What are the implications of PBL for engineering education?
- RQ3 What challenges are experienced during the implementation of PBL in engineering education?
- RQ4 How can PBL be contextualized to reinforce engineering education?

These workshops focused on real-life engineering problems, including those aligned with SDG-related issues. Data were collected through surveys, focus group discussions, and project evaluations, followed by both statistical analysis and qualitative narrative analysis. The findings indicate that, even within a controlled academic environment, PBL provides substantial benefits. It enables students to engage with practical challenges under the guidance of academic mentors and industry professionals.

The study also highlights that PBL reinforces student learning by helping them achieve key exit-level outcomes and graduate attributes, such as problem-solving, design, interdisciplinary collaboration, and entrepreneurial thinking. For maximum impact, PBL should be systematically embedded into undergraduate engineering curricula. This integration requires alignment between Intended Learning Outcomes (ILOs), teaching and learning activities, and assessment tasks. Furthermore, by incorporating PBL more comprehensively into engineering programs, students can acquire the practical, cognitive, and collaborative skills needed to navigate the complexities of modern engineering practice and contribute meaningfully to technological and societal advancement.

7.2 A Theoretical Perspective on PBL for Reinforcing Engineering Education: Relevance, Implications, Challenges, and Contextualization

PBL has emerged as a highly relevant pedagogical strategy in engineering education, addressing the limitations of traditional, lecture-centric methods. Rooted in learner-centred pedagogy, PBL emphasizes active engagement through real-world problem-solving, which not only enhances knowledge retention but also aligns educational objectives with the dynamic demands of the engineering profession (Abou-Hayt et al., 2020; Barrett, 2005; Chauhan, 2017; Riazifar & Khajehsaeid, 2024). The approach fosters critical thinking, creativity, and practical application of theory, enabling students to develop both technical and soft skills essential for modern engineering practice (Castro-Vargas & Cabana-Caceres, 2024; Naik et al., 2024). In contrast to rote memorization, PBL situates students in complex, ill-structured problems that mirror real-world engineering challenges, thereby bridging the gap between classroom learning and professional requirements (Das & Mishra, 2018; Schmidt, 1983, 1998; Ward & Lee, 2002; Yeung et al., 2003).

Moreover, PBL promotes self-directed learning—a crucial skill in the context of rapid technological change. Students learn to identify their knowledge gaps, formulate objectives, and synthesize information independently, thereby preparing for lifelong learning and adaptability (Chen, 2008; Havenga & Van Wyk, 2017; Wolf, 2007). As engineering increasingly demands interdisciplinary competence, collaboration, and innovation, PBL has become especially relevant in preparing graduates to meet these challenges (Delaney & Nagle, 2019; Tsalapatas et al., 2019).

The pedagogical implications of integrating PBL into engineering curricula are extensive. One of the most impactful outcomes is the enhancement of student engagement and learning outcomes. PBL empowers students to take ownership of their education, which leads to increased motivation, academic persistence, and deeper conceptual understanding (Naik et al., 2024; Riazifar & Khajehsaeid, 2024). In various contexts, including civil and computer engineering, PBL has demonstrated improvements in both technical proficiency and soft skills, such as leadership, teamwork, and project management (Noor et al., 2019; Striegel & Rover, 2002; Zeng & Xu, 2010).

By aligning curricula with real-world problems, PBL ensures that engineering graduates are better prepared to solve the complex, multidisciplinary issues faced by industry (Tsalapatas et al., 2019). Institutions like Aalborg University exemplify how large-scale interdisciplinary projects can build essential graduate attributes and reinforce academic learning (Kolmos & de Graaff, 2014). Furthermore, the integration of digital tools in PBL—such as simulations, collaborative platforms, and modelling software—has expanded its impact, making it scalable and accessible in diverse educational contexts (Lavado-Anguera et al., 2024; Patnawar, 2023).

Importantly, PBL supports the development of competencies tied to OBE, helping students meet specific learning outcomes and graduate attributes (Das & Mishra, 2018). When aligned with the SDGs, PBL also fosters a sense of social responsibility, allowing students to engage with global socio-economic and environmental challenges (Alavi, 1995; Chen, 2008).

Despite its advantages, PBL poses several implementation challenges that must be addressed to realize its full potential. Faculty resistance is a recurring issue, as many educators are accustomed to traditional pedagogical models and may lack the training to facilitate PBL effectively (Kjellberg et al., 2015; Rehman, 2023). Faculty development programs are essential to build instructional capacity and support the transition to a facilitator role (van Barneveld & Strobel, 2015).

Resource limitations also pose a barrier, especially in institutions with constrained infrastructure, technology, or administrative support (Rehman, 2023; Tuohi & Roslöf, 2007). PBL requires time-intensive planning, the design of meaningful problem scenarios, and often smaller class sizes to be effective, all of which can strain institutional resources.

Assessment and curriculum alignment remain particularly difficult. Traditional assessment methods may not adequately capture the depth of learning achieved in PBL settings. Developing holistic evaluation frameworks—including peer assessment, reflective journals, and formative feedback—is crucial for measuring both cognitive and process-based outcomes (Hunt et al., 2010; Servant-Miklos, 2018). Furthermore, balancing content coverage with PBL's open-ended structure within time-constrained curricula requires thoughtful course design and often institutional reform (Maladzhi et al., 2024).

Student resistance can also occur, especially in educational cultures where teacher-centred learning is the norm. Students unfamiliar with PBL may struggle with the autonomy and ambiguity it entails (Tsalapatas et al., 2019). Addressing this challenge requires scaffolding strategies that gradually increase complexity and independence (Hunt et al., 2010).

Effective contextualization of PBL involves tailoring its implementation to suit institutional, cultural, and disciplinary needs. Interdisciplinary collaboration is a vital strategy; engineering projects that span multiple domains provide a more holistic learning experience and better mirror industry challenges (Lavado-Anguera et al., 2024; Maladzhi et al., 2024). For instance, integrating environmental, economic, and social dimensions into PBL projects supports alignment with the SDGs and fosters responsible engineering practice.

Digital PBL environments, including virtual labs and simulation-based learning, expand accessibility and can enhance engagement in large or resource-limited programs (Patnawar, 2023; Tsalapatas et al., 2019). These tools also allow institutions to scale PBL while maintaining interactive and inquiry-driven learning.

Industry partnerships are another essential component of contextualization. Collaborating with engineering firms enables the co-design of authentic project experiences and exposes students to current practices and expectations (Kjellberg et al., 2015). These collaborations also facilitate the alignment of ILOs with real-world applications.

Faculty support remains a cornerstone of successful contextualization. Institutions must invest in ongoing professional development, provide access to shared resources, and cultivate communities of practice to help educators refine their approach (van Barneveld & Strobel, 2015).

Globally, institutions like those in the UK, Denmark, and South Africa have piloted various PBL integration models. These include hybrid approaches combining PBL with lecture-based teaching, modular PBL structures, and capstone projects embedded within the curriculum (Das & Mishra, 2018). These adaptations demonstrate that, with careful planning and stakeholder involvement, PBL can be effectively contextualized to reinforce engineering education across diverse settings.

Thus, PBL stands as a transformative approach in engineering education, offering a pathway to align academic learning with the complexities of real-world engineering. Its relevance lies in its capacity to prepare students for dynamic professional environments through active, interdisciplinary, and student-centred learning. While challenges in implementation persist—ranging from faculty resistance to resource constraints—strategic contextualization can overcome these barriers. Through collaborative efforts involving educators, institutions, and industry, PBL can be fully integrated into engineering curricula to cultivate the competencies required for sustainable innovation and global impact. However, to better understand these dynamics and assess their practical implications, this study was undertaken.

7.3 Approach of the Study

The research adopted a mixed-methods approach grounded in experiential inquiry and reflective pedagogical practice. Central to the study was the critical examination of industry-based PBL workshops conducted at a South African university in 2018. These workshops served both as an intervention and as a research setting, allowing for the real-time observation and evaluation of PBL processes and outcomes within undergraduate engineering education. The workshops were designed to simulate authentic learning environments and to align PBL with complex, real-world problems drawn from the United Nations SDGs. The participant cohort included a total of 89 undergraduate students and 17 facilitators, comprising lecturers from various engineering disciplines. The first workshop involved 21 students and seven academic staff members representing the disciplines of Civil Engineering, Electrical, Electronics and Computer Engineering, and Mechanical Engineering. The second workshop engaged 22 students and three facilitators, all drawn from various sub-disciplines within Civil Engineering. The third workshop was comprised of 24 students and three lecturers from the Civil Engineering and Bachelor of Science in Hydrology programmes. The fourth workshop was conducted with 22 students and four lecturers, representing Civil Engineering, Electrical, Electronics and Computer Engineering, Mechanical Engineering, as well as students from the Built Environment disciplines.

The dual objectives of the workshops were to introduce both the pedagogical philosophy and methodological underpinnings of PBL to students and academic staff, and to encourage active engagement in the teaching and learning process through problem-solving. A central component of the PBL method implemented in these workshops was the collaborative and interdisciplinary nature of the problem-solving task. Participants were divided into smaller groups and asked to identify a real-life or industry-based problem contextualized within the framework of the SDGs. They were then tasked with formulating conceptual engineering solutions to complex socio-economic and environmental challenges. Each group worked intensively within a defined timeframe—three days in the first workshop and four days in the second—under the facilitation of academic staff who assumed non-directive roles, consistent with the tenets of PBL.

7.3.1 Data Collection

To complement the qualitative insights gained through the implementation of the workshops, a purposive sampling technique was employed to gather both quantitative and qualitative data from the student and facilitator participants. At the end of the workshops, a pre-tested questionnaire was administered to all 89 student participants and 17 facilitators. The questionnaire included both closed-ended and open-ended questions. Closed-ended items captured quantitative data on participants' perceptions regarding the relevance, benefits, and challenges of PBL in engineering education. Open-ended items sought to elicit participants' reflections on the integration of PBL into existing engineering curricula and its alignment with sustainable development themes.

In addition to the survey, a focus group discussion was organized to further explore the opportunities, constraints, and implications of PBL in higher education contexts. The focus group was composed of lecturers, academic developers, academic leaders, and industry professionals with significant experience in engineering pedagogy and curriculum development. Participants were selected for their familiarity with educational innovations and for their roles in academic leadership and engagement with industry. The focus group aimed to generate deeper discourse on PBL implementation strategies and to capture institutional and professional insights on embedding PBL into learning programmes. The researcher served as the facilitator for the discussion to ensure continuity and focus in addressing the research themes.

7.3.2 *Data Analysis*

The study employed a combination of analytical techniques to ensure comprehensive and triangulated findings. Quantitative data collected through structured questionnaires were analysed statistically using the Perception Index (PI), calculated as the mean of responses on a Likert scale. To evaluate the consistency and reliability of the responses, additional statistical measures—including the Coefficient of Variation (CoV) and the Interquartile Range (IQR)—were applied. Furthermore, a nonparametric Chi-square test was conducted to assess the significance of the responses and validate their statistical relevance. Together, these analyses offered valuable insights into overall trends, participant satisfaction, and the perceived effectiveness of the PBL approach.

Qualitative data, which included student project outputs developed during the workshops, were analysed using content analysis. This involved a systematic coding and categorization process to identify recurring themes, application of engineering knowledge, and innovative elements within the projects. Additionally, narrative analysis was applied to interpret the open-ended questionnaire responses and the focus group discussion transcripts. This method allowed for the synthesis of participant narratives, enabling the identification of emergent patterns, underlying beliefs, and nuanced perspectives regarding PBL's integration into engineering education.

By drawing from multiple sources and analytical lenses, the study generated a robust understanding of the practical implications, effectiveness, and transformative potential of PBL within a real-world engineering education context.

The integrated methodological approach further underscored the pedagogical value of PBL as a means of addressing complex, interdisciplinary challenges aligned with the SDGs.

7.4 Study Outcomes on PBL to Reinforce Engineering Education

The role of PBL in engineering education, including that of civil engineering, was analysed based on three aspects: the relevancy of the pedagogical method; its implications on student learning and education; and challenges to its practice and implementation. For this purpose, the perceptions of the respondents from the survey, focus group discussions, and lessons learned from the workshops. A discussion on the contextualization of PBL was made to examine how PBL can be integrated into the learning programmes to reinforce the engineering education.

7.4.1 Relevancy of PBL to Engineering Education

The relevancy of PBL to engineering education was assessed premised upon different teaching and learning attributes such as whether it is related to exit level outcomes, has a combination of both theoretical knowledge and practical application, deals with multidisciplinary and interdisciplinary problems, assists in independent learning, enables group work, and applies engineering tools, techniques and models. The results presented in Table 7.1 illustrate a generally strong consensus among participants regarding the relevance of PBL to engineering education. The survey assessed various dimensions of PBL's alignment with educational objectives and pedagogical goals within the engineering discipline.

Table 7.1 Perceptions on the Relevance of PBL to Engineering Education

Relevance of PBL to Engineering Education	*Perception Index (PI)*	*SD*	*CoV*	*IQR*	*P-values*
Related to exit-level outcomes	3.89	0.771	198	1	<0.001
Combination of both theoretical knowledge and practical application	4.06	0.770	190	2	<0.001
Multidisciplinary and interdisciplinary problems	4.20	0.722	17.2	1	<0.001
Independent learning	3.52	0.836	238	1	<0.001
Group work	4.36	0.640	147	1	<0.001
Application of engineering tools, techniques and models	3.85	0.712	185	1	<0.001

The highest-rated item was group work, with a Perception Index (PI) of 4.36, accompanied by a relatively low Standard Deviation (SD) of 0.640, a Coefficient of Variation (CoV) of 14.7%, and an Interquartile Range (IQR) of 1. This indicates strong agreement on the value of collaborative learning promoted by PBL. It also suggests that participants found the PBL format effective in fostering team-based problem-solving, a skill crucial in engineering practice.

Following closely was the item related to multidisciplinary and interdisciplinary problems, which recorded a PI of 4.20, an SD of 0.722, a CoV of 17.2%, and an IQR of 1. This reflects the perception that PBL provides authentic exposure to complex challenges that transcend disciplinary boundaries, mirroring real-world engineering tasks. Similarly, the integration of theoretical knowledge with practical application was rated highly (PI = 4.06, SD = 0.770, CoV = 19.0%, IQR = 2), affirming that PBL serves as a pedagogical bridge between classroom learning and field practice.

Items related to application of engineering tools, techniques, and models (PI = 3.85, SD = 0.712, CoV = 18.5%, IQR = 1) and exit-level outcomes (PI = 3.89, SD = 0.771, CoV = 19.8%, IQR = 1) further reinforce the suitability of PBL in meeting educational standards and competency expectations within engineering programmes.

A relatively lower, though still positive, rating was observed for independent learning (PI = 3.52, SD = 0.836, CoV = 23.8%, IQR = 1), indicating that while PBL supports autonomy, it may challenge students unfamiliar with self-directed learning modes. Importantly, all items under this category reported p-values < 0.001, suggesting that the results are statistically significant and not due to random variation.

Moreover, the focus group discussion supported the perceptions of the respondents and the outcomes of the workshops. According to the focus group:

> ...students can use their cognitive knowledge to solve the complex problems belonging to practical or real-life issues in a multi-disciplinary group environment through their direct active intellectual engagements and doing it by themselves, which also refines and enhances their knowledge and skill.

Further, another member in the focus group argued that:

> .... the PBL process can enables student to learn how to deal with complex practical socio-economic-environmental (non-engineering wicked problems—having no specific solutions) problems, thereby making them to attain the exit level outcomes of undergraduate engineering education in some of the universities of South Africa because PBL demands both cognitive knowledge and follow[ing a] systematic learning process to deal with real life challenges.

Therefore, it is argued that PBL is highly relevant for engineering education and assists in offering the students the opportunities to attain critical exit-level outcomes such as problem-solving, design, the application different engineering tools, techniques, theoretical knowledge, principles, working in a multidisciplinary environment, working in groups while learning independently, etc.

7.4.2 Implications of PBL for Engineering Education

Table 7.2 captures participants' perceptions of the pedagogical and professional implications of integrating PBL into engineering education. The findings suggest that PBL is widely viewed as beneficial across multiple dimensions of learning and competency development (Figure 7.1).

A key implication identified was that PBL improves the application of knowledge to real-life or practical problems, which received a PI of 4.15, an SD of 0.712, a CoV of 17.2%, and an IQR of 1. This was matched by the perception that PBL enables students to link non-engineering problems to engineering solutions, which also recorded a PI of 4.15, but with a slightly higher SD of 0.842, CoV of 20.3%, and IQR of 1. These findings highlight PBL's potential to develop systems thinking and the ability to apply engineering knowledge within broader socio-economic and environmental contexts, especially relevant in relation to the UN SDGs.

Participants also noted that PBL assists in learning through hands-on training, with a PI of 4.10, SD of 0.720, CoV of 17.5%, and IQR of 1. This affirms that experiential learning, as facilitated by PBL, enhances technical competencies in ways that are not easily achieved through traditional teaching methods.

Moderately high ratings were observed for instilling innovative thinking (PI = 3.82, SD = 0.801, CoV = 21.0%, IQR = 1) and enabling students to eval-

Table 7.2 Implications of PBL for Engineering Education

Implications of PBL	*Perception Index (PI)*	*SD*	*CoV*	*IQR*	*P-values*
Assists in learning through hands-on training	4.10	0.720	17.5	1	<0.001
Improves application of knowledge to real-life or practical problems	4.15	0.712	17.2	1	<0.001
Instils innovative thinking	3.82	0.801	210	1	<0.001
Enable students to learn to link non-engineering problems to engineering solutions	4.15	0.842	20.3	1	<0.001
Enable the students to learn to evaluate different alternatives and context-specific solutions	3.83	0.782	20.4	2	<0.001
Imbibing of entrepreneurial attributes	3.62	0.942	26.0	1	<0.001
Better industry and academia interlinkage	3.44	1.005	29.2	1	<0.001

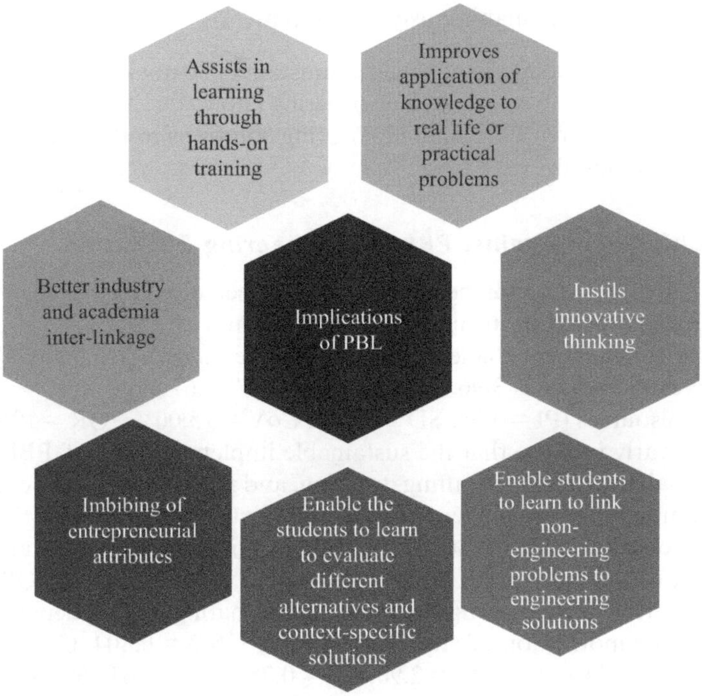

Figure 7.1 Implications of PBL for Engineering Education.

uate alternative and context-specific solutions (PI = 3.83, SD = 0.782, CoV = 20.4%, IQR = 2). These scores suggest that PBL nurtures critical reasoning and the exploration of diverse problem-solving strategies, although the variation consistently indicates room for enhancement in facilitating such outcomes.

Lower scores were found for PBL's contribution to imbibing entrepreneurial attributes (PI = 3.62, SD = 0.942, CoV = 26.0%, IQR = 1) and improving industry–academia interlinkage (PI = 3.44, SD = 1.005, CoV = 29.2%, IQR = 1). These results, while still positive, reflect variability in perceptions and suggest that institutional efforts may be needed to explicitly link PBL initiatives to entrepreneurial thinking and structured industry collaboration. Notably, all items returned p-values < 0.001, confirming the statistical significance of the responses.

In this context, the focus group observed that:

> ...usually engineering students remain aloof from the soft aspects of the projects such as selling their ideas and solutions and understanding the commercialisation prospects. The conventional education system does not really allow to learn those attributes. However, through PBL, they can acquire those attributes.

Also, the focus group unanimously agreed that.

> ...engineering students in normal circumstances do not link socio-economic-environmental problems to their engineering knowledge. However, PBL makes them think differently to find engineering solutions to the non-engineering problems at hand.

7.4.3 Challenges against PBL in Engineering Education

Table 7.3 provides a comprehensive view of the perceived challenges associated with implementing PBL in engineering education.

The most prominent challenges reported were those associated with human resource capacity (PI = 3.96, SD = 0.847, CoV = 67.44%, IQR = 21.4) and financial resources (PI = 3.94, SD = 0.755, CoV = 75.00%, IQR = 19.1). These findings clearly indicate that the sustainable implementation of PBL requires institutional investment in staffing, training, and materials and resources which may be limited in many higher education contexts.

Issues related to the formation of multidisciplinary groups were also viewed as a substantial challenge (PI = 3.49, SD = 0.522, CoV = 75.61%, IQR = 14.9), reflecting the logistical complexities of coordinating across departments or faculties. The motivation of students (PI = 2.93, SD = 0.804, CoV = 59.52%, IQR = 27.4) and lecturers (PI = 2.90, SD = 0.780, CoV = 63.63%, IQR = 26.9) were also identified as a moderate to significant concern. These findings indicate the psychological and cultural readiness required to shift from traditional teaching-learning models to more active, student-centred paradigms.

Challenges in industry participation (PI = 3.24, SD = 0.794, CoV = 71.05%, IQR = 24.5) and integrating PBL with the existing curriculum (PI = 2.60, SD = 0.857, CoV = 73.91%, IQR = 33.0) further illustrate structural barriers to implementation. The high variation and IQRs for these items suggest a wide

Table 7.3 Challenges against PBL in Engineering Education

Challenges	*Perception Index (PI)*	*SD*	*CoV*	*IQR*	*P-values*
Integrating with the existing curriculum	2.60	0.857	73.91	33.0	<0.001
Industry participation	3.24	0.794	71.05	24.5	<0.001
Motivation of the students[a]	2.93	0.804	59.52	27.4	<0.001
Motivation of the lecturers[a]	2.90	0.780	63.63	26.9	<0.001
Formation of multi-disciplinary groups	3.49	0.522	75.61	14.9	<0.001
Financial resources	3.94	0.755	75.00	19.1	<0.001
Human resource capacity of the universities	3.96	0.847	67.44	21.4	<0.001

[a] In the absence of being an integral part of the curriculum or syllabus.

range of opinions, possibly due to differing levels of exposure or institutional readiness.

Crucially, all items in this table also had p-values < 0.001, indicating that these perceptions are statistically significant and should be considered in any strategic planning for PBL integration.

These challenges were also corroborated in the focus group discussion. According to the focus group members:

> ...Although it might not be difficult and challenging to integrate PBL into the existing curriculum, the regulatory constraints and pedagogical method changes need to be considered, which might prove challenging. Moreover, implementation of such learning methods requires highly adept human resources, a significant amount of financial resources and industry participation.

In addition, according to students and lecturers, although motivation for PBL is not a problem, it will add to the existing workloads of the lecturers and students without much tangible benefits in terms of adding to the credit value unless it is a part of the curriculum. Furthermore, some academicians in the focus group showed their reservations, indicating that PBL might not assist students in acquiring the conceptual, theoretical knowledge and principles which are fundamental to engineering education, although they agreed that appropriate framing of the problems might be able to meet such challenges. Thus, the implementation of PBL in engineering education under the prevailing conditions might be challenging.

7.4.4 Contextualizing PBL to Reinforce Engineering Education

It is evident from the survey responses, workshop reflections, and focus group discussions that PBL is highly relevant to engineering education. Participants acknowledged its positive implications for students, particularly in enhancing critical thinking, problem-solving, and interdisciplinary learning. PBL not only enriches the educational experience but also introduces new dimensions to the learning process by bridging theory and practice. However, despite these benefits, several challenges continue to hinder its sustainable implementation. These include limitations in curriculum integration, resource constraints, and varying levels of stakeholder motivation and readiness.

Given both its potential advantages and practical obstacles, there is a need to contextualize PBL within the prevailing institutional and socio-economic conditions. While most focus group participants agreed on the necessity for a paradigm shift—from traditional, content-driven instruction to a competency-based, multidisciplinary and interdisciplinary model (Figure 7.2)—many also acknowledged that such a transformation remains aspirational and is unlikely to be fully realized in the immediate future. The current academic infrastructure and regulatory environment do not yet fully support this level of systemic change.

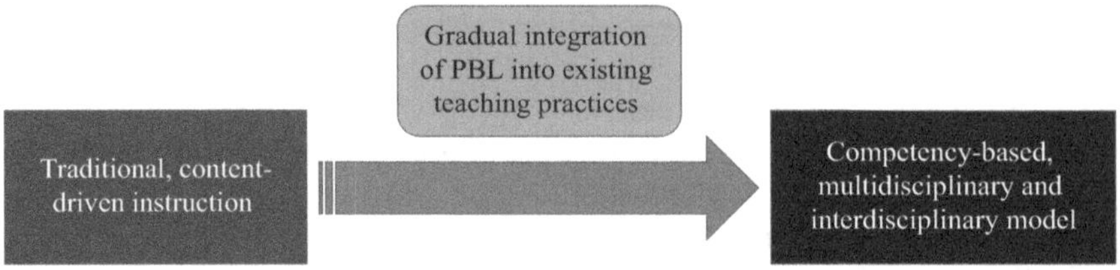

Figure 7.2 Paradigm Shift for Contextualizing PBL to Reinforce Engineering Education.

Within this context, several pragmatic suggestions emerged from the focus group discussions. One widely supported recommendation involves gradually integrating PBL into existing teaching practices by encouraging students to engage with real-life, practical problems in collaboration with industry partners, government agencies, and local communities. As one participant noted:

> Perhaps identifying real-life and practical problems in collaboration with industry partners and different social-solidarities (such as governance agencies and people) and assigning the students to solve these problems as part of their assignment and projects will assist integration of PBL into the current conventional form of teaching and learning gradually.

This incremental approach could sustain motivation among both lecturers and students while requiring minimal additional financial or human resources. With modest adjustments to existing curricula and teaching strategies, universities can begin to adopt PBL principles without overhauling their entire educational framework.

Another group of participants advocated for adapting successful international models of PBL implementation, particularly those from European universities. One notable example is the approach taken by University College London (UCL) in the United Kingdom, which integrates interdisciplinary PBL as a core component of its engineering programmes. Focus group members suggested that a similar model could be contextualized and implemented within South African Universities of Technology (UoTs), with careful consideration of local constraints and opportunities. As one participant explained:

> One compulsory module can be incorporated into the curriculum at the appropriate level, which should be inter- and multi-disciplinary. The students should be assigned practical/real-life problems to solve within a specified duration by following the process of PBL. This might be possible without significant changes in the regulations and policies of the universities.

These suggestions reflect a practical understanding of the current educational climate while acknowledging the transformative potential of PBL. Although full-scale implementation may be challenging, a contextualized and phased approach—anchored in existing teaching structures and enriched with problem-solving projects—can foster key graduate attributes. These include exit-level outcomes, such as independent learning, teamwork, innovation, design thinking, and the application of engineering tools and techniques. Additionally, PBL can contribute to the development of entrepreneurial competencies by exposing students to real-world challenges that demand creativity and strategic thinking.

Thus, while the widespread adoption of PBL in engineering education—particularly in universities from the Global South—may encounter structural and institutional challenges, as evidenced by this study conducted at a South African university, a gradual and context-sensitive implementation strategy can support meaningful transformation. When carefully adapted to local conditions, PBL has significant potential to enhance the quality, relevance, and impact of engineering education across the country.

7.5 Discussion and Implications

PBL is increasingly recognised as a transformative pedagogical model in engineering education. It addresses many of the limitations inherent in traditional, lecture-based approaches by shifting the focus from passive knowledge transmission to active inquiry, problem-solving, and collaboration. Rooted in constructivist and learner-centred theories (Barrows, 1988; Barrows & Tamblyn, 1980; Schmidt & Moust, 1995), PBL reframes the learning process as one where students construct knowledge through engagement with real-world, ill-structured problems. This shift is particularly relevant in engineering, a field that demands the ability to navigate complex, interdisciplinary, and evolving challenges.

The present study affirms the growing consensus in the literature that PBL effectively cultivates key competencies required in engineering practice, including critical thinking, collaboration, and the integration of theoretical knowledge with practical application (Das & Mishra, 2018; Delaney & Nagle, 2019). Both students and educators in the study characterized PBL as an approach that bridges the long-standing gap between academic instruction and professional engineering practice. This perception aligns with the graduate attributes articulated by many international accreditation frameworks, which emphasize problem-solving, communication, ethical reasoning, and the application of engineering principles in sociotechnical contexts.

PBL's relevance becomes even more pronounced when viewed through the lens of OBE—a framework that prioritizes demonstrable competencies over content memorization. The alignment between PBL and OBE is critical in regions such as the Global South, where engineering education is undergoing transformation. PBL's emphasis on self-directed learning, reflective inquiry, and contextual problem-solving complements OBE's objectives by fostering metacognition, adaptability, and systems thinking (Chen, 2008; Havenga & Van Wyk, 2017). Furthermore, when integrated with global frameworks like the United Nations SDGs, PBL supports the cultivation of socially responsive engineers equipped to address environmental and socio-economic challenges.

Despite its transformative potential, the study highlights several challenges associated with PBL implementation. Structural and institutional barriers—such as limited resources, rigid curricula, and faculty resistance—continue to constrain its broader adoption. Many educators are accustomed to traditional pedagogical roles and may lack training in facilitative teaching methods required for PBL (Kjellberg et al., 2015; van Barneveld & Strobel, 2015). Without sustained professional development and institutional support, PBL risks superficial adoption, which can dilute its pedagogical impact.

Concerns also persist regarding PBL's capacity to deliver core theoretical knowledge, particularly in foundational engineering sciences. As noted by Marcinauskas et al. (2024), without careful integration, students may struggle to acquire the disciplinary depth required for technical proficiency. In response, many scholars advocate for a blended approach, in which PBL complements rather than replaces traditional teaching methods. Embedding PBL activities within a strategically scaffolded curriculum allows for a balance between conceptual depth and practical engagement, thereby supporting a more holistic learning experience.

The study identifies several contextualization strategies that can mitigate these challenges and support effective PBL adoption. Incremental integration—beginning with capstone projects, interdisciplinary modules, or elective courses—enables institutions to pilot and adapt PBL without disrupting existing structures. Industry collaboration emerges as another critical enabler; authentic, practice-oriented projects co-designed with industry partners ensure alignment between educational outcomes and workforce expectations (Striegel & Rover, 2002). These partnerships also foster professional exposure and mutual learning between academia and industry.

The integration of digital tools and virtual platforms further expands the possibilities for PBL, especially in resource-constrained environments. Technologies such as simulation software, collaborative design tools, and online problem repositories allow for interactive and scalable learning experiences (Patnawar, 2023; Tsalapatas et al., 2019). These tools also facilitate innovative assessment practices and accommodate diverse learning needs, enhancing PBL's flexibility and reach.

In the Global South context, as evident from this study in a South African university, where engineering education must navigate persistent inequalities and structural constraints, PBL offers both a challenge and an opportunity. Its learner-centred orientation supports goals of educational equity and transformation by empowering students from varied backgrounds to engage in meaningful, socially relevant learning. Pilot initiatives embedded in community-based projects exemplify how PBL can simultaneously enhance student learning and contribute to local development. However, for such models to succeed, they must be underpinned by institutional investment in curriculum reform, faculty development, and policy alignment.

International models, such as those implemented at Aalborg University and University College London, provide compelling evidence of how PBL can be systematically integrated into engineering education. These institutions demonstrate that with flexible curricula, assessment innovation, and strong institutional culture, PBL can enhance both academic learning and employability outcomes. Importantly, the hybridization of PBL with conventional approaches is not a compromise but a pedagogical strength that supports diverse learning styles and curricular objectives.

Thus, PBL has demonstrated substantial potential to reinforce engineering education by aligning teaching and learning practices with the real-world demands of the profession. It cultivates not just technical expertise but also the collaborative, reflective, and adaptive capacities necessary for contemporary engineering challenges. While challenges remain—especially in terms of resources, curriculum restructuring, and faculty readiness—the benefits outweigh the limitations when carefully planned and contextually implemented. PBL, therefore, should be seen not merely as a pedagogical option but as a strategic tool for transforming engineering education towards greater relevance, impact, and inclusivity.

7.6 Conclusion

Reinforcing engineering education in the 21st century requires pedagogical approaches that transcend traditional content delivery and rote learning. Developing higher-order cognitive skills—such as application, analysis, evaluation, design, and creativity—alongside essential graduate attributes like problem-solving, interdisciplinary teamwork, communication, and entrepreneurial thinking, remains a significant challenge within conventional teaching frameworks. Although OBE is widely practised in many universities, particularly within engineering and technology programmes, the inconsistent integration of WIL raises legitimate concerns about whether students are graduating with the competencies required for workplace relevance and professional readiness.

In this context, PBL emerges not merely as an alternative but as a transformative pedagogical strategy. It promotes active, student-centred learning through engagement with real-world, open-ended problems. This study has demonstrated that PBL environments—such as the industry-based workshops conducted with undergraduate engineering students—provide meaningful opportunities for learners to apply both theoretical and practical knowledge to complex socio-economic and environmental challenges. Working collaboratively in multidisciplinary teams, students develop critical thinking, innovation, and context-responsive problem-solving skills.

Despite its promise, the literature remains limited in studies that critically explore the relevance, integration strategies, and broader implications of PBL specifically within the Global South, including South African engineering education. This study addressed this gap by examining the perceptions, implications, and challenges associated with PBL to reinforce engineering education, drawing on empirical data collected through workshops, a structured perception survey, and focus group discussions involving both academic and industry stakeholders.

The findings offer both insight and direction for enhancing pedagogical practice. The results reveal that PBL reinforces key dimensions of engineering education by cultivating graduate attributes that are central to the profession. These include technical competencies like design and the application of engineering tools and broader attributes such as systems thinking, contextual awareness, and the ability to respond to non-engineering problems through engineering solutions. Importantly, PBL also helps to bridge the theory–practice divide those conventional approaches often fail to address. Both students and academic staff perceive PBL as a powerful mechanism for achieving exit-level outcomes.

However, several challenges hinder its widespread adoption. Structural and operational barriers—including the integration of PBL into existing curricula, limited industry engagement, and resource constraints—remain significant. Additional concerns around sustaining lecturer and student motivation, particularly when PBL is not embedded as a credit-bearing component, must also be addressed. These challenges are consistent with international experiences but are compounded by the unique dynamics of South Africa's higher education system.

Nonetheless, the evidence strongly supports the argument that PBL can reinforce engineering education when implemented strategically and contextually. A phased approach—beginning with the alignment of student projects and assignments to real-world and industry problems—offers a feasible and low-risk entry point. Ensuring constructive alignment across ILOs, teaching and learning activities, and assessment strategies is essential to maintaining pedagogical coherence and academic rigour.

When supported institutionally and embedded with curricular flexibility, PBL serves as a high-impact educational strategy. It equips graduates to be not only technically proficient but also innovative, adaptable, and capable of addressing interdisciplinary and real-world challenges. By institutionalizing PBL in a way that is inclusive, flexible, and aligned with national development goals, engineering education can be transformed into a more responsive, future-ready, and socially impactful endeavour.

References

Abou-Hayt, I., Dahl, B., & Rump, C. Ø. (2020). A problem-based approach to teaching a course in engineering mechanics (pp. 499–509). https://vbn.aau.dk/ws/files/344818442/A_Problem_Based_Approach_to_Teaching_a_Course_in_Engineering_Mechanics.pdf

Alavi, C. (1995). *Problem-based learning in a health science curriculum*. Routledge.

Barrett, T. (2005). Understanding problem-based learning. In T. Barrett, I. Mac Labhrainn, & H. Fallon (Eds.), *Handbook of enquiry and problem-based learning* (pp. 13–25). CELT.

Barrows, H. S. (1988). *The tutorial process*. Southern Illinois University School of Medicine.

Barrows, H. S., & Tamblyn, R. (1980). *Problem-based learning: An approach to medical education*. Springer.

Boud, D., & Feletti, G. (Eds.). (1991). *The challenge of problem-based learning*. Kogan Page.

Brady, L. (1996). Outcome-based education: A critique. *The Curriculum Journal*, *7*(1), 5–16. https://doi.org/10.1080/0958517960070102

Buentello-Montoya, D. (2022). Problem-based learning as a tool to foster electricity and electromagnetism-related concepts in an engineering course. *International Conference on Teaching, Assessment, and Learning for Engineering*, 174–178. https://doi.org/10.1109/TALE54877.2022.00035

Castro-Vargas, C., & Cabana-Caceres, M. (2024). Problem-based learning and knowledge of digital electronics among engineering students. *International Journal of Engineering Pedagogy (IJEP)*. https://doi.org/10.3991/ijep.v14i4.45075

Chauhan, A. (2017). An overview of problem-based learning in engineering education. *Journal of Emerging Technologies and Innovative Research*, *4*(10), 225–228. https://www.jetir.org/view?paper=JETIR1710035

Chen, N.C. (2008). An educational approach to problem-based learning. *Kaohsiung Journal of Medical Sciences*, *24*(3 Suppl), S23–S30.

Das, D., & Mishra, B. (2018). Exploring the complementarity of problem-based learning with outcomes-based education in engineering education: A case study in South Africa. In W. Sunyu, A. Kolmos, A. Guerra, & Q. Weifeng (Eds.), *7th international research symposium on PBL: Innovation, PBL and competences in engineering education* (pp. 297–307). Aalborg Universitetsforlag.

Delaney, K., & Nagle, G. (2019). Engineering education through problem-based learning: A case study of teaching mechanical engineering design in TU Dublin. *Engineers Journal, Special Education Issue*. https://doi.org/10.21427/0rsq-4x20

Duch, B. J., Groh, S. E., & Allen, D. E. (2001). Why problem-based learning? In B. J. Duch, S. E. Groh, & D. E. Allen (Eds.), *The power of problem-based learning* (pp. 3–13). Stylus Publishing.

Havenga, M., & Van Wyk, M. (2017). A conceptualised design-based research framework for a problem-based learning approach in a first-year engineering course. *Global Journal of Engineering Education*, *19*, 200–206.

Hidayat, T., & Osei Kwabena, J. (2023). Problem based learning in engineering course in Malaysia. *Asia Pacific Global Academy Journal*, *2*(2), 95–105. https://doi.org/10.53623/apga.v2i2.234

Hmelo-Silver, C. E. (2004). Problem-based learning: What and how do students learn? *Educational Psychology Review*, *16*(3), 235–266.

Hmelo-Silver, C. E., & Barrows, H. S. (2006). Goals and strategies of a problem-based learning facilitator. *Interdisciplinary Journal of Problem-Based Learning*, *1*(1), 21–39.

Hunt, E. M., Lockwood-Cooke, P., & Kelley, J. (2010). Linked-class problem-based learning in engineering: Method and evaluation. *American Journal of Engineering Education*, *1*(1), 79–88. https://doi.org/10.19030/AJEE.V1I1.794

Kjellberg, M., Adawi, T., & Brolin, K. (2015). Challenges in implementing PBL: Chalmers Formula Student as a case. https://research.chalmers.se/en/publication/226614

Kolmos, A., & de Graaff, E. (2014). *Problem-based and project-based learning in engineering education: Merging models* (pp. 141–161). Cambridge University Press. https://doi.org/10.1017/CBO9781139013451.012

Lavado-Anguera, S., Velasco-Quintana, P.-J., & Terrón-López, M.-J. (2024). Project-Based Learning (PBL) as an experiential pedagogical methodology in engineering education: A review of the literature. *Education Sciences*, *14*(6), 617. https://doi.org/10.3390/educsci14060617

Maladzhi, R., Nemavhola, F., Ntunka, M. G., Moloi, K., Mthombeni, N. H., & Tsoe, M. S. (2024). Addressing challenges and solutions for scaling up project-based learning initiatives in large engineering programs. In *2024 World Engineering Education Forum - Global Engineering Deans Council (WEEF-GEDC)*, Sydney, Australia, 2024, pp. 1–9. https://doi.org/10.1109/weef-gedc63419.2024.10854922

Marcinauskas, L., Iljinas, A., Čyvienė, J., & Stankus, V. (2024). Problem-based learning versus traditional learning in physics education for engineering program students. *Nevelėstudomány*. https://doi.org/10.3390/educsci14020154

McCarthy, M. M. (1993). Challenges to the public-school curriculum: New targets and strategies. *The Phi Delta Kappan*, *75*(1), 58–60.

Naik, S. M., Bandi, S., & Reddy, C. (2024). Enhancing engineering education: Exploring the impact of problem-based learning on freshman students' skills and engagement – A case study. *Journal of Engineering Education Transformations*, *38*(is1), 133–139. https://doi.org/10.16920/jeet/2024/v38is1/24222

Noor, N. A., Alias, A., Ariffin, K., Bhkari, N. M., & Hashim, A. H. (2019). Employing the problem-based learning approach in civil engineering education: The highway engineering experience. In *Proceedings of the International Conference on Civil Engineering Education* (pp. 519–528). Springer. https://doi.org/10.1007/978-981-10-8730-1_52

Patnawar, S. T. (2023). A comprehensive review on PBL and digital PBL in engineering education – Status, challenges and future prospects. *Journal of Engineering Education Transformations*, *37*(2), 142–157. https://doi.org/10.16920/jeet/2023/v37i2/23157

Rehman, S. U. (2023). Trends and challenges of project-based learning in computer science and engineering education. https://doi.org/10.1145/3629296.3629360

Riazifar, N., & Khajehsaeid, H. (2024). Enhancing engineering education through problem-based learning. In *UK and Ireland Engineering Education Research Network Conference Proceedings 2023*. https://doi.org/10.31273/10.31273/9781911675167/1636

Schmidt, H. G. (1983). Problem-based learning: rationale and description. *Medical Education*, *17*, 11–16.

Schmidt, H. G. (1998). Problem-based learning: Does it prepare medical students to become better doctors? *Medical Journal of Australia*, *168*, 429–430.

Schmidt, H. G., & Moust, J.H.C. (1995). What makes a tutor effective? A structural equation modelling approach to learning in problem-based learning. *Academic Medicine*, *70*(8), 708–714.

Servant-Miklos, V.F.C. (2018). Fifty years on: A retrospective on the world's first problem-based learning programme at McMaster University Medical School. *Health Professions Education*. https://doi.org/10.1016/j.hpe.2018.04.002

Stoyanov, I., Ivanov, V., & Koşunalp, S. (2023). Applying a problem-based learning in the topic of modelling of the power and improvement the voltage stability of wind turbines. https://doi.org/10.1109/elma58392.2023.10202556

Striegel, A., & Rover, D. T. (2002). Problem-based learning in an introductory computer engineering course. *Frontiers in Education Conference*, *2*. https://doi.org/10.1109/FIE.2002.1158138

Tsalapatas, H., Vaz de Carvalho, C., Heidmann, O., & Houstis, E. N. (2019). Active problem-based learning for engineering higher education. *International Conference on Computer Supported Education*, *2*, 347–351. https://doi.org/10.5220/0007720403470351

Tuohi, R., & Roslöf, J. (2007). Practical challenges with a PBL implementation in engineering education. http://icee2007.dei.uc.pt/proceedings/papers/180.pdf

Utama, N. Y., Mindarta, E. K., Komarudin, K., & Suhartadi, S. (2024). Implementasi problem-based learning untuk meningkatkan kompetensi perbaikan sistem bahan bakar siswa pusdiklat. *Jurnal Pendidikan Vokasi Otomotif*, *7*(1), 168–183. https://doi.org/10.21831/jpvo.v7i1.79356

van Barneveld, A., & Strobel, J. (2015). Implementation of PBL in engineering education: Conceptualisation and management of tensions. https://doi.org/10.24908/PCEEA.V0I0.5790

Ward, J. D., & Lee, C. L. (2002). A review of problem-based learning. *Journal of Family and Consumer Sciences Education*, *20*(1), 16–26.

Wolf, A. (2007). Tutoring PBL: A model for student-centred teaching. In L.E. Young & B.L. Paterson (Eds.), *Teaching nursing*. Lippincott.

Yeung, E., Au-Yeung, S., Chiu, T. et al. (2003). Problem design in problem-based learning: Evaluating students' learning and self-directed learning practice. *Innovations in Education and Teaching International*, *40*(3), 237–244.

Zeng, L., & Xu, C. (2010). Problem-based learning in civil engineering education. *International Conference on Education Technology and Computer*, *3*. https://doi.org/10.1109/ICETC.2010.5529601

Chapter 8

Problem-Based Learning as a Means for Embedding Entrepreneurial Attributes among Engineering Students

8.1 Introduction

Fostering entrepreneurial attributes among university students—particularly within engineering education—has become a strategic imperative in the higher education system. Amidst rising unemployment and the growing demand for innovation-driven economic development, universities are under increasing pressure to produce graduates who are not only technically competent but also entrepreneurial in mindset and capable of creating value in uncertain and dynamic environments. Although entrepreneurship education has gained traction in recent years, particularly through standalone undergraduate courses or specialized master's programmes in management sciences, these initiatives often serve a narrow cohort of students with a predefined interest in entrepreneurship (Garavan & O'Cineide, 1994; Hytti & O'Gorman, 2004; Kolvereid & Moen, 1997; Zhao et al., 2005). As a result, a significant portion of students in technically focused disciplines—such as engineering—are left with limited exposure to the competencies, skills, behaviours, and mindsets needed to thrive in entrepreneurial contexts.

To address this shortfall, Problem-Based Learning (PBL) has emerged as a promising pedagogical strategy that can embed entrepreneurial development within the core structure of engineering education. PBL aligns closely with entrepreneurial learning outcomes, particularly those associated with real-world engagement, innovation, and autonomous decision-making. Rather than teaching entrepreneurship as a detached theory or through traditional lecture formats, PBL allows students to experience entrepreneurship in action—working through complex, ill-structured problems that demand creativity, risk-taking, initiative, and teamwork (Dewiyani & Kosasih, 2017; Hasanah & Malik, 2019).

In the context of engineering education, PBL encourages students to think beyond technical problem-solving and engage with broader entrepreneurial processes such as identifying market needs, evaluating business models, managing constraints, and interacting with stakeholders. These experiences mirror the realities of startup ventures and innovation projects, offering students a low-risk but high-impact space to develop entrepreneurial competencies.

DOI: 10.1201/9781003529743-11

Projects might involve, for example, designing a low-cost water purification system for rural communities or developing a mobility solution for a congested urban centre—scenarios that require both engineering proficiency and entrepreneurial insight.

Studies indicate that PBL improves students' critical thinking, adaptability, and creative problem-solving abilities—all of which are vital for entrepreneurship, particularly in the fast-evolving technological landscape of Industry Revolution (IR) 4.0 (Hasanah & Malik, 2019; Riazifar & Khajehsaeid, 2024). These attributes are further reinforced when students are required to work in interdisciplinary teams, navigate uncertain project outcomes, and iteratively refine their ideas based on stakeholder feedback. Such processes promote the kind of effectual reasoning (Gerhart & Melton, 2016) and value creation orientation that lie at the heart of entrepreneurial thinking.

Moreover, PBL encourages entrepreneurial behaviour by shifting the locus of control from the instructor to the student, thereby cultivating self-direction, initiative, and ownership over learning outcomes. The collaborative nature of PBL also mirrors the dynamics of entrepreneurial teams, where diverse perspectives and complementary skills are essential for success (Bezerra de Araújo et al., 2020; Machado-Toledo et al., 2018). Through these team-based, experiential learning environments, students learn to manage conflict, negotiate roles, and leverage group dynamics—skills that are just as critical in startups and innovation hubs as they are in engineering projects.

Importantly, PBL can be woven directly into core engineering modules such as Mechatronics, Fluid Mechanics, or Structural Design, allowing students to develop technical acumen and entrepreneurial capability in tandem (Gerhart & Melton, 2016; Liu et al., 2015). This dual focus ensures that entrepreneurship is not seen as an optional add-on but as a fundamental aspect of professional engineering practice. When done effectively, this integration leads to more holistic graduates—engineers who can solve problems, identify opportunities, and mobilize resources to turn ideas into a tangible impact.

However, the successful integration of PBL for entrepreneurial development requires careful curriculum design and institutional support. Educators must ensure that PBL tasks are purposefully designed to incorporate entrepreneurial challenges and are assessed in a manner that values both technical and entrepreneurial contributions (Morselli, 2019; Tan & Ng, 2006). Faculty development, access to real-world project opportunities, and the creation of interdisciplinary collaboration spaces are also critical to realizing the full potential of this approach.

While implementing PBL is undoubtedly resource-intensive and may pose challenges for institutions unaccustomed to such pedagogical shifts, its long-term benefits are significant. As the demands on engineering graduates continue to evolve, PBL offers a robust and adaptable framework for embedding entrepreneurship directly into the heart of technical education. Doing so enables universities to produce graduates who are not only capable engineers but also confident, innovative, and resilient entrepreneurs, ready to contribute meaningfully to any nation's development and beyond.

However, the linkage between PBL activities and entrepreneurial attributes has not been explored explicitly. Therefore, this study investigates the relationship between PBL and the development of entrepreneurial attributes among engineering students. In other words, as higher education increasingly aims to equip graduates with the skills necessary for innovation and value creation, this research explores how PBL, as an experiential and student-centred pedagogical approach, may contribute to entrepreneurial capability in engineering education. In this context, the following research questions were investigated.

RQ1 What attributes do engineering students develop through participation in PBL?

RQ2 How are specific PBL activities linked to developing entrepreneurial attributes?

To address these questions, a mixed-methods approach was employed. A cross-sectional survey was administered to 120 purposively selected engineering students using a pre-tested self-completion questionnaire. In addition, focus group discussions were conducted with key stakeholders—including lecturers, academic leaders, industry professionals, entrepreneurs, and students—to gain deeper insights into the perceived connection between PBL and entrepreneurship.

Findings indicate that entrepreneurial learning is an inherent outcome of PBL. Activities such as preparing design briefs based on real-world industry challenges, developing conceptual products, engaging with external stakeholders, and pitching business ideas facilitate the development of key entrepreneurial attributes. These include innovation, opportunity recognition, stakeholder engagement, risk analysis, and basic marketing skills. Through these hands-on experiences, students not only strengthen their technical abilities but also learn how to translate ideas into viable products and attract investment, skills that are essential for entrepreneurial success.

The study concludes that PBL serves as an effective pedagogical method for cultivating entrepreneurial attributes in engineering students. As such, integrating entrepreneurship-focused PBL activities into engineering curricula is recommended to support the broader goal of producing graduates equipped for innovation-driven economies.

8.2 Problem-Based Learning and Entrepreneurship: A Theoretical Perspective

PBL is widely recognised as an effective pedagogical approach for developing competencies essential to entrepreneurial success, particularly in professional and technical fields like engineering (Servant-Miklos, 2019). Grounded in constructivist learning theory, PBL positions students as active participants in their learning, engaging with real-world, open-ended problems in collaborative settings (Barrows & Tamblyn, 1980; Boud & Feletti, 1991; Schmidt, 1998). This student-driven process fosters critical skills aligned with entrepreneurial thinking, such as independence, effective communication, and problem-solving (Chen, 2008; Duch et al., 2001). Learning activities—including case analysis, stakeholder engagement, peer discussions, and presentations—closely mirror the stages of entrepreneurial ventures, making PBL particularly suitable for cultivating entrepreneurial attributes (Barrett, 2005; Hmelo-Silver, 2004; Luryi et al., 2007; Williams & Paltridge, 2017).

A distinguishing feature of PBL is its emphasis on formative, competency-based assessment, which evaluates students' ability to perform in complex and unpredictable contexts rather than their ability to recall static knowledge (Das & Mishra, 2018; Servant-Miklos, 2019). This aligns closely with the experiential and innovation-driven nature of entrepreneurship, where ambiguity, iteration, and adaptability are essential. PBL fosters the development of entrepreneurial capabilities such as self-directed learning, critical reflection, creative ideation, and the ability to conceptualize and pitch innovative solutions (Biley & Smith, 1998; Moore, 2009). By integrating cognitive and behavioural competencies, PBL provides a powerful framework for entrepreneurship that promotes autonomy, resilience, collaboration, and lifelong learning (Kehinde, 2018; Mubuuke et al., 2017; Ward & Lee, 2002).

PBL's relevance to entrepreneurship lies in its ability to simulate the conditions and demands of real entrepreneurial activity. As entrepreneurship is increasingly recognised as a driver of economic development, higher education has responded by introducing various forms of entrepreneurship training. These range from standalone modules and embedded curriculum elements to full postgraduate programmes in entrepreneurship, particularly across engineering, science, and business faculties (Doboli et al., 2010; Kriewall & Mekemson, 2010; Lumsdaine & Binks, 2003; Swart, 2014). However, the effectiveness of these programmes in fostering entrepreneurial attributes remains under-evaluated and inconsistently understood (Garavan & O'Cineide, 1994; Hytti & O'Gorman, 2004; Kolvereid & Moen, 1997; Zhao et al., 2005).

Increasingly, as discussed in previous chapters, PBL has been integrated into engineering education as an interactive, learner-centred method that fosters entrepreneurial capabilities. By engaging students with authentic problems, PBL enhances critical thinking, promotes creativity, and supports the application of theoretical knowledge in practical contexts (Dolmans et al., 2005; Duch et al., 2001; Hasanah & Malik, 2019; Riazifar & Khajehsaeid, 2024). Its alignment with the interdisciplinary and iterative nature of entrepreneurship makes it a pedagogically coherent method for teaching students how to approach uncertainty, identify opportunities, and devise innovative solutions. Numerous studies have confirmed PBL's role in fostering an entrepreneurial mindset and enhancing the relevant skill set for entrepreneurial activities (Bezerra de Araújo et al., 2020; Gerhart & Melton, 2016; Machado-Toledo et al., 2018).

Notably, PBL is effective in cultivating creativity and innovation, which are key dimensions of entrepreneurial behaviour. In STEM education, PBL has enabled students to engage in innovation-focused tasks that reflect entrepreneurs' challenges (Hynes, 1996; Zhou & Frisdahl, 2024). In engineering specifically, PBL scenarios often replicate entrepreneurial situations, encouraging students to develop, test, and apply knowledge in realistic, project-based settings. These experiences help students build expertise in areas such as project management, team leadership, and risk assessment—elements central to entrepreneurial competence (Machado-Toledo et al., 2018). Furthermore, PBL's constructivist structure contributes to developing Individual Entrepreneurial Orientation (IEO), including proactive behaviour, self-efficacy, and opportunity recognition (Surjanti & Nugrohoseno, 2018).

PBL has been successfully applied across diverse educational contexts to enhance entrepreneurship education, from secondary schools to university-level programmes. Activities like ideation workshops and hackathons have been integrated into engineering programmes to replicate entrepreneurial ecosystems and inspire entrepreneurial thinking (Kabilan, 2025). These experiential learning opportunities have led to measurable gains in students' critical thinking and decision-making abilities, vital for successful entrepreneurship (Hasanah & Malik, 2019). Moreover, PBL promotes practical entrepreneurial competencies such as business model development and value proposition design through hands-on, interdisciplinary projects that require teamwork and real-time problem-solving (Gómez, 2024).

Despite its benefits, implementing PBL to inculcate entrepreneurship attributes poses certain challenges. Effective application requires thoughtful instructional design, well-structured facilitation, and a supportive learning environment that encourages experimentation and tolerates failure, both essential to fostering entrepreneurial capacity (Juan & Lingling, 2013; Tan & Ng, 2006). Institutions must also provide adequate training and support for educators and develop assessment frameworks that capture the nuanced, developmental outcomes of entrepreneurial learning. Furthermore, while short-term impacts have been documented, there remains a need for longitudinal research to assess the sustained influence of PBL on students' entrepreneurial development and to refine pedagogical strategies accordingly (Raimundo & Rosário, 2024).

Thus, PBL presents a promising approach to entrepreneurship education by providing a framework that encourages active learning and the practical application of entrepreneurial concepts. However, its effectiveness depends on the thoughtful design and implementation of the learning environment, as well as continuous research to optimize its application in various educational contexts. This approach not only enhances students' entrepreneurial skills but also prepares them to tackle complex challenges in their professional lives. While alternative pedagogies continue to be explored, PBL provides engineering students with unique opportunities to develop entrepreneurial attributes through structured engagement with real-world challenges. Nonetheless, further empirical research is required to more precisely understand the extent and mechanisms through which PBL fosters entrepreneurial capabilities in engineering education.

8.3 Research Methods

The study adopted a mixed-methods approach, incorporating survey research and focus group discussions to gather both quantitative and qualitative insights from key stakeholders. The research aimed to examine the effectiveness of PBL in fostering entrepreneurial competencies among engineering students, as well as to explore stakeholder perspectives on the integration of PBL into engineering education.

8.3.1 PBL Workshops and Student Participation

A series of PBL workshops were organized at a university in South Africa, involving approximately 150 students and 27 facilitators from diverse disciplines, including Built Environment, Civil Engineering, Electrical Engineering, Mechanical Engineering, and Information Technology. These workshops were designed to immerse students in experiential learning, where they tackled real-world challenges aligned with the United Nations' Sustainable Development Goals (SDGs).

Students were required to analyse socio-economic and environmental issues, applying engineering knowledge to propose innovative, interdisciplinary solutions. The workshops were structured to simulate real-world problem-solving environments, encouraging students to collaborate, think critically, and engage in hands-on learning experiences.

Throughout the process, students were mentored by trained facilitators, academic staff, and industry professionals, ensuring they received guidance on both technical and strategic aspects of problem-solving. This industry–academia collaboration enhanced students' exposure to practical challenges and entrepreneurial thinking in engineering.

8.3.2 Assessment of Student Learning

The evaluation of student learning and project outcomes was conducted using a multi-stage assessment framework:

1. Formative Assessment: At different stages of the workshops, facilitators conducted ongoing assessments to track students' progress, providing continuous feedback to refine their problem-solving approaches.
2. Summative Assessment: The final showcasing of proposed solutions was assessed by a jury panel composed of industry experts, faculty members, and interdisciplinary facilitators to ensure a comprehensive and multi-perspective

evaluation. The jurors assessed students' ability to apply technical knowledge, critical thinking, creativity, teamwork, and entrepreneurial reasoning in solving the given challenges.

8.3.3 Survey Design and Implementation

To quantify students' perceptions of the effectiveness of PBL in enhancing entrepreneurial attributes, a cross-sectional survey was conducted among the workshop participants. A purposive sampling method was used, targeting students who actively engaged in the workshops. In total, 120 students completed the survey.

A pre-tested, self-completion questionnaire was developed based on:

- Findings from the literature review on PBL pedagogy and entrepreneurial learning.
- Case studies from universities with well-established PBL models, such as University College London (UCL), UK.
- Practical experiences from previous PBL implementations in engineering education.

The questionnaire explored key themes, including:

- The skills and attributes students developed through PBL.
- The linkage between PBL activities and entrepreneurial learning.
- Students' perceptions of how PBL influenced their creativity, problem-solving, teamwork, and business acumen.

A five-point Likert scale was used to gauge student perceptions, with response options ranging from:

1 = Not significant,
2 = Significant to a certain extent,
3 = Fairly significant,
4 = Significant,
5 = Highly significant.

This structured approach enabled the obtaining of quantifiable insights into the perceived benefits of PBL in fostering entrepreneurial competencies within an engineering education context.

8.3.4 Focus Group Discussions with Stakeholders

Beyond the student survey, focus group discussions were conducted with diverse stakeholders, including:

- Lecturers and academic leaders involved in PBL curriculum design and facilitation.
- Industry professionals and entrepreneurs who mentor students and assess final solutions.
- Students who participated in the workshops.

The discussions aimed to gather qualitative insights on:

- The perceived strengths and challenges of integrating PBL with entrepreneurial education.
- The effectiveness of mentorship and industry engagement.
- Opportunities to enhance PBL implementation to better align with industry needs and entrepreneurial skill development.

By combining quantitative survey analysis with qualitative insights from focus group discussions, this study provided a comprehensive understanding of the role of PBL in fostering entrepreneurial attributes among engineering students. The findings offer valuable recommendations for refining PBL pedagogy, enhancing industry–academia collaboration, and strengthening the entrepreneurial dimensions of engineering education in future iterations.

8.3.5 Data Analysis

The survey data was analysed using multiple statistical methods to ensure robustness and validity. A Perception Index (PI) was computed as the mean of all Likert scale responses to quantify participants' overall perceptions. To assess the variability and consistency of responses, the Coefficient of Variation (CoV) was calculated, providing a relative measure of dispersion. Additionally, the Interquartile Range (IQR) was used to evaluate the spread of responses and identify potential outliers within the dataset.

To test the statistical significance of the responses and validate the findings, a Chi-square test was conducted as a nonparametric significance test, determining whether observed distributions significantly differed from expected distributions. This test was particularly useful for analyzing categorical data and identifying relationships between different variables.

For the qualitative data obtained from discussions, journal entries, and stakeholder opinions, an interpretative analysis was applied to uncover underlying themes and meanings. A narrative analysis was also employed to examine how participants structured their experiences and perspectives, providing a deeper contextual understanding of their viewpoints.

The integration of both quantitative and qualitative findings, along with feedback from instructors, technical experts, and students, generated valuable insights for refining the PBL approach. Furthermore, the perspectives and insights obtained from the focus group discussions were systematically analyzed using narrative analysis, ensuring a comprehensive interpretation of the discourse and its implications for improving PBL implementation in future iterations.

8.4 Results and Discussions

The linkage between entrepreneurial education for engineering students and PBL was examined based on survey respondents' perceptions in two key areas: (1) the attributes students develop through PBL; and (2) how PBL activities contribute to the development of entrepreneurial attributes. Additionally, findings from the narrative analysis of stakeholder discussions were used to compare, corroborate, or challenge the results from the perception survey.

8.4.1 Attributes Students Learn through PBL

The results presented in Table 8.1 provide a detailed overview of students' perceptions regarding the entrepreneurial attributes developed through PBL. According to the students' perceptions, based on indicators such as the Perception Index (PI), Standard Deviation (SD), Coefficient of Variation (CoV), Interquartile Range (IQR), and statistical significance from the Chi-square test. The student learning was classified into three categories: high, mid and low. The data is supported by measures.

8.4.1.1 High Perceived Learning Outcomes

Attributes in this category scored above 4.00 on the Perception Index, indicating that students strongly agreed that PBL enhanced their skills in these areas. All attributes here also had low CoV values (≤18.8%), IQR values and were statistically significant ($p < 0.05$), reflecting high consensus and reliability in students' responses.

- **Teamwork** (PI = 4.30, SD = 0.643, CoV = 15.0%, IQR = 1.0, $p < 0.001$): Rated highest among all attributes, teamwork was clearly perceived as a strength of the PBL approach. The low variability suggests widespread agreement that PBL supports collaborative learning.
- **Conceptualization** (PI = 4.19, SD = 0.737, CoV = 17.6%, IQR = 1.0, $p < 0.001$): This indicates strong support for PBL's effectiveness in helping students generate and develop ideas conceptually.
- **Design, Creativity, and Innovation** (PI = 4.09, SD = 0.756, CoV = 18.5%, IQR = 1.0, $p < 0.001$): Students acknowledged that PBL stimulates creative problem-solving and innovation, reinforcing its value in entrepreneurial learning.
- **Analytical Skills** (PI = 4.04, SD = 0.760, CoV = 18.8%, IQR = 1.0, $p = 0.049$): Although slightly lower in mean and higher in variability, this still falls

Table 8.1 The Attributes Students Learn through PBL

Attributes of PBL	*Perceptions of Respondents Regarding the Learning of Entrepreneurship Attributes*				
	PI	*SD*	*CoV (%)*	*IQR*	*Chi-square Test (p-value)*
Theories and principles	3.93	0.719	18.3	1.0	<0.001
Multidisciplinary knowledge	3.58	0.693	18.3	1.0	<0.001
Analytical skills	4.04	0.760	18.8	1.0	0.049
Conceptualization	4.19	0.737	17.6	1.0	<0.001
Design, creativity and innovation	4.09	0.756	18.5	1.0	<0.001
Practical work	2.93	0.796	27.1	2.0	<0.001
Communication and presentation skills	3.86	0.652	16.9	1.0	<0.001
Engagement with stakeholders	2.96	0.715	24.2	1.0	<0.001
Teamwork	4.30	0.643	15.0	1.0	<0.001
Contribute individually	3.72	0.688	18.5	1.0	<0.001
Evaluation of the outcomes and risk analysis	2.78	0.692	24.9	1.0	<0.001
N	120				

within the high range. The statistical significance affirms its positive reception in supporting critical thinking.

The consistent significance across all these high-scoring attributes confirms that PBL is highly effective in fostering essential entrepreneurial skills—especially collaboration, conceptual reasoning, and creative analysis—making it a strong pedagogical tool in these domains.

8.4.1.2 Moderate Perceptions

These attributes had Perception Index values between 3.50 and 3.99, indicating moderate agreement among students regarding the effectiveness of PBL in these areas. Variability remained low (CoV between 16.9% and 18.5%), and also low IQR values, and all results were statistically significant ($p < 0.001$), reflecting a dependable, though less enthusiastic, consensus.

- **Theories and Principles** (LI = 3.93, SD = 0.719, CoV = 18.3%, IQR = 1.0, $p < 0.001$): Students agreed that theoretical grounding is addressed, although it may not be as strongly emphasized as practical or teamwork skills.
- **Communication and Presentation Skills** (LI = 3.86, SD = 0.652, CoV = 16.9%, IQR = 1.0, $p < 0.001$): These skills are moderately developed, possibly reflecting variations in how PBL is facilitated or assessed.

- **Individual Contribution** (LI = 3.72, SD = 0.688, CoV = 18.5%, IQR = 1.0, $p < 0.001$): While teamwork is well supported, individual accountability may require more explicit emphasis within group-based projects.
- **Multidisciplinary Knowledge** (LI = 3.58, SD = 0.693, CoV = 18.3%, IQR = 1.0, $p < 0.001$): Students perceive some benefit in integrating knowledge across disciplines, though the implementation of this dimension might vary depending on the problem scenarios provided.

These attributes, while positively rated, highlight areas for enhancement. Educators could strengthen content delivery and individual responsibility within group settings, as well as better scaffold communication and cross-disciplinary engagement.

8.4.1.3 Low Perceived Learning Outcomes

This group comprises attributes with Likert Index values below 3.50, indicating that students felt PBL was less effective in fostering these competencies. These attributes also showed higher CoV values (≥24.2%) and a wider spread of responses, although all were still statistically significant ($p < 0.001$), suggesting the results are reliable but reflect critical gaps.

- **Engagement with Stakeholders** (LI = 2.96, SD = 0.715, CoV = 24.2%, IQR = 1.0, $p < 0.001$): Students reported limited exposure to real-world external stakeholders such as clients, industry mentors, or community partners.
- **Practical Work** (LI = 2.93, SD = 0.796, CoV = 27.1%, IQR = 2.0, $p < 0.001$): This had the highest variability (IQR = 2.0), showing inconsistent experiences with hands-on or field-based activities across PBL projects.
- **Evaluation of Outcomes and Risk Analysis** (LI = 2.78, SD = 0.692, CoV = 24.9%, IQR = 1.0, $p < 0.001$): The lowest-rated attribute suggests students lack sufficient experience in assessing results or navigating risk, key entrepreneurial capabilities.

Despite high statistical significance, these low scores call attention to limitations in the practical, real-world components of PBL implementation. Enhancing stakeholder interaction, incorporating structured opportunities for practical application, and teaching risk evaluation strategies are essential steps to address these shortcomings.

The results illustrate that students perceive PBL as a powerful pedagogical approach for developing soft skills such as teamwork, creativity, and analytical thinking, which are essential for entrepreneurship. However, it also reveals critical areas for pedagogical improvement, especially concerning experiential learning, stakeholder engagement, and risk management—core elements of practical entrepreneurship. The uniformly significant Chi-square p-values across all attributes reinforce the reliability of these perceptions across the respondents.

Furthermore, these findings were corroborated by the stakeholders' discussions. On this account, according to some of the stakeholders discussed, it was noted that:

> ...PBL forces students to work in a group environment and at the same time allows them to contribute individually according to their disciplines and specializations, which are generally discussed by the team, and, accordingly, decisions are made. Thus, PBL assists in both teamwork and individual contributions simultaneously...

Some of the facilitators said that:

> ...It is amazing to learn how students learn quickly in the company of each other and appreciate multidisciplinary knowledge. The students are open to the multidisciplinary and interdisciplinary knowledge, which is generally not seen in the conventional learning systems...

According to some of the assessors and jury members:

> ...the quality of outputs shows the level of knowledge, analytical skills, ingenuity, ability of design, creativity, innovation and communication the students have gained. PBL must have definitely assisted them...

Thus, the findings of the perceptions of surveys such as PBL could able to assist in learning the attributes such as theories and principles, multidisciplinary knowledge, analytical skills, conceptualisation, design, creativity, innovation, communication and presentation skills, teamwork, and contributing individually, are corroborated by the stakeholders.

8.4.2 Linkage of Entrepreneurial Attributes with PBL Activities

An attempt was made to map or link the entrepreneurial aspects or attributes with the teaching, learning, and assessment activities conducted in PBL. The entrepreneurial attributes were selected from the findings of the scholars evidenced in the mainstream literature in entrepreneurship education. However, the mapping was conducted based on the perceptions of the respondents and corroborated by the opinions of stakeholders that emanated from the discussions with them. Table 8.2 presents the linkage between entrepreneurial attributes and the teaching, Learning, and assessment activities undertaken in PBL. The entrepreneurial attributes were categorized under five specific categories: personal, intellectual, resource, technical, and communication and engagement.

The analysis of students' perceptions of entrepreneurial attributes developed through PBL activities reveals important insights into the strengths and

Table 8.2 The Linkage between Entrepreneurial Attributes and PBL Activities

Entrepreneurial Aspects	*Activities Embedded in PBL Linking Entrepreneurial Aspects*	*Perception of Students on the Learning of Entrepreneurship Attributes Through PBL*				
		PI	*SD*	*CoV (%)*	*IQR*	*Chi-square Test (p-value)*
Personal						
Motivation	Exploring the problems and undertaking projects for problem-solving	4.13	0.705	17.1	1.0	<0.001
Attitude	Working in groups trying to solve problems, and exposing to constructive feedback and criticisms	2.68	0.661	24.6	1.0	<0.001
Work experience	Practical work/ Hands-on experience	4.19	0.702	16.7	1.0	<0.001
Intellectual						
Prior knowledge base	Selection of a problem. Exploring the availability of discipline-based knowledge	3.38	0.536	15.9	1.0	<0.001
Idea generation	Brainstorming among the group and exchange of ideas	4.28	0.663	15.5	1.0	<0.001
Articulation of the idea and writing of the project brief	Writing the design brief	4.17	0.737	17.7	1.0	<0.001
Evaluation of the feasibility of the idea	Presenting the design brief to the peers and stakeholders and understanding the demands, Potentials, and costs	4.02	0.799	19.9	2.0	0.723
Resource						
Generating resources (human, know-how, and skill)	Formation of groups, exploring the knowledge, skills, and competencies available in the group and among the peers	3.12	0.735	23.9	1.0	<0.001
Findings investors/ venture capitalists to commercialize the product	Engagement with the stakeholders Presentation of the design brief and the final product to the potential investors	2.75	0.664	24.2	1.0	<0.001

Technical						
Product/design conceptualization and innovation	Transforming the ideas into concepts	4.02	0.820	20.4	2.0	<0.975
Alternate designs/ solutions	Analysis/generating alternate products/design/ solutions	3.62	0.768	21.2	1.0	<0.001
Final product/design/solution and risk analysis	Evaluation of alternate products/solutions. Examining the product/ design or solution valueEvaluation of the risks of the final product and its acceptance	3.43	0.589	17.2	1.0	<0.001
Communication and Engagement						
Stakeholders' engagement	Engagement with the stakeholders, such as potential buyers, consumers, investors, technical partners, engineers, technologists, etc.	2.88	0.740	25.4	1.0	<0.001
Selling the idea/product/design	Presentation, showcasing, and competition	3.82	0.698	18.3	1.0	<0.001
N = 120						

Source of entrepreneurial aspects: Charney & Libecap, 2000; Karima, 2016; Kriewall and Mekemson, 2010; Lumsdaine & Binks, 2003.

limitations of this pedagogical approach. The results, as measured using a five-point Likert scale, are interpreted based on mean scores (Perception Index), dispersion measures (SD, CoV, IQR), and Chi-square significance tests.

Under the category of personal attributes, students rated work experience (PI = 4.19, SD = 0.702, CoV = 16.7%, IQR = 1.0) and motivation (PI = 4.13, SD = 0.705, CoV = 17.1%, IQR = 1.0) as the most positively influenced by PBL activities, particularly through hands-on tasks and problem-solving. These attributes also recorded statistically significant Chi-square values ($p < .001$), indicating consistent agreement across the student cohort. In contrast, attitude received a lower mean score (PI = 2.68, SD = 0.661, CoV = 24.6%, IQR = 1.0), suggesting that group work and classroom feedback mechanisms may not sufficiently foster entrepreneurial mindsets. Despite its lower score, the result was statistically significant ($p < 0.001$), underscoring a shared perception of limited impact in this area.

In the domain of intellectual attributes, idea generation (PI = 4.28, SD = 0.663, CoV = 15.5%, IQR = 1.0) and articulation of ideas (PI = 4.17, SD = 0.737, CoV = 17.7%, IQR = 1.0) received high ratings, with p-values less than 0.001, reflecting strong and consistent student perceptions that brainstorming activities and project briefs effectively enhance cognitive entrepreneurial competencies. The prior knowledge base was moderately rated (PI = 3.38) but had the lowest CoV (15.9%), IQR = 1.0, and SD (0.536), indicating relatively homogeneous views. However, evaluation of feasibility (PI = 4.02) showed a wider spread in responses (SD = 0.799, IQR = 2.0). It did not yield a statistically significant result ($p = 0.723$), suggesting inconsistencies in how students experience feasibility assessment within PBL contexts.

The resource management category presented more mixed results. Resource mobilization, which is a critical component of any entrepreneurial venture, appeared to be only partially addressed within the PBL framework. It was observed that PBL activities—such as the formation of student groups and the exploration of the knowledge, skills, and competencies within these groups and among peers—provided some opportunity for students to learn how to identify and gather human resources. This process also helped them begin to map available competencies to specific project needs. While students moderately acknowledged their ability to generate internal resources within their teams (PI = 3.12, SD = 0.735, CoV = 23.9%, IQR = 1.0, $p < .001$), the attribute related to finding external investors received a notably lower rating (PI = 2.75, SD = 0.664, CoV = 24.2%, IQR = 1.0, $p < .001$). These findings suggest a significant gap in the integration of external entrepreneurial elements—such as accessing funding and engaging with stakeholders—within the current PBL activities, even though the perceptions were statistically significant and consistently shared among respondents.

Regarding technical development, product/design conceptualization received a high mean score (PI = 4.02). However, it was associated with high variability (SD = 0.820, IQR = 2.0) and a non-significant p-value ($p = 0.975$), indicating inconsistent student experiences in this area. Similarly, alternate designs and solutions were moderately rated (PI = 3.62, $p < .001$), reflecting a fair perception of technical exploration through iterative design. The evaluation of the final product/design and risk analysis was somewhat positive (PI = 3.43), with low SD (0.589) and CoV (17.2%), and IQR = 1.0 value supported by a significant p-value ($p < .001$), showing moderate consensus about its relevance in PBL projects.

Finally, under communication and engagement, the ability to sell the idea/product was relatively well-rated (PI = 3.82, SD = 0.698, CoV = 18.3%, IQR = 1.0, $p < .001$), indicating that activities such as project showcases and competitions support the development of entrepreneurial presentation skills. In other words, respondents opine that the skill of selling the idea/product/design, which is one of the most important aspects of successful entrepreneurship, is learned fairly through PBL activities such as presentations. showcasing and competition, which the students have to carry out within a very limited time, which is often limited to a few minutes. However, stakeholder engagement received a low score (PI = 2.88, SD = 0.740, CoV = 25.4%, IQR = 1.0), highlighting a consistent gap in real-world exposure through industry or community collaboration, though the result remained statistically significant ($p < .001$).

Overall, the results indicate that PBL is particularly effective in fostering entrepreneurial attributes such as personal motivation, ideation, articulation of ideas, and experiential learning—elements that are foundational to entrepreneurial thinking. These strengths suggest that PBL provides a supportive environment for the development of critical soft and cognitive skills through active, student-centred learning. The Chi-square significance tests affirm the reliability of these perceptions, with most attributes yielding p-values below the 0.05 threshold, thereby lending statistical support to the representativeness of the findings.

However, significant gaps persist in areas such as stakeholder engagement, external resource mobilization (e.g., attracting investors or venture capitalists), and structured risk assessment. These dimensions are crucial for holistic entrepreneurial development but appear to be underrepresented or insufficiently addressed in current PBL practices. Thus, while a majority of respondents agreed that key entrepreneurial attributes across the five broad categories—personal, intellectual, resource, technical, and communication and engagement—are embedded within PBL activities and can be effectively learned through them, certain attributes such as cultivating entrepreneurial attitudes, accessing external funding, and engaging with stakeholders were not conclusively linked to PBL experiences.

8.4.3 Discussion

Entrepreneurship education has gained significant traction in higher education, particularly within engineering disciplines, due to the growing demand for graduates who can innovate, adapt, and contribute to socio-economic development. However, there is considerable debate over how such education should be designed and integrated into the curriculum. Some scholars argue that entrepreneurship should be offered as a standalone qualification or comprehensive course, typically at the postgraduate diploma or Master's level, where it often forms part of management education or is delivered independently with a well-structured curriculum covering topics such as marketing, finance, competitive analysis, and business plan development (Charney & Libecap, 2000). In some institutions, entrepreneurship is embedded within undergraduate engineering or management programs, while some scholars support the delivery of focused modules—such as new business development or business plan preparation—as a means of imparting entrepreneurial skills (Charney & Libecap, 2000). More recently, thought leaders in engineering education have called for the integration of entrepreneurial thinking into the very culture of engineering pedagogy itself (Karima, 2016; Kriewall & Mekemson, 2010). Despite the lack of consensus on the delivery model, there is widespread agreement that engineering students should graduate with entrepreneurial knowledge and capabilities—for instance, the ability to identify problems with market potential and to design solutions from the customer's perspective (Karima, 2016; Lumsdaine & Binks, 2003).

The findings from this study offer evidence of the transformative potential of PBL in fostering such entrepreneurial attributes among engineering students in South Africa. The results highlight how PBL—anchored in constructivist and experiential learning theories—serves as a catalyst for entrepreneurial thinking and behaviour in the engineering education context. Based on the findings, a conceptual framework linking the PBL activities and entrepreneurial attributes is presented in Figure 8.1.

Survey responses reveal that a substantial proportion of students perceive PBL as instrumental in developing critical entrepreneurial competencies, including opportunity identification, innovative problem-solving, teamwork, stakeholder engagement, and project management. These competencies align with the literature on entrepreneurship education, which emphasizes real-world learning, resilience, creativity, and effective communication as central to entrepreneurial development (Fayolle & Benoît, 2008; Nabi et al., 2017).

Theoretically, these findings are consistent with Barrows and Tamblyn's (1980) framework of PBL as a learner-centred methodology that immerses students in complex, authentic problems, mirroring the uncertainty and multidisciplinary challenges faced by entrepreneurs. Through such learning processes, students are not only applying technical knowledge but also developing vital entrepreneurial capabilities like decision-making under uncertainty, strategic planning, and resource allocation (Schmidt, 1998; Servant-Miklos, 2019).

Qualitative insights from focus group discussions further illustrate how entrepreneurial mindsets are cultivated through PBL. Students shared that

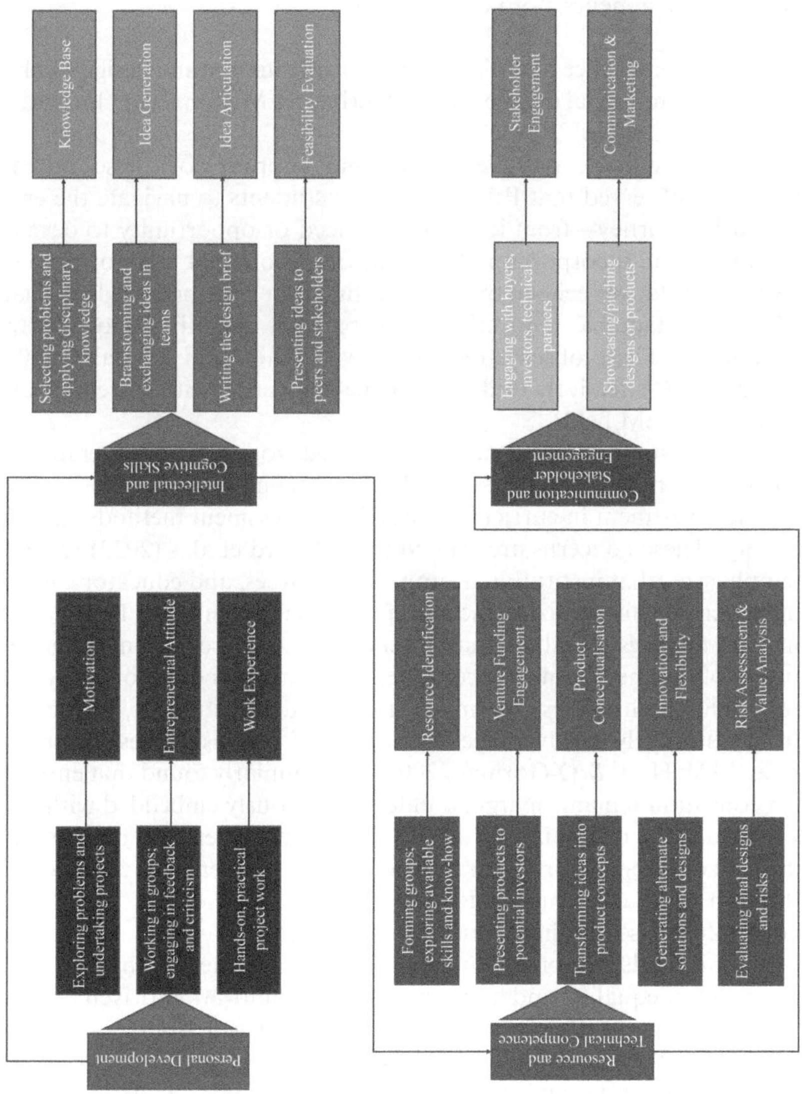

Figure 8.1 Conceptual Framework Linking PBL Activities to Entrepreneurial Attributes.

working on real-world problems, especially in community-based or industry-linked projects, required them to think beyond traditional engineering roles. One student remarked,

> When we worked on real-life problems with actual users in mind, we had to think like entrepreneurs, not just engineers.

This reflects the significance of authenticity, user orientation, and design thinking in building entrepreneurial competencies (Gerhart & Melton, 2016; Pittaway & Cope, 2007).

Stakeholder feedback provided additional support for these findings. Academic staff observed that PBL encourages students to navigate the entire entrepreneurial journey—from identifying a need or opportunity to developing a prototype and incorporating feedback from potential users or partners. Industry representatives echoed this view, noting that PBL-prepared graduates tend to be more adaptable, innovative, and capable of operating across disciplinary boundaries. These observations align with Zhou and Frisdahl's (2024) assertion that PBL effectively bridges technical education with entrepreneurial competence in STEM fields.

However, the study also surfaced several pedagogical and systemic challenges. Lecturers noted that the successful implementation of PBL demands considerable investment in curriculum redesign, assessment methods, and faculty training. These concerns are echoed in Holgaard et al.'s (2022) research, which emphasizes that institutional support, resources, and educator capacity building are critical for realizing the entrepreneurial potential of PBL.

Moreover, findings reveal that entrepreneurial exposure within engineering programs is often uneven and concentrated in elective modules or postgraduate tracks. This limits access for many undergraduate students, particularly those from historically disadvantaged institutions. Previous studies (Garavan & O'Cineide, 1994; Hytti & O'Gorman, 2004) have similarly found that entrepreneurship education remains marginal unless consciously embedded within the core curriculum. To maximize inclusivity and impact, there is a clear need to institutionalize entrepreneurial learning across the broader engineering curriculum through approaches such as PBL.

Importantly, these findings must be contextualized within the broader socio-economic landscape of a country or region, characterized by high youth unemployment, inequality, and an urgent need for innovation-driven development. In this setting, PBL offers not only a pedagogical strategy but also a developmental intervention—empowering engineering graduates to become job creators, innovators, and responsive problem-solvers. Thus, integrating entrepreneurial learning through PBL supports national priorities and contributes to long-term economic transformation (Fatoki & Chindoga, 2011; Ndedi, 2009).

This study thus reinforces the argument for leveraging PBL as an effective framework for entrepreneurship education in engineering. It confirms that when thoughtfully designed and contextually relevant, PBL can move students from passive recipients of knowledge to active, entrepreneurial change-makers. To fully harness this potential, higher education institutions must commit to curriculum transformation, cross-disciplinary collaboration, and sustained investment in teaching capacity. Future research could explore the long-term entrepreneurial trajectories of engineering graduates exposed to PBL and examine comparative outcomes across diverse institutional settings.

8.5 Conclusion

There are various viewpoints and ongoing debates regarding how entrepreneurship education should be delivered in higher education generally, and in engineering education in particular. Drawing from the recent successes of PBL implementations across universities worldwide, an emerging argument is that PBL can effectively foster essential entrepreneurial attributes among engineering students. In this context, the present study sought to examine the linkage between the development of entrepreneurial attributes and the application of PBL in engineering education.

The study was grounded in the perceptions of engineering students, collected through a perception survey using a pre-tested questionnaire, as well as qualitative discussions with key stakeholders, including lecturers, facilitators, and industry partners. Findings suggest that PBL is highly relevant for engineering education, as it encourages students to explore engineering solutions to complex social, economic, and environmental problems—many of which lie outside the traditional scope of engineering yet have clear commercial demand.

Throughout the PBL process, students engage in a variety of tasks and activities that help them develop a wide range of cognitive and entrepreneurial competencies. These include the ability to understand the relevance of multidisciplinary knowledge, apply analytical and conceptual skills, engage in design thinking, foster creativity and innovation, communicate effectively, collaborate in teams, engage with stakeholders, and evaluate outcomes, including conducting risk analysis. These competencies align with five key dimensions of entrepreneurial attributes: personal, intellectual, resource, technical, and communication and engagement. Thus, PBL offers students a valuable platform to cultivate these traits within a structured, problem-oriented learning environment.

The findings from this study suggest that PBL can serve as a viable pedagogical approach for embedding entrepreneurial attributes in engineering education. However, the study is not without limitations. It is based on a relatively small-scale perception survey and qualitative discussions, which may not fully capture the broader context or long-term impact.

Nevertheless, the evidence indicates that incorporating PBL into engineering curricula has the potential not only to support the development of cognitive skills and exit-level competencies (such as problem identification, analysis, and idea generation, conceptualization, design, innovation, teamwork, and communication) but also to enhance students' entrepreneurial readiness. In doing so, PBL can contribute meaningfully to producing graduates who are not only technically competent but also entrepreneurial in mindset and practice.

References

Barrett, T. (2005). Understanding problem-based learning. In T. Barrett, I. Mac Labhrainn, & H. Fallon (Eds.), *Handbook of enquiry and problem-based learning* (pp. 13–25). CELT.

Barrows, H. S., & Tamblyn, R. (1980). *Problem-based learning: An approach to medical education*. Springer.

Bezerra de Araújo, R. G., Americano da Costa, M. V., Joseph, B., & Sánchez, J. (2020). Developing professional and entrepreneurship skills of engineering students through problem-based learning: A case study in Brazil. *International Journal of Engineering Education, 36*(1), 155–169. https://dialnet.unirioja.es/servlet/articulo?codigo=7339013

Biley, F., & Smith, K. (1998). Exploring the potential of problem-based learning in nurse education. *Nurse Education Today, 18*(5), 353–361.

Boud, D., & Feletti, G. (Eds.). (1991). *The challenge of problem-based learning*. Kogan Page.

Charney, A., & Libecap, G. (2000). *Impact of entrepreneurship education*. Insights. A Kauffman Research Series. Kauffman Centre for Entrepreneurial Leadership.

Chen, N.C. (2008). An educational approach to problem-based learning. *Kaohsiung Journal of Medical Sciences, 24*(3 Suppl), S23–S30.

Das, D., & Mishra, B. (2018). Exploring the complementarity of problem-based learning with outcomes-based education in engineering education: A case study in South Africa. In W. Sunyu, A. Kolmos, A. Guerra, & Q. Weifeng (Eds.), *Proceedings of the 7th International Research Symposium on PBL* (pp. 297–307). Aalborg University Press.

Dewiyani, L., & Kosasih, M. (2017). Development of an entrepreneur learning model based on problem-based learning to increase competency, independence and creativity in students of industrial engineering. *International Journal of Scientific & Technology Research, 6*(10), 170–175. https://www.ijstr.org/paper-references.php?ref=IJSTR-1017-17975

Doboli, S., Kamberova, G. L., Impagliazzo, J., Xiang Fu, & Currie, E. H. (2010, October 27–30). A model of entrepreneurship education for computer science and computer engineering students [Paper presentation]. In *40th ASEE/IEEE Frontiers in Education Conference*, Washington, DC.

Dolmans, D. H. J. M., De Grave, W., Wolfhagen, I. H. A. P., & van der Vleuten, C. P. M. (2005). Problem-based learning: Future challenges for educational practice and research. *Medical Education, 39*(7), 732–741. https://doi.org/10.1111/j.1365-2929.2005.02205.x

Duch, B. J., Groh, S. E., & Allen, D. E. (2001). Why problem-based learning? A case study of institutional change in undergraduate education. In B. J. Duch, S. E. Groh, & D. E. Allen (Eds.), *The power of problem-based learning* (pp. 3–13). Stylus Publishing.

Fatoki, O., & Chindoga, L. (2011). An investigation into the obstacles to youth entrepreneurship in South Africa. *International Business Research*, *4*(2), 161–169. https://doi.org/10.5539/ibr.v4n2p161

Fayolle, A., & Benoît, G. (2008). From craft to science: Teaching models and learning processes in entrepreneurship education. *Journal of European Industrial Training*, *32*(7), 569–593. https://doi.org/10.1108/03090590810899838

Garavan, T. N., & O'Cineide, B. (1994). Entrepreneurship education and training programs: A review and evaluation—Part 1. *Journal of European Industrial Training*, *18*(8), 3–12.

Gerhart, A. L., & Melton, D. E. (2016). *Entrepreneurially minded learning: Incorporating stakeholders, discovery, opportunity identification, and value creation into problem-based learning modules with examples and assessment specific to fluid mechanics*. American Society for Engineering Education. https://peer.asee.org/entrepreneurially-minded-learning-incorporating-stakeholders-discovery-opportunity-identification-and-value-creation-into-problem-based-learning-modules-with-examples-and-assessment-specific-to-fluid-mechanics

Gómez, C. V. (2024). Teaching entrepreneurship in higher education: The application active based learning activities to environmental protection. *Thinking Skills and Creativity*, *52*, 101502. https://doi.org/10.1016/j.tsc.2024.101502

Hasanah, & Malik, M. N. (2019). Implementation of problem-based learning to improve critical thinking skills in entrepreneurs' learning. *Journal of Physics: Conference Series*, *1387*(1), 012037. https://doi.org/10.1088/1742-6596/1387/1/012037

Hmelo-Silver, C. (2004). Problem-based learning: What and how do students learn? *Educational Psychology Review*, *16*(3), 235–266.

Holgaard, J. E., Du, X., & Guerra, A. (2022). *When problem-based learning becomes entrepreneurial—a facilitator's view on student challenges*. Universitat Politècnica de València. https://doi.org/10.5821/conference-9788412322262.1266

Hynes, B. (1996). Entrepreneurship education and training—Introducing entrepreneurship into non-business disciplines. *Journal of European Industrial Training*, *20*(8), 10–17.

Hytti, U., & O'Gorman, C. (2004). What is enterprise education? An analysis of the objectives and methods of enterprise education programmes in four European countries. *Education and Training*, *46*(1), 11–23.

Juan, S., & Lingling, Z. (2013). An analysis of the development of student entrepreneurship abilities based on PBL mode of instruction. In *2013 Third International Conference on Intelligent System Design and Engineering Applications*, 392–396. https://doi.org/10.1109/ICIII.2013.6703168

Kabilan, S. J. (2025). A problem-based learning (PBL) approach in effective implementation of innovation and entrepreneurship education for engineering undergraduates. *Journal of Engineering Education Transformations*, *38* (Special Issue), 120–126.

Karima, M. S. A. (2016). Entrepreneurship education in an engineering curriculum. *Procedia Economics and Finance*, *35*, 379–387.

Kehinde, E. E. (2018). Effects of problem-based learning on polytechnic students' academic performance in entrepreneurship education in Ogun State. *Nigerian Journal of Business Education*, *4*(1), 311–318. https://www.nigjbed.com.ng/index.php/nigjbed/article/download/95/95

Kolvereid, L., & Moen, O. (1997). Entrepreneurship among business graduates: Does a major in entrepreneurship make a difference? *Journal of European Industrial Training*, *21*(4), 154–160.

Kriewall, T. J., & Mekemson, K. (2010). Instilling an entrepreneurial mindset into engineering. *The Journal of Engineering Entrepreneurship*, *1*(1), 5–19. http://www.jeenonline.com/Vol1/Num1/Vol1No1P1.pdf

Liu, L., Mynderse, J. A., Gerhart, A. L., & Arslan, S. (2015). Fostering the entrepreneurial mindset in the junior and senior mechanical engineering curriculum with a multi-course problem-based learning experience. In *2015 IEEE Frontiers in Education Conference (FIE)*, 1–5. https://doi.org/10.1109/FIE.2015.7344040

Lumsdaine, E., & Binks, M. (2003). Teaching entrepreneurship to engineers. In *Proceedings of the 2003 American Society for Engineering Education Annual Conference & Exposition*. http://ipmall.info/hosted_resources/teaching_ip/edward_lumsdaine_2003.pdf

Luryi, S., Tang, W., Lifshitz, N., Wolf, G., Doboli, S., Betz, J. A., Maritato, P., & Shamash, Y. (2007). Entrepreneurship in engineering education. In *37th Annual Frontiers in Education Conference – Global Engineering: Knowledge Without Borders, Opportunities Without Passports* (pp. T2E-10–T2E-15). IEEE. https://doi.org/10.1109/FIE.2007.4418174

Machado-Toledo, J., Mora, C. E., Añorbe-Díaz, B., González-Marrero, A., & Martín-Gutiérrez, J. (2018). Project-oriented problem-based learning for an entrepreneurial vision in engineering education. In *Re-engineering and automating the job of the future* (pp. 483–491). Springer, Cham. https://doi.org/10.1007/978-3-319-91152-6_37

Moore, J. (2009). An exploration of the lecturer as a facilitator within the context of problem-based learning. *Nurse Education Today*, *29*(2), 150–156.

Morselli, D. (2019). Teaching a sense of initiative and entrepreneurship through problem-based learning. *Form@re—Open Journal per La Formazione in Rete*, *19*(2), 149–160. https://doi.org/10.13128/FORMARE-25176

Mubuuke, G., Louw, A. J. N., & Van Schalkwyk, S. (2017). Cognitive and social factors influencing students' response and utilization of facilitator feedback in a problem-based learning context. *Health Professions Education*, *3*(2), 85–98.

Nabi, G., Liñán, F., Fayolle, A., Krueger, N., & Walmsley, A. (2017). The impact of entrepreneurship education in higher education: A systematic review and research agenda. *Academy of Management Learning & Education*, *16*(2), 277–299. https://doi.org/10.5465/amle.2015.0026

Ndedi, A. A. (2009). *Entrepreneurship and wealth creation at the African bottom of the pyramid.* SSRN. https://ssrn.com/abstract=1799125

Pittaway, L., & Cope, J. (2007). Entrepreneurship education: A systematic review of the evidence. *International Small Business Journal*, *25*(5), 479–510. https://doi.org/10.1177/0266242607080656

Raimundo, R., & Rosário, A. T. (2024). A bibliometric systematic literature review on the relationship between problem-based learning methodology and entrepreneurship. *Businesses*, *4*(4), 765–790. https://doi.org/10.3390/businesses4040042

Riazifar, N., & Khajehsaeid, H. (2024). Enhancing engineering education through problem-based learning. In *Proceedings of UK and Ireland Engineering Education Research Network Conference 2023*. University of Warwick. https://journals.warwick.ac.uk/index.php/eern/article/view/1636

Schmidt, H. G. (1998). Problem-based learning: Does it prepare medical students to become better doctors? *The Medical Journal of Australia*, *168*(9), 429–430.

Servant-Miklos, V. F. C. (2019). Fifty years on: A retrospective on the world's first problem-based learning programme at McMaster University Medical School. *Health Professions Education*, *5*(1), 3–12. https://doi.org/10.1016/j.hpe.2018.04.002

Surjanti, J., & Nugrohoseno, D. (2018). Antecedents of individual entrepreneurial orientation (IEO): Problem-based learning (PBL). *Journal of Business and Retail Management Research*, *13*(2). https://doi.org/10.24052/JBRMR/V13IS02/ART-12

Swart, A. J. (2014). Using problem-based learning to stimulate entrepreneurial awareness among senior African undergraduate students. *Eurasia Journal of Mathematics, Science and Technology Education*, *10*(2), 125–134. https://doi.org/10.12973/EURASIA.2014.1023A

Tan, S. S., & Ng, C. K. F. (2006). A problem-based learning approach to entrepreneurship education. *Education + Training*, *48*(6), 416–428. https://doi.org/10.1108/00400910610692606

Ward, J. D., & Lee, C. L. (2002). A review of problem-based learning. *Journal of Family and Consumer Sciences Education*, *20*(1), 16–26.

Williams, J. C., & Paltridge, D. J. (2017). What we think we know about the tutor in problem-based learning. *Health Professions Education*, *3*(1), 26–31.

Zhao, H., Seibert, S. E., & Hills, G. E. (2005). The mediating role of self-efficacy in the development of entrepreneurial intentions. *Journal of Applied Psychology*, *90*(6), 1265–1272.

Zhou, C., & Frisdahl, L. (2024). Entrepreneurship and problem-based learning (PBL) in STEM teacher education in high schools: An everyday creativity perspective. In S. Salman & A. Bhaumik (Eds.), *Entrepreneurship Innovation and Education for Performance Improvement* (pp. 30–52). IGI Global Scientific Publishing. https://doi.org/10.4018/979-8-3693-7903-5.ch002

Chapter 9

Decolonisation and Internationalisation of Engineering Education through Problem-Based Learning

9.1 Introduction

9.1.1 Context and Significance of Transforming Engineering Education

Engineering education globally is undergoing a significant paradigm shift as institutions grapple with the imperative to both decolonise curricula and promote internationalisation. These dual objectives—once perceived as divergent—are now recognised as complementary pathways towards more inclusive, relevant, and socially just forms of education. At the heart of this transformation lies the critical question: how can engineering education simultaneously honour local knowledge systems, cultures, and contexts while equipping graduates with global competencies to thrive in an interconnected world?

The transformation of engineering education is a multidimensional process propelled by the need to prepare students for the complexities of an increasingly interconnected and technologically advanced world. This transformation extends beyond the enhancement of technical competencies; it emphasises the cultivation of engineers who can respond innovatively and ethically to societal challenges. Central to this shift is the adoption of Problem-Based Learning (PBL)—a pedagogical approach that situates students at the core of the learning process through active engagement with real-world problems. PBL fosters a learning environment where theoretical knowledge is applied in practice, encouraging critical thinking, creativity, and the development of solutions grounded in local and global contexts (Lavi et al., 2023). It also promotes interdisciplinary collaboration and learner autonomy, equipping students with essential transversal or "21st century" skills such as communication, teamwork, and adaptive problem-solving, which are increasingly valued in modern engineering practice.

Moreover, the integration of contextual learning and sensitivity to cultural frameworks is fundamental to engineering education's transformation. Learning that is rooted in the social, environmental, and cultural contexts of learners ensures relevance and applicability, allowing engineering solutions to

DOI: 10.1201/9781003529743-12

be responsive to specific community needs (Crawley et al., 2008). For instance, models like Measure–Model–Manipulate–Make in biological engineering demonstrate how contextual frameworks (Ideker et al., 2006) can be embedded even in disciplines that are less product-oriented, reinforcing the importance of situating technical education within meaningful, real-world narratives.

The rapid pace of technological advancement and the pressures of globalisation further necessitate pedagogical innovation. Technology-enhanced learning—through online platforms, blended learning, and e-learning—has made engineering education more accessible and flexible, accommodating diverse learning preferences and improving retention and completion rates (Samaka & Ally, 2016). Simultaneously, globalisation compels the development of engineers who are not only technically adept but also innovative and entrepreneurial, capable of contributing to global value chains and sustainable development (Tryggvason & Apelian, 2011). PBL supports this shift by exposing students to interdisciplinary and international collaboration, preparing them to operate in transnational teams and navigate sociotechnical complexities.

In addition, the integration of social justice principles into engineering curricula is reshaping the discipline's identity. By questioning who gets to become an engineer and how engineering is practised, educators are beginning to address issues of equity, inclusion, and systemic bias within the field (Leydens & Lucena, 2018). PBL, with its emphasis on collaborative inquiry and stakeholder engagement, serves as an ideal framework for embedding these values, allowing students to explore the ethical dimensions of engineering challenges in ways that are informed by diverse experiences and perspectives.

Finally, curriculum reform efforts grounded in local contexts illustrate how PBL can be adapted to meet specific regional needs. Comparative studies of curriculum change in engineering departments across Kenya, Tanzania, and South Africa, for example, highlight the influence of sociopolitical, institutional, and economic factors on reform processes (Case et al., 2016). These examples reinforce the importance of a flexible and responsive educational model—such as PBL—that can accommodate varying local realities while aligning with global standards of engineering education.

Despite its transformative potential, the integration of PBL and other innovative approaches poses challenges. These include ensuring equitable access to digital and pedagogical resources, maintaining quality across diverse institutional settings, and overcoming resistance from entrenched traditional structures. Yet the momentum towards transformation reflects a growing recognition of the need to align engineering education with the evolving demands of the profession and society at large. As such, PBL stands not only as a method but as a catalyst for redefining what it means to educate an engineer in the 21st century—one who is technically competent, socially conscious, globally engaged, and locally responsive.

9.1.2 Definitions and Tensions: Decolonisation vs. Internationalisation

The concepts of decolonisation and internationalisation in higher education are often perceived as being in tension, yet they also hold potential as complementary processes when approached with critical intentionality. Decolonisation centres on dismantling enduring colonial structures and ideologies within educational systems, advocating for epistemic justice through the recognition and valorisation of diverse knowledge systems, particularly those originating from the Global South (Byrne & Alexander, 2023; Hall et al., 2021; Kohn & McBride, 2014; Wildcat, 2022). This process moves beyond political independence to challenge the deeper entanglements of the coloniality of power, being, and knowledge (Byrne & Alexander, 2023; Hall et al., 2021; Wildcat, 2022).

In contrast, internationalisation is typically defined as the intentional integration of international, intercultural, or global dimensions into education to improve educational quality and foster societal contributions (Ozyonum, 2023). However, it is frequently critiqued for inadvertently advancing Eurocentric paradigms, thereby reinforcing Western dominance in the production and dissemination of knowledge (Ozyonum, 2023).

A core tension between decolonisation and internationalisation lies in their epistemological orientations. While decolonisation explicitly challenges the supremacy of Western epistemologies and calls for pluralistic recognition of knowledge traditions, internationalisation efforts often perpetuate the notion of Western models as universal standards (Arif et al., 2022; Ozyonum, 2023). Additionally, the power dynamics embedded in internationalisation processes—such as the unidirectional flow of students and academics from the Global South to the Global North—risk reinforcing hierarchical structures of knowledge and privilege (Ozyonum, 2023). In contrast, decolonial approaches seek to disrupt these asymmetries by emphasising a pluriversal world in which multiple ways of knowing are equally validated (Byrne & Alexander, 2023).

Despite these tensions, there is significant potential for synergy between the two paradigms. For instance, internationalisation efforts that are grounded in decolonial perspectives can catalyse meaningful curriculum transformation, fostering more inclusive, reflective, and diverse educational environments (du Preez, 2018). This involves not only incorporating diverse epistemologies but also critically interrogating the ontological, epistemological, and ethical assumptions embedded in curricular design (du Preez, 2018). Moreover, virtual engagement platforms—such as Collaborative Online International Learning (COIL)—offer practical avenues for reimagining global learning partnerships. These platforms enable cross-cultural collaboration without physical mobility, reducing asymmetries and providing space for equitable knowledge exchange that respects local contexts (Cooke et al., 2019).

In a broader sense, decolonisation and internationalisation need not be viewed as mutually exclusive. Rather, when reframed through a critical lens, each can inform and enrich the other. Decolonial scholarship calls for a decentring of Western hegemony, which, if embraced by internationalisation initiatives, can result in more equitable global education practices (Byrne & Alexander, 2023). Conversely, internationalisation offers a vehicle for amplifying decolonial discourses and facilitating intercultural dialogue that broadens collective understanding (Kramm et al., 2024). This interplay suggests that a transformative, inclusive, and globally relevant education system is possible when decolonisation and internationalisation are engaged as dynamic, intersecting forces rather than oppositional paradigms.

9.1.3 Rationale for Using Problem-Based Learning (PBL) as a Mediating Approach

PBL is an increasingly adopted pedagogical approach across various disciplines due to its strong emphasis on active, student-centred learning. The core rationale for employing PBL lies in its ability to immerse students in real-world problems, thereby fostering critical thinking, collaboration, and self-directed learning. Unlike traditional lecture-based methods, PBL shifts the focus towards interactive and practical experiences, aligning more closely with the skills required in professional environments. Grounded in established educational theories, PBL has been shown to enhance knowledge retention and application, making it a valuable strategy in contemporary education.

One of the defining features of PBL is its engagement with authentic, real-world problems, which allows students to recognise the relevance of their academic work to real-life situations. This aspect is particularly impactful in professional fields such as medicine, law, and business, where the ability to apply theoretical knowledge in practice is essential (Hunter et al., 2025; Leong, 2009; Pepper, 2014). Through engagement with complex, often ill-structured problems, students are encouraged to think critically and creatively, exploring multiple solutions and developing essential problem-solving competencies (Meifitri, 2020; Pepper, 2014).

PBL also fosters the development of higher-order thinking skills, requiring students to analyse, synthesise, and evaluate information—skills that are often underdeveloped in more traditional, rote-based education systems (Hawamdeh & Adamu, 2021; Leong, 2009). Rather than passively receiving information, students are actively involved in constructing their understanding by drawing on prior knowledge and applying it in new and varied contexts, which promotes deeper learning (Hawamdeh & Adamu, 2021; Leong, 2009).

A critical component of PBL is its focus on collaborative and self-directed learning. Students typically work in small, cooperative groups where they share diverse perspectives, communicate effectively, and refine their teamwork skills (Lalwani, 2024; Leong, 2009; Meifitri, 2020). Simultaneously, PBL encourages learners to take ownership of their education, nurturing habits of inquiry, autonomy, and lifelong learning (Movafegh & Azimaraghi, 2015; Srilatha et al., 2022).

Moreover, PBL has been linked to enhanced retention and application of knowledge, with research showing that students engaged in active, problem-based scenarios tend to remember content more effectively and apply it with greater confidence (Finucane et al., 1998; Lalwani, 2024). This experiential approach also helps students develop systematic problem-solving strategies that are transferable across academic and professional domains (Hawamdeh & Adamu, 2021).

Despite its many advantages, PBL is not without its challenges. Implementation can be resource-intensive, demanding more time, planning, and effort from both instructors and students. The effectiveness of the method also hinges on the quality of the problems posed and the ability of educators to facilitate rather than instruct (Finucane et al., 1998). Nevertheless, the educational benefits of PBL—particularly in cultivating critical thinking, collaborative learning, and practical skill application—make it a compelling and transformative approach within modern educational contexts.

However, despite the potential of PBL as a transformative pedagogical approach, there remains a significant research gap regarding its role in advancing decolonisation and internationalisation within academic contexts.

9.1.4 Research Aims/Objectives and Questions

Based on the discussion of PBL as a transformative approach in the context of decolonisation and internationalisation of engineering education, the objectives of the study are:

Research Objectives

1. To investigate how PBL can serve as a mediating framework for the transformation of engineering education in a globalised context.
2. To explore the role of PBL in fostering decolonial and inclusive pedagogies within engineering curricula.
3. To assess the potential of PBL to bridge the tensions between decolonisation and internationalisation in engineering education.

For this purpose, the Research Questions investigated include:

- RQ1 To what extent can PBL help reconcile the goals of decolonisation and internationalisation in engineering education, particularly in the Global South?
- RQ2 In what ways can PBL support the decolonisation of engineering curricula by integrating diverse epistemologies and local contexts?
- RQ3 What are the challenges, opportunities, and impacts of implementing PBL as a means to promote inclusive, socially just, and globally relevant engineering education?

9.2 Theoretical and Conceptual Framework

9.2.1 Decolonisation of Engineering Education

The decolonisation of engineering education is a transformative process that critically examines and reshapes curricula to address both historical legacies and ongoing colonial influences. Rooted in a commitment to inclusivity and justice, this approach challenges the dominance of Eurocentric perspectives and seeks to integrate diverse epistemologies, particularly those of Indigenous and marginalised communities (Byrne & Alexander, 2023; Kohn & McBride, 2014). The ultimate goal is to create an equitable educational environment that not only acknowledges past injustices but also fosters cultural relevance, innovation, and empowerment.

Historically, engineering education has been shaped by colonial powers, embedding Western ideologies, values, and methodologies into global curricula. This dominance has systematically excluded local knowledge systems and cultural practices, narrowing the scope of what is considered valid knowledge (Dumbuya, 2025; Fomunyam, 2020). The continued prioritisation of Eurocentric standards limits the discipline's ability to address complex global challenges and alienates students from non-Western backgrounds (Muller, 2018; Winter et al., 2022).

A key element of decolonising engineering education involves recognising the myth of neutrality and objectivity in the field. Engineering is often viewed as apolitical and purely technical, yet this perception masks embedded colonial assumptions and power dynamics (Lord et al., 2019). Many programmes still rely on curricula that reflect technocratic and positivist models, marginalising Indigenous knowledge systems and communal, spiritual, or oral traditions (Adam, 2024).

In response, several institutions are undertaking efforts to reimagine engineering education. Initiatives in countries like Canada and Nigeria are redesigning curricula to include culturally relevant content and local problem-solving practices (Cicek et al., 2024; Fomunyam, 2020). Similarly, in South Africa, as an example, students have identified indigenous practices such as *umfuso* (sun-drying vegetables) and traditional brewing as engineering-relevant and worthy of curricular integration (Kondratyev, 2023; Mathaba, 2023). In Canada, the use of "threshold concepts" is helping educators and students cross epistemological divides and engage meaningfully with Indigenous perspectives (Wyllie-Runner et al., 2024).

A decolonial praxis in engineering also promotes radical pluralism and the inclusion of subaltern epistemologies, challenging hierarchical knowledge structures and fostering global justice and ecological sustainability (Moleka, 2024). Alongside curricular reforms, attention must also be paid to educational technology, which often reproduces colonial logics under the guise of neutrality. A justice-oriented framework is needed to expose and address the material and epistemic injustices embedded in EdTech systems (Adam, 2024).

Despite these promising developments, challenges remain. Integrating indigenous knowledge into existing frameworks is complex, and there is no single path forward. The process may evoke trauma for Indigenous students and faculty, and institutional resistance can impede progress (Cicek et al., 2024; Mathaba, 2023). Nevertheless, sustained dialogue, critical reflection, and collaborative action offer pathways towards a more inclusive, culturally responsive, and socially embedded vision of engineering education.

9.2.2 Internationalisation in Higher Education

The internationalisation of higher education is a multifaceted and dynamic process shaped by globalisation, technological advancements, and the growing interconnectedness of economies. Its goals extend beyond academic mobility to include the development of global competencies, intercultural understanding, and ethical engagement. However, this process also entails significant risks, including the potential reinforcement of neocolonial structures. To navigate these complexities, institutions have adopted two key approaches, "internationalisation at home" and "internationalisation abroad," each offering distinct pathways toward global engagement.

"Internationalisation at home" refers to efforts aimed at embedding international and intercultural dimensions within the curriculum and campus life without requiring students to travel. This approach ensures that all students, including those who cannot participate in mobility programmes, develop global competencies through diverse course content, inclusive teaching practices, and virtual exchanges (de Wit, 2020; Lazari & Matsoukas, 2025). In contrast, "internationalisation abroad" involves physical mobility through study abroad programmes, faculty exchanges, and international partnerships. Traditionally, the dominant form of this model is often motivated by institutional goals such as improving global rankings and increasing revenue through international student enrollment (de Wit, 2019, 2020).

One of the primary aims of internationalisation is to cultivate global competencies in students. These include skills such as cross-cultural communication, critical thinking, and adaptability—traits essential for thriving in an increasingly interconnected world. Institutions achieve this through curriculum internationalisation and by fostering inclusive campus environments that embrace diversity (Kraja et al., 2024; Lazari & Matsoukas, 2025). Alongside skill development, ethical engagement is a critical dimension. Institutions are encouraged to pursue equitable access to educational resources, ensure respectful engagement with diverse cultures, and address systemic power imbalances. Ethical internationalisation avoids reducing education to a market commodity and instead promotes fairness, mutual respect, and social responsibility (de Wit, 2020; Kraja et al., 2024).

Despite its potential, internationalisation also presents risks that warrant critical attention. One such risk is cultural hegemony, where dominant cultural norms and academic paradigms overshadow local traditions and knowledge systems. This can lead to a homogenisation of educational practices, marginalising alternative worldviews (Kraja et al., 2024). Power imbalances are also prevalent in international collaborations, especially between institutions in the Global North and the South. These asymmetries can result in unequal partnerships where benefits are disproportionately skewed towards more affluent institutions (Kraja et al., 2024). Additionally, the commodification of education—fuelled by the pursuit of international student revenue—can undermine educational quality and ethical commitments, shifting focus from learning outcomes to economic returns (de Wit, 2019, 2020).

While internationalisation offers meaningful opportunities for enriching higher education, it must be approached with a commitment to ethical principles and cultural sensitivity. To avoid reinforcing colonial dynamics, institutions must critically assess their practices, foster equitable partnerships, and prioritise diversity and inclusion. By doing so, they can contribute to a more just, balanced, and genuinely global educational landscape.

9.2.3 Problem-Based Learning as a Transformative Pedagogy for Local Relevance with Global Perspectives

PBL has evolved into a powerful student-centred methodology (Loyens et al., 2011; Storey & Schiavo, 2013). Its pedagogical strength lies in its grounding in constructivist learning theory, which posits that learners actively construct knowledge through meaningful experiences and social interaction (Loyens et al., 2011). Within PBL environments, collaboration, inquiry, and self-direction are central, reflecting core tenets of constructivist pedagogy (Rehmat et al., 2018). Furthermore, PBL aligns with the principles of critical pedagogy as conceptualised by Paulo Freire, by engaging students in the critical examination of real-world social issues. This pedagogical alignment facilitates not only deep learning but also transformative engagement, empowering learners to become reflective thinkers and socially responsible actors (Cavanagh et al., 2019).

A distinctive strength of PBL is its capacity to integrate local relevance with global perspectives. PBL scenarios frequently embed global challenges, allowing students to engage with interdisciplinary problems that transcend national borders while remaining grounded in their own sociocultural contexts. This dual focus enables students to connect global concerns—such as sustainability, inequality, or technological change—with locally meaningful actions. In doing so, they develop critical competencies such as interdisciplinary reasoning, holistic analysis, and adaptive problem-solving, which are essential in navigating the complexities of an interconnected world (Thomas, 2009). Moreover, engaging with culturally diverse, real-world contexts enhances students' understanding of global interdependence and cultivates a sense of civic responsibility. In other words, PBL positions it as a transformative pedagogy for local relevance with global perspectives (Figure 9.1).

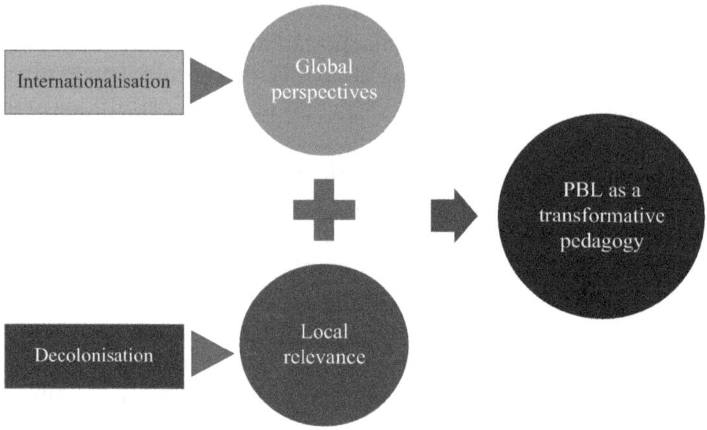

Figure 9.1 PBL as a Transformative Pedagogy for Local Relevance with Global Perspectives.

Despite these advantages, PBL also presents certain challenges. Research suggests that students in traditional, lecture-based environments may initially perform better on assessments of factual recall. However, these disparities tend to diminish as PBL students gain deeper conceptual understanding and more versatile skills over time (Loyens et al., 2011). The integration of critical pedagogy within PBL further enhances its impact by encouraging learners to reflect on the sociocultural and ethical dimensions of the issues they investigate (Cavanagh et al., 2019). Consequently, PBL emerges not only as an effective strategy for academic development but also as a vital approach for cultivating global citizenship and socially responsive learning.

9.3 PBL in Practice: Bridging Decolonisation and Internationalisation

Three important aspects can be considered to bridge decolonisation and internalisation in engineering education. They are designing contextually relevant engineering problems, embedding diverse knowledge systems and cultivating global competencies through PBL (Figure 9.2).

9.3.1 Designing Contextually Relevant Engineering Problems

Designing contextually relevant engineering problems involves the integration of local sociotechnical challenges and the co-creation of problems with communities and diverse stakeholders. This approach enhances the educational experience for engineering students while promoting the development of sustainable and culturally appropriate solutions. By engaging with real-world

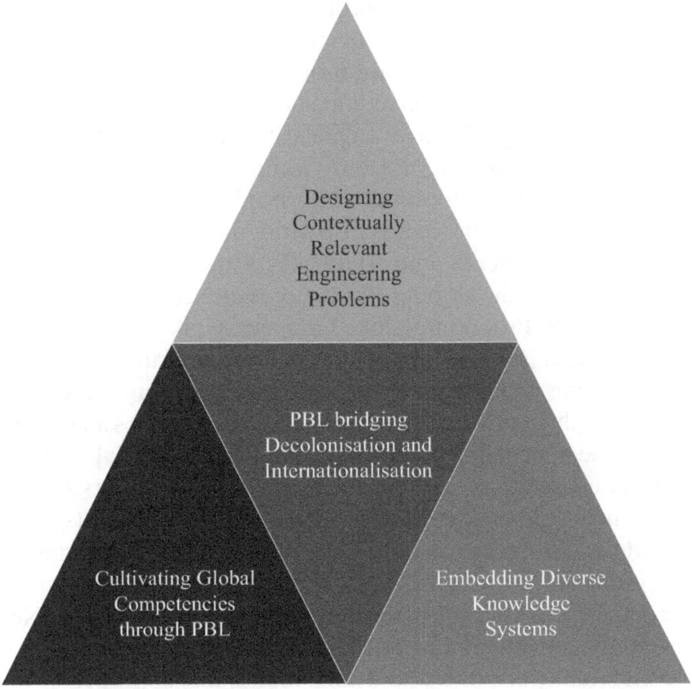

Figure 9.2 Bridging Decolonisation and Internationalisation in Engineering Education through PBL.

issues and collaborating directly with local communities, engineering education can produce graduates who are better prepared to address complex global challenges.

A key aspect of this process is incorporating local sociotechnical challenges into engineering curricula (Nieusma, 2015). Engineering problems often possess both technical and social dimensions. For instance, a course on homelessness combined engineering design with social issues by tasking students with developing solar water heater prototypes for mobile shower units used by a non-profit organisation. This project underscored the importance of understanding the sociotechnical context to create effective and impactful solutions (Chen et al., 2020). Similarly, the use of sociotechnical design challenges in a study involving first-year civil engineering students, such as addressing acid mine drainage, demonstrated an increase in student engagement and self-efficacy. These challenges required students to explore both technical solutions and social implications, thereby fostering a deeper understanding of engineering's societal role. Projects centred on sustainable development, such as the creation of mercury-free mineral processing technologies for small-scale mining communities, also exemplify this approach. These initiatives highlight the importance of context-specific solutions and stakeholder engagement in understanding the local sociotechnical environment (Smith et al., 2018).

Equally important is the co-creation of engineering problems in collaboration with communities and diverse stakeholders. Community engagement is essential for developing effective engineering solutions. This collaboration ensures that the solutions were not only technically sound but also culturally appropriate and responsive to community needs. Involving a broad range of stakeholders in the design process is also vital, as it allows for a deeper understanding of the values, needs, and technical competencies of different groups. This inclusion helps in developing more representative and inclusive solutions (Mahyar et al., 2020). Programmes like Lafayette College's Engineering Studies degree further emphasise the importance of interdisciplinary approaches by bridging engineering with liberal arts, thereby encouraging students to consider the complex interplay between technology and society (Rossmann et al., 2020).

Despite the clear benefits, integrating sociotechnical challenges and fostering community co-creation in engineering education presents significant challenges. These include the need for sensitivity to cultural contexts, the mitigation of biases, and the consideration of ethical implications. Furthermore, adapting engineering curricula to embrace these methods requires substantial changes in teaching practices, institutional priorities, and resource allocation. Nevertheless, the potential to cultivate more socially responsible engineers and produce effective, contextually grounded solutions makes this approach both necessary and highly valuable.

9.3.2 *Embedding Diverse Knowledge Systems*

Integrating diverse knowledge systems—particularly indigenous and community-based knowledge—into engineering solutions and education is essential for fostering sustainable and inclusive development. This integration not only enriches the engineering field with a broader range of perspectives but also addresses historical injustices and promotes intercultural understanding. By valuing multiple epistemologies, engineering can move beyond its traditionally Western-centric foundation and become more responsive to local and global needs. The role of interdisciplinary and intercultural learning, particularly within PBL frameworks, is central to bridging the goals of decolonisation and internationalisation. Such approaches encourage collaboration, respect, and mutual learning among diverse knowledge systems, ultimately leading to more holistic and effective solutions.

Indigenous knowledge systems offer distinct insights into sustainability and community-centred innovation. These systems are grounded in relationality and reciprocity, standing in contrast to the often-extractive and utilitarian approaches common in conventional engineering practice (Schaefer et al., 2021). Including indigenous perspectives in engineering education deepens students' understanding of science and technology by positioning these knowledge systems as valid and analogous to Western engineering sciences in both rigour and applicability (Kondratyev, 2023). At institutions like the University of Manitoba, efforts to decolonise and Indigenise engineering curricula illustrate how collaboration with Indigenous communities can prepare students for culturally sensitive, ethically grounded professional practice. These initiatives actively embed indigenous worldviews, practices, and priorities into teaching and learning environments (Cicek et al., 2022; Ruta et al., 2022).

PBL serves as an effective pedagogical framework to foster interdisciplinary and intercultural learning. PBL encourages students to engage with complex, real-world challenges that demand the integration of diverse knowledge systems for meaningful solutions (Cicek et al., 2022). This approach supports the incorporation of indigenous perspectives into science and technology education, promoting intercultural learning and advancing sustainability goals while respecting Indigenous values and worldviews (Rendón, 2023). Moreover, collaborations between Indigenous and non-Indigenous scholars have led to the emergence of transdisciplinary frameworks that integrate various types of knowledge. These frameworks promote the co-design and co-development of community-owned systems that are both locally relevant and globally informed (Schaefer et al., 2021).

However, the integration of Indigenous and Western knowledge systems is not without challenges. Deep-seated colonial structures and epistemological biases can hinder genuinely integrative practices (Castleden et al., 2017). Overcoming these obstacles requires a conscious effort to abandon the "arrogance of expertise" and instead cultivate co-learning environments where all forms of knowledge are acknowledged as valuable and legitimate (Castleden et al., 2017). The Truth and Reconciliation Commission of Canada's Calls to Action further emphasise the urgency of educational reforms that incorporate Indigenous knowledge systems and pedagogical methods. Such reforms aim to build student capacity for intercultural understanding, empathy, and reconciliation (Ruta et al., 2022).

Although complex, the integration of Indigenous knowledge into engineering and education offers significant opportunities for innovation, equity, and sustainability. Achieving this integration demands a paradigm shift—from a traditional, positivist view of science and engineering to a more inclusive model that recognises the value of diverse perspectives. This shift can lead to more effective and equitable solutions that benefit both Indigenous and non-Indigenous communities. Through interdisciplinary and intercultural learning, educational institutions have the potential to serve as key agents in bridging decolonisation and internationalisation, ultimately contributing to a more just, inclusive, and sustainable future.

9.3.3 Cultivating Global Competencies through PBL

Cultivating global competencies through PBL is increasingly recognised as essential for preparing engineering students to thrive in a highly interconnected world. PBL provides a dynamic and experiential platform that enables students to engage with real-world problems, fostering not only technical expertise but also vital soft skills such as collaboration, cultural awareness, and ethical decision-making. This approach is particularly impactful in international contexts, where students work in diverse teams and are exposed to different cultural perspectives. Through these experiences, engineering education evolves into a space that not only sharpens technical acumen but also builds the interpersonal and intercultural competencies necessary for global practice.

International PBL collaborations are a cornerstone of this globalised approach. Virtual teams and global studios enable students from different countries to collaborate on complex projects, providing authentic opportunities for cross-cultural communication and teamwork. The Collaborative Online International Learning (COIL) model, for example, uses information technology to create virtual environments where students can engage asynchronously across time zones. This setup simulates the realities of global engineering work and helps students build digital collaboration skills and cultural fluency (Yamamoto et al., 2018). In response to these trends, universities are increasingly incorporating global PBL strategies into their curricula to enhance future work skills, particularly in areas like cross-cultural competence and virtual team collaboration (Yamamoto et al., 2018). Additionally, the use of case studies and real-world scenarios in PBL introduces students to the complexities of international engineering projects, offering critical insights into the challenges and opportunities of working across diverse cultural and geographic contexts.

Incorporating reflexivity, ethics, and cultural humility into PBL further deepens its impact. A strong foundation in engineering ethics and professionalism is essential for navigating the social, human, and contextual dimensions of global engineering practice. Engineers must be prepared to adapt to unforeseen challenges and lead ethically grounded innovations in diverse settings (Barakat, 2015). Cultural humility is especially vital—it involves recognising one's cultural limitations, respecting others' values and practices, and maintaining an openness to learning. This mindset is crucial for engineers working in international or cross-cultural environments, where adaptability and sensitivity are as important as technical know-how (Johnston, 2001).

Moreover, bridging the goals of decolonisation and internationalisation in engineering education requires a critical examination of the values and assumptions that underpin engineering decisions. By encouraging students to reflect on the societal impacts of their work and question dominant narratives in engineering practice, educators can foster a more inclusive and equitable approach. This intersection of decolonisation and internationalisation promotes a deeper understanding of global engineering's responsibilities and potential for transformative change (Lord et al., 2019).

While the benefits of PBL in cultivating global competencies are significant, its implementation also presents challenges. The effectiveness of international virtual teams, for example, can be hindered by technological limitations, communication barriers, and time zone differences. Additionally, fostering cultural humility and reflexivity demands continuous learning and self-awareness—an ongoing process that may be difficult to sustain in traditional classroom settings. Nonetheless, these challenges do not diminish the value of PBL. Rather, they highlight the need for thoughtful design, institutional support, and pedagogical innovation. When implemented with care and intentionality, PBL remains a powerful tool for preparing engineering students to navigate and contribute meaningfully to a multicultural, interconnected world.

9.4 Case Studies or Empirical Insights

9.4.1 Case 1: A South African University's Curriculum Transformation through PBL

The transformation of the curriculum at South African universities, particularly through PBL, reflects a dynamic response to the dual imperatives of decolonisation and internationalisation. This multifaceted process involves reimagining educational frameworks to integrate indigenous knowledge systems alongside global perspectives. It seeks to redress the historical dominance of Eurocentric educational models by promoting inclusivity and relevance in diverse cultural contexts. The experience of institutions like the Durban University of Technology (DUT) provides a practical lens through which to explore these curricular shifts, along with the attendant challenges and opportunities.

In the context of decolonisation, universities such as DUT are actively engaging in efforts to move away from Eurocentric paradigms by embracing African epistemologies and pedagogies. Rather than adopting a prescriptive approach, the process encourages critical dialogue and collaboration among students, academic staff, and communities to reshape educational content and methodologies in a locally meaningful way, as discussed by Mheta et al. (2018). At the University of South Africa (UNISA), curriculum transformation is integrated into the institution's strategic vision. This vision emphasises the incorporation of African knowledge systems, especially within disciplines like anthropology that are historically rooted in colonial worldviews. The ultimate goal is to cultivate graduates who are both responsive to contemporary challenges and firmly grounded in African intellectual traditions, as highlighted by Menegatti (2023). More broadly, the national response to decolonial imperatives underscores the importance of policy development and execution to ensure that curricula genuinely reflect African students' lived experiences and advance epistemic justice for historically marginalised communities, according to Govender and Naidoo (2023).

Simultaneously, the internationalisation of the curriculum (IoC) is being pursued as a complementary strategy aimed at diversifying knowledge systems and advancing the decolonial agenda. In South Africa, IoC is seen not merely as a means of aligning with global academic trends but as a way to support students in becoming globally competent professionals who remain connected to their cultural and linguistic roots. This perspective is articulated by Pap (2024), who emphasises that true internationalisation must be grounded in local realities and follow a bottom-up approach. Rather than imposing foreign standards, the goal is to create curricula that are relevant, inclusive, and contextually appropriate for each institution's unique geographic and cultural setting.

Despite these promising developments, the implementation of a decolonised and internationalised curriculum faces several challenges (Jansen & Walters, 2022). A key barrier is resistance from academic staff who may be deeply invested in traditional Eurocentric approaches to teaching and learning. As Govender and Naidoo (2023) point out, fostering a sense of ownership and openness to change among faculty members is critical to sustaining meaningful transformation. Additionally, there is often a misalignment between existing pedagogical practices and the diverse learning needs of African students, signalling the need for a curriculum redesign that effectively bridges local and global knowledge systems, as Gobingca (2024) argues. Further complexities arise from identity-related concerns and the necessity of ensuring that curricula remain responsive to the sociocultural contexts of students (Jansen & Walters, 2022). Mendy and Madiope (2020) offer insights into how these issues are being addressed within open distance and e-learning frameworks in South Africa.

Ongoing debates continue to shape the discourse around curriculum transformation. While some critics argue that decolonisation may compromise academic freedom, others contend that these efforts can, in fact, enhance academic integrity, particularly when informed by globally recognised principles, such as those outlined by UNESCO. Olsson (2023) asserts that decolonisation and internationalisation need not be mutually exclusive but can instead function as mutually reinforcing strategies. Ultimately, the process of curriculum transformation in South African higher education requires sustained collaboration among a wide range of stakeholders and a long-term commitment to fostering educational systems that are both locally resonant and globally connected.

9.4.2 Case 2: International PBL Project between African and European Institutions

The processes of internationalisation and decolonisation in higher education are deeply intertwined yet complex, requiring a fundamental rethinking of traditional academic paradigms and the cultivation of equitable global partnerships. The cases of the University of Nairobi and the Platform for African-European Studies serve as important illustrations of these dynamics, highlighting both the potential and the persistent challenges involved in reconciling international collaboration with the imperative to decolonise curricula and knowledge production.

The University of Nairobi, as Kenya's first national public university, exemplifies the struggle of post-colonial institutions to assert their agency within the global academic landscape. Its efforts at internationalisation must be understood within a historical context marked by the legacy of colonialism. Although the university has made strides in gaining control over its curriculum and forging international partnerships, it continues to face systemic challenges, including limited financial and human resources and enduring structural influences from its colonial foundations (Otieno, 2012). The motivations behind the university's internationalisation initiatives are multifaceted. On one hand, academic rationales—such as expanding research opportunities and enhancing professional development—drive collaboration. On the other hand, economic considerations significantly shape the nature and extent of these engagements (Otieno, 2012). Parallel to its internationalisation agenda, the University of Nairobi is also actively pursuing curriculum decolonisation. This entails confronting colonial legacies and redefining its global academic identity by fostering more equitable and reciprocal relationships with Western institutions (Otieno, 2012).

The Platform for African-European Studies, a transcontinental initiative involving 22 universities, offers a broader perspective on international academic partnerships through the lens of science diplomacy and collaborative knowledge production. This initiative centres on rebalancing scholarly relationships between Africa and Europe by integrating decolonial principles into curriculum development and research design (Nshimbi et al., 2023). At the heart of this project lies a commitment to dismantling hegemonic knowledge structures and fostering equitable academic collaborations. The platform promotes the inclusion of African epistemologies, aiming to counteract the persistent dominance of Western paradigms in higher education (Nshimbi et al., 2023). Nonetheless, while the project provides a promising model for cross-continental engagement, it also encounters significant hurdles, including entrenched power imbalances and the institutional inertia that often impedes meaningful change (Nshimbi et al., 2023).

Expanding the view beyond individual institutions, the broader discourse on decolonial internationalisation calls for a critical re-evaluation of the Eurocentric norms that continue to shape global higher education. This approach advocates for the recognition and integration of diverse knowledge systems and challenges the structural patterns of exclusion embedded in conventional models of academic exchange (Ozyonum, 2023). However, efforts to decolonise the curriculum are not without complications. While many institutions organise public lectures, seminars, and workshops under the banner of decolonisation, these activities sometimes result in symbolic rather than substantive change—a phenomenon referred to as "decolonial-washing" (Le Grange et al., 2020). Virtual platforms present both new challenges and opportunities in this landscape. Projects such as Personalised Engineering Education in Southern Africa (PEESA) III demonstrate how virtual collaboration can facilitate internationalisation while also advancing decolonial goals. By enabling knowledge exchange across borders without reproducing the physical hierarchies of global academia, such initiatives can support more equitable forms of engagement (Cooke et al., 2019).

Taken together, the experiences of the University of Nairobi and the African-European PBL project illustrate the ongoing, multifaceted efforts required to harmonise decolonisation and internationalisation in higher education. These cases reveal that while important progress has been made, significant work remains. A critical interrogation of institutional power dynamics and a deliberate integration of pluralistic knowledge traditions are essential. As higher education institutions continue to navigate these processes, sustained commitment, reflexivity, and innovative approaches will be necessary to build genuinely equitable and decolonised global academic partnerships.

9.4.3 Lessons Learned: Challenges, Opportunities, and Impacts

9.4.3.1 Challenges

Across both South African and Kenyan higher education contexts, a key challenge in the pursuit of decolonisation and internationalisation is the persistent influence of colonial legacies on institutional structures, curricula, and epistemological foundations. At the University of Nairobi, for instance, post-colonial constraints continue to affect academic autonomy, particularly in curriculum design and faculty capacity (Otieno, 2012). Similarly, South African universities such as DUT and UNISA face resistance from faculty accustomed to Eurocentric models, which hinders transformation efforts (Govender & Naidoo, 2023).

Another widespread challenge is the risk of superficial engagement, often described as “decolonial-washing,” where symbolic gestures (e.g., public lectures or branding) replace deep structural change (Le Grange et al., 2020). Moreover, imbalanced global partnerships, especially between African and European institutions, continue to reproduce historical asymmetries, making it difficult to achieve genuine epistemic equity (Nshimbi et al., 2023).

Resource limitations—financial, infrastructural, and human—further constrain both local reform efforts and transnational collaborations. These limitations particularly affect institutions trying to implement curriculum redesign or sustain meaningful international networks while maintaining their local relevance.

9.4.3.2 Opportunities

Despite these challenges, significant opportunities have emerged. The growing discourse on decolonial internationalisation presents a chance to rethink global academic partnerships in more equitable and context-sensitive ways.

Initiatives like the African–European PBL project are at the forefront of this, demonstrating how science diplomacy and co-created curricula can foster mutual respect and knowledge exchange (Nshimbi et al., 2023).

PBL, adopted by South African institutions, has shown promise in enabling student-centred, context-aware learning environments that incorporate both indigenous and global knowledge systems. This pedagogical shift not only enhances relevance but also encourages critical engagement with diverse epistemologies (Mheta et al., 2018).

Virtual platforms also offer a key opportunity. Projects such as PEESA III demonstrate how digital collaboration can overcome geographic and economic barriers, enabling broader participation in international education while supporting decolonial aims through democratised access (Cooke et al., 2019).

9.4.3.3 Impacts

The impacts of these transformation efforts are still unfolding, but several are already evident. First, there is an increasing awareness within African universities of the need to redefine institutional identities and academic missions in alignment with African realities and priorities. This has led to more deliberate policies and strategic plans aimed at decolonising knowledge and enhancing global relevance.

Second, the integration of African epistemologies into the curriculum—though uneven—has helped shift the narrative from dependency on Western models to one of knowledge pluralism. This is not only academically enriching but also empowers students by validating their cultural and intellectual heritage (Menegatti, 2023).

Finally, while challenges remain, efforts to foster more equitable international partnerships have the potential to reshape global knowledge production, moving away from extractive practices to ones rooted in reciprocity, respect, and shared ownership (Nshimbi et al., 2023; Ozyonum, 2023).

9.5 Implications for Curriculum and Policy Reform

Decolonisation and internationalisation in engineering education, particularly through PBL, carry significant implications for curriculum and policy reform (Figure 9.3). This approach challenges traditional Eurocentric paradigms and seeks to integrate diverse perspectives and knowledge systems, thereby fostering a more inclusive and globally relevant educational framework.

Central to this transformation is the reimagining of accreditation and assessment standards. Conventional models often reflect Eurocentric values that marginalise non-Western epistemologies. A decolonial critique advocates for the inclusion of diverse knowledge systems and the recognition of local contexts in engineering standards (Dumbuya, 2025; Heleta & Chasi, 2023). PBL serves as a powerful pedagogical tool in this process by engaging students

Figure 9.3 Implications of Lessons Learned from Case Studies on PBL for Decolonisation and Internationalisation.

with real-world problems that demand multifaceted, culturally sensitive solutions. This approach promotes holistic assessments that prioritise creativity and critical thinking over standardised testing (Abraham et al., 2024; Graham, 2012).

Developing staff capacity for decolonial and global pedagogies is also essential. Faculty development programmes focused on cultural competence, critical pedagogy, and the integration of indigenous knowledge are key to equipping educators for this shift (Cicek et al., 2024; Lord et al., 2019). Additionally, fostering collaborative learning environments that enable dialogue among educators from diverse backgrounds can encourage the exchange of ideas and lead to more innovative and inclusive teaching practices (De Lissovoy, 2010; Shahjahan et al., 2021).

On an institutional level, strategies for establishing equitable international partnerships must emphasise mutual benefit and reciprocity. Such partnerships should value the contributions of all collaborators and avoid perpetuating one-sided agendas (Clarke, 2024; Morreira et al., 2020). To truly decolonise, institutions must also work to decentre Western dominance by amplifying voices from the Global South and supporting joint curriculum development, research initiatives, and exchange programmes that foreground local knowledge and expertise (Dumbuya, 2025; Heleta & Chasi, 2023).

Despite the clear opportunities, challenges remain. Resistance to change, entrenched institutional practices, and the complexities of integrating diverse epistemologies can hinder progress. Moreover, the successful implementation of decolonial pedagogies requires significant institutional commitment and resources, which may not always be readily available. Nevertheless, the potential benefits of a decolonised and internationalised engineering education system are substantial, offering a more equitable, inclusive, and globally attuned foundation for educating future engineers.

9.6 Critical Reflections

Decolonisation and internationalisation in engineering education through PBL face significant challenges rooted in historical and structural inequities within higher education systems. These challenges include institutional inertia, entrenched power dynamics in knowledge validation, and the superficial or tokenistic adoption of decolonial frameworks. Addressing these issues requires a deep, critical examination of current practices and a sustained commitment to genuine transformation.

Institutional inertia is a major barrier, as many universities remain grounded in Eurocentric traditions that resist change. This is reflected in the continued dominance of Western epistemologies and methodologies in engineering education, often at the expense of alternative knowledge systems (Camfield, 2024; Ozyonum, 2023). Structural barriers such as rigid accreditation policies and risk-averse administrative cultures further hinder the adoption of decolonial practices by prioritising established methods over innovative, inclusive approaches (Hall et al., 2021). Cultural resistance also plays a role, as there is often a lack of awareness or willingness among faculty and administrators to meaningfully engage with decolonial initiatives, resulting in tokenistic rather than transformative efforts (Agrawal & Heydenrych, 2024; Lord et al., 2019).

Power dynamics in knowledge validation present another challenge. The validation of knowledge in engineering education is often biased towards Western paradigms, creating hierarchies that marginalise non-Western epistemologies (Mitra et al., 2023; Ozyonum, 2023). This imbalance is evident in collaborative projects where partnerships between Western institutions and Indigenous or non-Western communities often remain unequal, with Western perspectives dominating outcomes (Camfield, 2024; Dei & Karanja, 2022). Such dynamics perpetuate epistemic injustice, systematically disadvantaging groups whose knowledge systems are excluded from academic validation and discourse (Porter & Siddiqui, 2024; Wyllie-Runner et al., 2024).

Moreover, there is a risk that decolonial frameworks are adopted in ways that are superficial or symbolic. Tokenism in curriculum design can occur when Indigenous or local content is added without addressing underlying structural inequalities (Cicek et al., 2024; Porter & Siddiqui, 2024). Without critical reflection, these efforts may become performative, serving more as a compliance exercise than as a driver of meaningful change (Agrawal & Heydenrych, 2024; Hall et al., 2021). In some cases, decolonial methods are even co-opted by dominant power structures, reducing their transformative impact and reinforcing existing hierarchies (Camfield, 2024).

Despite these challenges, the movement to decolonise and internationalise engineering education presents important opportunities for transformative change. To realise this potential, institutions must be willing to dismantle entrenched power structures and genuinely value diverse knowledge systems. This involves not only revising curricula and policies but also fostering authentic, equitable collaborations with marginalised communities. Ultimately, the success of decolonial initiatives depends on a sustained commitment to change and a readiness to critically engage with the structural roots of inequality in education.

9.7 Conclusion

The study presents a case for reimagining engineering education at the intersection of decolonisation and internationalisation. It argues that these two paradigms—traditionally viewed as conflicting—can, when approached critically, function as mutually reinforcing strategies to create inclusive, contextually relevant, and globally engaged education systems. Across case studies in South Africa, Kenya, and broader African-European collaborations, the study highlights how colonial legacies continue to shape curricula, knowledge hierarchies, and global academic relations. However, it also demonstrates how institutions are increasingly taking ownership of transformation, integrating African epistemologies, and promoting ethical international engagement through intentional pedagogy and policy reform.

At the heart of this educational shift can be PBL, positioned as a powerful mediating framework. PBL promotes transformative, inclusive, and socially embedded learning by actively engaging students with real-world problems grounded in both local and global contexts. Its pedagogical strength lies in cultivating higher-order thinking, interdisciplinarity, reflexivity, and cultural humility. PBL's flexible, student-centred design fosters collaboration, self-direction, and ethical awareness—skills vital for addressing today's complex engineering challenges. Moreover, its application in diverse settings, including international virtual teams and community-engaged learning, underscores its capacity to bridge global and local knowledge systems. By embedding indigenous and community-based knowledge into the curriculum, PBL disrupts traditional hierarchies and opens pathways to epistemic justice.

Despite its promise, the implementation of PBL also reveals structural and cultural challenges, ranging from institutional inertia and resource constraints to power asymmetries in global collaborations. Addressing these barriers requires intentional curriculum redesign, faculty development, and policy interventions that prioritise equity and mutuality. PBL cannot function as a panacea, but, when aligned with decolonial and inclusive values, it becomes a transformative force for reconfiguring both teaching and learning in engineering education.

Looking ahead, future research and practice should focus on three strategic areas. First, there is a need for longitudinal studies that assess the long-term impact of PBL on student competencies, institutional change, and social outcomes across varied contexts. Second, methodological innovation is essential to explore how indigenous and community knowledge can be systematically integrated into engineering curricula without co-optation or dilution.

Third, future work should explore how digital and virtual learning platforms, such as Collaborative Online International Learning (COIL), can be harnessed to scale PBL while ensuring equitable access and intercultural integrity.

Moreover, engineering education must evolve to prepare graduates who are not only technically skilled but also globally conscious and socially responsible. By engaging with both decolonisation and internationalisation through the lens of PBL, higher education institutions can cultivate engineers capable of navigating—and transforming—the complex realities of our interconnected world.

References

Abraham, S., Vurkaç, M., Miguel, A., Takach, M., Ferré, E., Singh, S., & Louie, H. (2024). Work in Progress: Reimagining the ECE Curriculum: Bridging Technical Preparation, Professional Formation, and University Mission for a Holistic Education. https://peer.asee.org/48511

Adam, T. (2024). A justice-oriented conceptual and analytical framework for decolonising and desecularising the field of educational technology. *Education Sciences*, *14*(9), 962. https://doi.org/10.3390/educsci14090962

Agrawal, A., & Heydenrych, H. (2024). An epistemic shift: Exploring students' learning experiences with a decolonisation initiative in an engineering course. *European Journal of Engineering Education*. https://doi.org/10.1080/03043797.2024.2303023

Arif, A., Lumb, P., Mohebali, M., & Khan, A. I. (2022). Exploring tensions in decolonisation of internationalisation of higher education. *Critical Internationalisation Studies Review*, *1*(1), 18–21. https://scholarship.shu.edu/cisr/vol1/iss1/9

Barakat, N. (2015). Engineering ethics and professionalism education for a global practice. QScience Proceedings (Engineering Leaders Conference 2014), *2015*(4), 5. https://doi.org/10.5339/QPROC.2015.ELC2014.5

Byrne, D. C., & Alexander, J. O. (2023). Special section editorial. Decolonising speculative fiction. *Staffrider*, *37*. https://doi.org/10.17159/2617-3255/2023/n37a28

Camfield, L. (2024). What challenges for global development research are posed by a more decolonial approach? Colonial genealogies and responses. *The European Journal of Development Research*. https://doi.org/10.1057/s41287-024-00676-3

Case, J. M., Fraser, D. M., Kumar, A., & Itika, A. (2016). The significance of context for curriculum development in engineering education: A case study across three African countries. *European Journal of Engineering Education*, *41*(3), 279–292. https://doi.org/10.1080/03043797.2015.1056103

Castleden, H., Martin, D., Cunsolo, A., Harper, S. L., Hart, C., Sylvestre, P., Stefanelli, R., Day, L., & Lauridsen, K. (2017). Implementing Indigenous and western knowledge systems (Part 2): "You have to take a backseat" and abandon the arrogance of expertise. *International Indigenous Policy Journal*, *8*(4), 8. https://doi.org/10.18584/IIPJ.2017.8.4.8

Cavanagh, A., Vanstone, M., & Ritz, S. A. (2019). Problems of problem-based learning: Towards transformative critical pedagogy in medical education. *Perspectives on Medical Education*, *8*(1), 38–42. https://doi.org/10.1007/S40037-018-0489-7

Chen, D. A., Chapman, M. A., & Mejia, J. A. (2020). Balancing complex social and technical aspects of design: Exposing engineering students to homelessness issues. *Sustainability*, *12*(15), 5917. https://doi.org/10.3390/SU12155917

Cicek, J. S., Herrmann, R., Forrest, R., & Monkman, K. (2022). Decolonising and indigenizing engineering: The design & implementation of a new course. *Proceedings of the … CEEA Conference*. https://doi.org/10.24908/pceea.vi.15886

Cicek, J. S., Morris, E. C., Coleman, J. P., Cook, C. R., Sagan, C., Wastasecoot, H., Wilson, S. L., & Herrmann, R. (2024). Decolonising and indigenizing engineering: The trauma of learning the truth about Canada's colonial present & history in curricula. https://doi.org/10.24908/pceea.2024.18518

Clarke, H. (2024). Toward pedagogies of decoloniality: Evaluating teaching practice and syllabus design in IR undergraduate modules. *International Studies Perspectives*. https://doi.org/10.1093/isp/ekad026

Cooke, L. A., Lungu, B. N., & Samuels, L. (2019). Virtual engagement: A nexus between internationalisation and decolonisation of the curriculum in the PEESA III project? *Balkan Region Conference on Engineering and Business Education, Sciendo*, (1), 35–43.

Crawley, E. F., Jianzhong, C., Malmqvist, J., & Brodeur, D. R. (2008). The context in engineering education. http://www.cdio.org/files/document/file/m1-crawley2008.pdf

De Lissovoy, N. (2010). Decolonial pedagogy and the ethics of the global. *Discourse: Studies in The Cultural Politics of Education*, *31*(3), 279–293. https://doi.org/10.1080/01596301003786886

De Wit, H. (2019). Internationalisation in higher education, a critical review. *SFU Educational Review*, *12*(3), 9–17. https://doi.org/10.21810/SFUER.V12I3.1036

De Wit, H. (2020). Internationalisation of higher education: The need for a more ethical and qualitative approach. *Journal of International Students*, *10*(1). https://doi.org/10.32674/JIS.V10I1.1893

Dei, G. J. S., & Karanja, W. W. (2022). Responding to the epistemic challenge—A decolonial project (pp. 79–111). https://doi.org/10.1007/978-3-030-84201-7_4

Du Preez, P. (2018). On decolonisation and internationalisation of university curricula: What can we learn from Rosi Braidotti? *Journal of Education*, *74*, 19–31. https://doi.org/10.17159/2520-9868/I74A02

Dumbuya, E. (2025). Decolonising the curriculum: Rethinking colonial influences in African Education for an empowered future. https://doi.org/10.2139/ssrn.5018488

Finucane, P., Johnson, S. M., & Prideaux, D. J. (1998). Problem-based learning: Its rationale and efficacy. *The Medical Journal of Australia*, *168*(9), 445–448. https://doi.org/10.5694/J.1326-5377.1998.TB139025.X

Fomunyam, K. G. (2020). Decolonising engineering education in Nigerian higher education. *International Journal of Recent Technology and Engineering*, *9*(1), 1887–1893. https://doi.org/10.35940/ijrte.a1568.059120

Gobingca, B. Z. (2024). Academics' views on Africanisation and decolonisation of teacher education curricula: A case study of a selected higher education institution in South Africa. *E-Journal of Humanities, Art and Social Sciences*, 2930–2940. https://doi.org/10.38159/ehass.202451620

Govender, L., & Naidoo, D. (2023). Decolonial insights for transforming the higher education curriculum in South Africa. *Curriculum Perspectives*. https://doi.org/10.1007/s41297-023-00200-3

Graham, R. (2012). The one less travelled by: The road to lasting systemic change in engineering education. *Journal of Engineering Education, 101*(4), 596–600. https://doi.org/10.1002/J.2168-9830.2012.TB01120.X

Hall, R., Ansley, L. J. C., Connolly, P., Loonat, S., Patel, K., & Whitham, B. (2021). Struggling for the anti-racist university: Learning from an institution-wide response to curriculum decolonisation. *Teaching in Higher Education*, 1–18. https://doi.org/10.1080/13562517.2021.1911987

Hawamdeh, M., & Adamu, I. (2021). Problem-based learning (PBL): A deep approach to learning in the 21st century. In G. Kurubacak-Meric & S. Sisman-Ugur (Eds.), *Improving Scientific Communication for Lifelong Learners* (pp. 38–56). IGI Global Scientific Publishing. https://doi.org/10.4018/978-1-7998-4534-8.ch003

Heleta, S., & Chasi, S. (2023). Curriculum decolonisation and internationalisation. *Journal of International Students*. https://doi.org/10.32674/jis.v14i2.6383

Hunter, C., Archer, S., & Illingworth, C. (2025). Problem-based learning. In F. Cownie, A. Bradney, & E. Jones (Eds.), *Elgar concise encyclopedia of legal education* (pp. 302–305). Elgar Online. https://doi.org/10.4337/9781035302932.ch83

Ideker, T., Winslow, L. R., & Lauffenburger, D. A. (2006). Bioengineering and systems biology. *Annals of Biomedical Engineering, 34*, 1226–1233. https://doi.org/10.1007/s10439-006-9119-3

Jansen, J. D., & Walters, C. A. (2022). The uptake of decolonisation: The case of the sciences and engineering. In: *The decolonisation of knowledge: Radical ideas and the shaping of institutions in South Africa and beyond* (pp. 172–202). Cambridge University Press.

Johnston, S. F. (2001). Towards culturally inclusive global engineering. *European Journal of Engineering Education, 26*(1), 77–89. https://doi.org/10.1080/03043790010008771

Kohn, M., & McBride, K. (2014). Decolonization. *The Encyclopedia of Political Thought*, 833–836. https://doi.org/10.1002/9781118474396.wbept0245

Kondratyev, N. (2023). Engineering knowledge as indigenous knowledge. *Contemporary Issues in Technology Education*, 33–42. https://doi.org/10.1007/978-981-99-1396-1_3

Kraja, G., Dedej, D., & Spahija, S. (2024). Navigating the global landscape: Dynamics and impacts of internationalisation in higher education. *Interdisciplinary Journal of Research and Development, 11*(1 S1), 7. https://doi.org/10.56345/ijrdv11n1s102

Kramm, M., Ludwig, D., Ngosso, T., Mosima, P., & Boogaard, B. K. (2024). Confrontation or dialogue? Productive tensions between decolonial and intercultural scholarship. *Ergo, 11*(0). https://doi.org/10.3998/ergo.6163

Lalwani, K. (2024). How to design and moderate an effective problem-based learning discussion. In K. Lalwani et al. (Eds.), *Professional, ethical, legal, and educational lessons in medicine: A problem-based learning approach*. Oxford Academic. https://doi.org/10.1093/med/9780197655979.003.0052

Lavi, R., Bertel, L. B., & Du, X. (2023). Guest editorial special issue on transforming engineering education. *IEEE Transactions on Education, 66*, 404–409. https://doi.org/10.1109/te.2023.3308668

Lazari, E., & Matsoukas, I. G. (2025). Bridging divides: The bold new era of university internationalisation. *Frontiers in Education, 9*. https://doi.org/10.3389/feduc.2024.1473045

Le Grange, L., du Preez, P., Ramrathan, L., & Blignaut, S. (2020). Decolonising the university curriculum or decolonial-washing? A multiple case study. *Journal of Education, 80*, 25–48. https://doi.org/10.17159/2520-9868/I80A02

Leong, P. N. C. (2009). The power of problem-based learning (PBL) in the EFL classroom. *Polyglossia*, *16*, 41–48. https://www.apu.ac.jp/rcaps/uploads/fckeditor/publications/polyglossia/Polyglossia_V16_Ng.pdf

Leydens, J. A., & Lucena, J. C. (2018). *Transforming engineering education and practice* (pp. 197–241). Wiley-IEEE Press. https://doi.org/10.1002/9781118757369.CH5

Lord, S. M., Mejia, J. A., Hoople, G. D., & Chen, D. A. (2019). Special session: Starting a dialogue on decolonisation in engineering education. *Frontiers in Education Conference*. https://doi.org/10.1109/FIE43999.2019.9028477

Loyens, S. M. M., Kirschner, P., & Paas, F. (2011). Problem-based learning. https://ro.uow.edu.au/edu.apers/1235

Mahyar, N., Jasim, M., Sarvghad, A., Tory, M., & Keefe, D. F. (2020). Designing technology for sociotechnical problems: Challenges and considerations. *IEEE Computer Graphics and Applications*, *40*(6), 76–87. https://doi.org/10.1109/MCG.2020.3017405

Mathaba, M. (2023). Perceptions of chemical engineering students on decolonising the curriculum by incorporating indigenous knowledge practices. *Scholarship of Teaching and Learning in the South*, *7*(2), 106–121. https://doi.org/10.36615/sotls.v7i2.260

Meifitri. (2020). Problem-Solving Ability and Communication Skills Through a Problem-Based Learning Approach. In *Proceedings of the Eighth International Conference on Languages and Arts (ICLA-2019)*. https://doi.org/10.2991/assehr.k.200819.028

Mendy, J., & Madiope, M. (2020). Curriculum transformation: A case in South Africa. *Perspectives in Education*, *38*(2). https://doi.org/10.18820/2519593X/PIE.V38.I2.01

Menegatti, E. (2023). *Decolonising the curriculum at Unisa* (pp. 257–272). BRILL eBooks. https://doi.org/10.1163/9789004540811_016

Mheta, G., Lungu, B. N., & Govender, T. (2018). Decolonisation of the curriculum: A case study of the Durban University of technology in South Africa. *South African Journal of Education*, *38*(4), 1–7. https://doi.org/10.15700/SAJE.V38N4A1635

Mitra, S., Sarkar, S., & Ganguli-Mitra, A. (2023). On the need for an anticolonial perspective in engineering education and practice. *Nature Communications*, *14*. https://doi.org/10.1038/s41467-023-43952-2

Moleka, P. (2024). Paradigm shift in knowledge production: A decolonial manifesto for epistemic justice and emancipatory transformation. https://doi.org/10.20944/preprints202411.0785.v1

Morreira, S., Luckett, K., Kumalo, S. H., & Ramgotra, M. (2020). Confronting the complexities of decolonising curricula and pedagogy in higher education. *Critical Studies in Teaching and Learning*, *5*, 1–18. https://doi.org/10.1080/23802014.2020.1798278

Movafegh, A., & Azimaraghi, O. (2015). Problem-based learning. *Academic Journal of Surgery*, *2*(1–2). https://ajs.tums.ac.ir/index.php/ajs/article/view/17

Muller, M. (2018). Decolonising engineering in South Africa – Experience to date and some emerging challenges. *South African Journal of Science*, *114*(5/6), 1–6. https://doi.org/10.17159/sajs.2018/a0270

Nieusma, D. (2015). *Analysing context by design: Engineering education reform via social-technical integration* (pp. 415–434). Springer. https://doi.org/10.1007/978-3-319-16169-3_20

Nshimbi, C. C., Develtere, P., & Debela, B. K. (2023). Rethinking African–European scientific cooperation: The case of the platform for African–European studies. *Journal of Contemporary European Research*, *19*(2). https://doi.org/10.30950/jcer.v19i2.1299

Olsson, E. J. (2023). Academic freedom and the decolonisation of knowledge: Curriculum transformation in South Africa from a UNESCO perspective. *Studies in Higher Education, 48*, 1172–1182. https://doi.org/10.1080/03075079.2023.2186389

Otieno, I. A. (2012). Internationalisation of an African university in the post-colonial era: a case study of the University of Nairobi. https://uknowledge.uky.edu/cgi/viewcontent.cgi?article=1002&context=epe_etds

Ozyonum, E. (2023). Let's talk about decolonial internationalisation. *Yu-Write, 2*(1). https://doi.org/10.25071/28169344.17

Pap, A. (2024). University administrators, leadership, and faculty views on the internationalisation of curriculum (A perspective through decolonial lenses from the Global South). *Journal of Higher Education, Theory, and Practice, 24*(6). https://doi.org/10.33423/jhetp.v24i6.7059

Pepper, C. (2014). *Problem-based learning (PBL)* (pp. 1–3). Springer. https://doi.org/10.1007/978-94-007-6165-0_128-2

Porter, S., & Siddiqui, Y. (2024). Addressing colonial and militarised themes in STEM education. *Frontiers in Education*. https://doi.org/10.3389/feduc.2024.1480199

Rehmat, A. P., Krista, G., & Hmelo-Silver, C. E. (2018). Contextualising problem-based learning: An overview of research and practice. In A. O'Donnell, N. C. Barnes, & J. Reeve (Eds.), *The Oxford handbook of educational psychology*. Oxford Academic. https://doi.org/10.1093/oxfordhb/9780199841332.013.24

Rendón, L. I. (2023). Indigenous knowledge and science and technology education (pp. 165–179). https://doi.org/10.1007/978-3-031-24259-5_12

Rossmann, J. S., Sanford, K. L., Nicodemus, J., & Cohen, B. (2020). The sociotechnical core curriculum: An interdisciplinary engineering studies degree program. https://peer.asee.org/the-sociotechnical-core-curriculum-an-interdisciplinary-engineering-studies-degree-program

Ruta, A., Cicek, J. S., Mante, A. A., Speare, M. E., & Herrmann, R. (2022). Ten calls to action to integrate indigenous knowledge and perspectives into the biosystems engineering program at the University of Manitoba. *Canadian Biosystems Engineering, 63*(1), 9.01–9.17. https://doi.org/10.7451/cbe.2021.63.9.1

Samaka, M., & Ally, M. (2016). *Model for transforming engineering education using technology-enhanced learning* (pp. 117–138). Springer. https://doi.org/10.1007/978-3-319-15323-0_5

Schaefer, M., Schmitt Olabisi, L., Arola, K. L., Poitra, C. M., Matz, E., Seigel, M., Schelly, C., Adesanya, A. A., & Bessette, D. (2021). Understanding socio-technological systems change through an indigenous community-based participatory framework. *Sustainability, 13*(4), 2257. https://doi.org/10.3390/SU13042257

Shahjahan, R. A., Estera, A. L., Surla, K. L., & Edwards, K. T. (2021). "Decolonising" curriculum and pedagogy: A comparative review across disciplines and global higher education contexts. *Review of Educational Research, 92*(1), 1–45. https://doi.org/10.3102/00346543211042423

Smith, N. M., Teschner, B. A., & Bullock, R. (2018). Bringing sustainable development challenges into the engineering classroom: Applying human centred design protocols to artisanal and small-scale mining. https://peer.asee.org/30159.pdf

Srilatha, C., Sucharitha, N., & Kumar, K. S. (2022). Problem-based learning (PBL) in nursing education. 13–17. https://doi.org/10.52711/ijnmr.2022.04

Storey, V. A. & Schiavo, M. A. (2013). Redefining the higher education landscape through problem-based learning. In V. Wang (Ed.), *Handbook of research on teaching and learning in K-20 education*, pp. 517–534. IGI Global Scientific Publishing. https://doi.org/10.4018/978-1-4666-4249-2.ch030

Thomas, I. (2009). Critical thinking, transformative learning, sustainable education, and problem-based learning in universities. *Journal of Transformative Education*, *7*(3), 245–264. https://doi.org/10.1177/1541344610385753

Tryggvason, G., & Apelian, D. (2011). *Meeting new challenges: Transforming engineering education* (pp. 1–17). John Wiley & Sons, Inc. https://doi.org/10.1002/9781118138267.CH1

Wildcat, M. (2022). *Decolonisation* (pp. 35–38). Routledge eBooks. https://doi.org/10.4324/9781003216575-21

Winter, J., Webb, O. J., & Turner, R. (2022). Decolonising the curriculum: A survey of current practice in a modern UK university. *Innovations in Education and Teaching International*, *60*(3), 1–12. https://doi.org/10.1080/14703297.2022.2121305

Wyllie-Runner, J., Cicek, S. J., & Zacharias, K. (2024). Towards defining threshold concepts for decolonising engineering. *Proceedings of the … CEEA Conference*. https://doi.org/10.24908/pceea.2023.17072

Yamamoto, T., Liao, A. Y. H., Wu, W.C. V., Shih, M., Shih, J. L., & Chu, H.C. (2018). A proposal for the global and collaborative PBL learning environment where all global members on different campuses are "on the same page" throughout the process of learning in the project. *International Conference on Technologies and Applications of Artificial Intelligence*, 90–95. https://doi.org/10.1109/TAAI.2018.00029

Part IV

Case Studies, Plausible Examples, and Reflections and Future Direction

Chapter 10

Problem-Based Learning in Engineering

Case Studies across Disciplines

10.1 Introduction

Case studies play a pivotal role in problem-based learning (PBL) within engineering education by presenting realistic scenarios that mirror the complexities of professional practice. This learner-centred approach enables students to apply theoretical knowledge to practical problems, thereby enhancing their understanding of engineering concepts and promoting the development of key professional competencies (Davis & Yadav, 2014; Jackson et al., 2015).

This chapter examines the implementation of PBL across a range of engineering disciplines—including Civil, Mechanical, Electrical, Chemical, and Software Engineering—through a comparative analysis of selected case studies. By analysing how PBL is designed, contextualised, and operationalised in different fields, the chapter highlights both shared strategies and discipline-specific adaptations that respond to the unique epistemic and professional demands of each domain.

10.1.1 Purpose of Case Studies in PBL

The purpose of the case studies in PBL includes understanding the real-world applications, skills development and holistic understanding of the challenges. They are discussed below:

- **Real-World Application**: Case studies expose students to real-life engineering challenges, enabling them to practice problem-solving in authentic contexts. This experience builds practical understanding and prepares students for workplace demands (Davis & Yadav, 2014; Jackson et al., 2015).
- **Skill Development**: Engaging with case studies enhances essential skills such as teamwork, communication, critical thinking, and self-directed learning—competencies vital for engineering professionals (Butan, 2007; Warnock & Mohammadi-Aragh, 2016).
- **Holistic Understanding**: By encouraging students to analyse complex systems, case studies foster an integrated view of engineering processes, helping students understand the interdependencies among technical, social, and contextual factors (Gandhi et al., 2012).

DOI: 10.1201/9781003529743-14

10.1.2 Significance of Case Studies in Engineering Education

The significance of this approach is underscored by global shifts in engineering education that demand graduates who are not only technically competent but also capable of critical thinking, interdisciplinary collaboration, and adaptive problem-solving. Case studies, as integral components of PBL, support these outcomes in the following ways:

- **Enhanced Engagement and Knowledge Retention**: Real-world relevance increases student engagement and promotes deeper learning and longer retention of engineering principles (Delaney & Nagle, 2019; Riazifar & Khajehsaeid, 2024).
- **Professional Preparedness**: Simulating workplace scenarios through case studies equips students with the mindset and problem-solving capabilities needed for seamless transitions into engineering practice (Azam et al., 2024; Delaney & Nagle, 2019)
- **Critical Thinking and Innovation**: The open-ended nature of case-based learning cultivates critical thinking and encourages students to explore multiple perspectives and innovative solutions to complex problems (Davis & Yadav, 2014; Jackson et al., 2015).

10.1.3 Rationale for the Case Study Approach Across Disciplines

The case study approach in PBL is a powerful pedagogical strategy in engineering education. It enhances student engagement, fosters critical thinking, and supports multidisciplinary learning. Its effectiveness lies in simulating real-world scenarios that allow students to apply theoretical knowledge to practical challenges. The rationale for employing case studies across engineering disciplines is supported by several core principles: the development of essential professional skills; the promotion of interdisciplinary collaboration; and the enhancement of student motivation and learning outcomes.

10.1.3.1 Development of Essential Skills

Case studies serve as catalysts for developing competencies that are central to engineering practice.

- **Critical Thinking and Problem-Solving**: Engaging with authentic, complex problems enables students to explore diverse solutions, fostering analytical reasoning and creativity—skills crucial across all engineering fields (Davis & Yadav, 2014; Naik et al., 2024).
- **Technical and Multidisciplinary Knowledge**: Case-based PBL encourages students to integrate knowledge from various subfields. For example, understanding electric vehicle design involves elements from mechanical, electrical, and software engineering (Syed et al., 2024).

- **Leadership and Teamwork**: Collaborative case study projects help students develop interpersonal and leadership skills essential for functioning in diverse, interdisciplinary teams (Delaney & Nagle, 2019; Naik et al., 2024).

10.1.3.2 Promotion of Interdisciplinary Collaboration

Modern engineering problems are seldom confined to a single discipline.

- **Integration of Multiple Disciplines**: Case studies promote holistic understanding by requiring students to draw upon knowledge from several engineering areas. This aligns with frameworks like India's National Education Policy (NEP) 2020, which underscores the importance of multidisciplinary learning (Syed et al., 2024; Taajamaa et al., 2014).
- **Systems Thinking**: Interdisciplinary cases, such as those found in Chemical or Environmental Engineering, help students understand how technical components interact within broader economic, environmental, and societal systems.

10.1.3.3 Enhancement of Student Motivation and Engagement

The interactive nature of case studies creates a dynamic learning environment that enhances student engagement.

- **Active Learning**: Students report increased motivation, better class attendance, and preference for PBL over traditional lectures due to the active role they play in the learning process (Naik et al., 2024; Riazifar & Khajehsaeid, 2024).
- **Ownership and Responsibility**: Solving real-world problems fosters a sense of ownership over learning, encouraging deeper engagement and better retention of engineering concepts (Gartner et al., 2022; Riazifar & Khajehsaeid, 2024).

10.1.3.4 Alignment with the Nature of Engineering Practice

Engineering challenges are inherently complex, context-specific, and open-ended. Case studies replicate this environment, training students to handle ambiguity and make informed decisions.

- **Civil Engineering** cases might involve trade-offs between structural safety, environmental sustainability, and budget constraints.
- **Electrical Engineering** cases may focus on integrating emerging technologies like smart grids, which intersect with policy and infrastructure.
- **Software Engineering** cases often involve managing evolving user needs, scalability, and cybersecurity.

This situated learning approach supports authentic, context-rich engagement, aiding knowledge transfer and retention (Loyens, Jones, Mikkers, & van Gog, 2015).

10.1.3.5 Pedagogical Flexibility and Contextual Adaptability

The case study method can be tailored to reflect the specific challenges and standards of each engineering discipline.

- In Mechanical Engineering, cases may involve physical prototyping and design iteration.
- In Software Engineering, they may focus on agile development, interface design, or project management.

This adaptability ensures relevance and effectiveness across diverse educational and cultural contexts, thereby maximising learning impact (Kolmos et al., 2009).

10.1.3.6 Development of Transferable Professional Competencies

Case-based PBL cultivates not only technical know-how but also vital professional attributes:

- **Communication**: Students learn to present complex ideas clearly to technical and non-technical audiences.
- **Collaboration**: Case studies promote teamwork and shared problem-solving.
- **Ethical and Social Awareness**: Students are exposed to ethical dilemmas and sustainability issues.
- **Lifelong Learning**: Open-ended problems encourage independent research and self-directed learning.

These competencies are valued across all branches of engineering and by employers globally.

10.1.3.7 Empirical Support and Scholarly Endorsement

There is robust empirical evidence supporting the effectiveness of case-based PBL in engineering education:

Improved Problem-Solving and Decision-Making (Davis & Yadav, 2014): Davis and Yadav (2014) emphasise the educational value of case studies in engineering, which use real-life scenarios to teach problem-solving in a controlled setting. These narratives expose students to complex, ill-structured problems similar to those they may face professionally. While case-based learning has a long history in fields like business, law, and medicine, it is increasingly adopted in engineering to enhance learning outcomes.

Case studies provide a safe space for students to develop problem-solving skills without real-world risks (Davis & Yadav, 2014). They foster active learning and deeper understanding through analysis and reflection. Research also shows improvements in student motivation and performance, including higher exam scores and positive feedback (Daun et al., 2014).

Higher Levels of Motivation and Engagement (Delaney & Nagle, 2019): Delaney and Nagle (2019) report that implementing PBL in TU Dublin's Mechanical Engineering programme significantly improved student motivation and engagement. Through hands-on projects—such as designing and building machines—students felt better prepared for their careers, despite the increased workload. These findings align with broader research supporting PBL's effectiveness in engineering education. PBL fosters active participation by involving students in real-world projects like robot construction, increasing engagement and the practical application of theory (Abu-Aisheh et al., 2016; Delaney & Nagle, 2019; Riazifar & Khajehsaeid, 2024). Students also reported higher motivation, finding PBL more relevant to their future careers (Delaney & Nagle, 2019). Research confirms that PBL boosts intrinsic motivation by encouraging student ownership of learning (Khalid & Rahman, 2014; Yusof et al., 2016). Feedback from TU Dublin students was overwhelmingly positive, with many preferring PBL over traditional methods—a sentiment echoed in other studies (Abu-Aisheh et al., 2016; Naik et al., 2024).

Better Preparedness for Real-World Engineering Practice (Warnock & Mohammadi-Aragh, 2016): PBL helps students develop key professional skills, including teamwork, time management, and problem-solving (Delaney & Nagle, 2019; Naik et al., 2024). The study by Warnock and Mohammadi-Aragh (2016) demonstrates that PBL significantly enhances both technical and professional skills in biomedical engineering students. Their research found substantial improvements in students' problem-solving abilities, written and oral communication, self-directed learning, and teamwork. These findings offer strong evidence for the effectiveness of PBL in developing core engineering competencies and align with broader educational research. For instance, similar gains in problem-solving skills have been reported by Baluarte-Araya and Bedregal-Alpaca (2022), while improvements in communication have been noted in other PBL-based programmes (Filho et al., 2024). PBL also supports self-directed learning, an essential skill for lifelong development, as emphasised by Riazifar and Khajehsaeid (2024). The ability to work effectively in teams—crucial in engineering practice—was another area of growth confirmed by Bozic et al. (2018).

Additionally, related studies have shown that PBL increases student engagement and satisfaction, which contributes to deeper understanding and better retention of engineering concepts (Riazifar & Khajehsaeid, 2024). Through real-world projects like the scissor lift design, students apply theoretical knowledge in practical settings, reinforcing learning and skill development (Filho et al., 2024). Overall, PBL fosters both generic and discipline-specific competencies, such as systems thinking and practical application, helping close the gap between academic preparation and industry expectations (Bozic et al., 2018; Baluarte-Araya & Bedregal-Alpaca, 2022).

10.1.4 Challenges in Implementing Case Studies in Engineering Education

While the pedagogical benefits of case studies in PBL are well-documented, their integration into engineering curricula presents several challenges. Effective implementation requires **careful planning**, **curricular alignment**, and **faculty expertise** to ensure that case studies complement learning outcomes meaningfully (Azam et al., 2024; Tham Yew Hua & Breslow, 2017).

Key challenges include:

- **Case Design**: Cases must strike a balance between structure and openness, providing enough guidance to support learning while remaining sufficiently complex to encourage critical thinking and problem-solving.
- **Instructor Preparedness**: Successful facilitation of PBL requires a shift in teaching approach. Instructors must move from being knowledge providers to facilitators of student-driven inquiry, which may require targeted professional development (Davis & Yadav, 2014).
- **Time and Resource Demands**: Developing and conducting case-based activities is time-intensive for both educators and students. Institutional support is often necessary to provide the time, resources, and flexibility needed to sustain this approach (Azam et al., 2024).
- **Student Adaptation**: Students unfamiliar with open-ended, inquiry-based learning may initially resist the approach. However, this can be mitigated through structured guidance, clear expectations, and opportunities for reflection (Ameta et al., 2020).

Despite the associated challenges, thoughtful implementation and institutional support can enable the case study method to significantly enhance the quality and relevance of engineering education (Ameta et al., 2020; Davis & Yadav, 2014).

Case studies are a foundational element of PBL in engineering. They serve as a vital bridge between theory and practice, engaging students in authentic, real-world problem-solving while cultivating essential skills such as critical thinking, collaboration, and interdisciplinary reasoning. Although their integration requires careful design and commitment, the pedagogical value of case studies in preparing students for the complexities of professional engineering practice makes them an indispensable tool in contemporary education.

The rationale for adopting case studies across engineering disciplines is both pedagogically robust and practically essential. By fostering transferable skills, supporting interdisciplinary learning, and grounding theory in application, case-based PBL equips students to navigate the complex, evolving demands of modern engineering fields—whether Civil, Mechanical, Electrical, Chemical, or Software Engineering.

Moreover, this approach contributes meaningfully to ongoing engineering education reform, particularly in regions seeking to align curricula with industry expectations and sustainable development goals. In these contexts, PBL—anchored in well-crafted case studies—enhances student agency, boosts employability, and nurtures lifelong learning, ultimately supporting a more responsive, innovative, and future-ready engineering workforce.

10.2 Methodology and Rationale for Case Study Selection in Engineering PBL

Selecting appropriate case studies for PBL in engineering disciplines—such as Civil, Mechanical, Electrical, Chemical and Computer/Software Engineering—required a strategic and pedagogically sound approach to ensure both relevance and educational impact. The methodology for case selection was guided by alignment with programme learning outcomes, the need to foster analytical and critical thinking, and the opportunity to develop self-directed, collaborative learning habits. The rationale for selecting case studies in each discipline was grounded in their real-world applicability, interdisciplinary scope, and their ability to facilitate the application of theoretical knowledge to practical, open-ended problems.

Case studies were chosen from existing literature based on their relevance to the curriculum, ensuring alignment with core disciplinary content and intended graduate attributes. In Civil Engineering, the projects involved fluid mechanics or soil liquefaction, directly engaging with foundational civil systems and geotechnical concerns. In Mechanical Engineering, problems included designing machines to clear street-gutter debris in developing countries, addressing critical public health concerns and developing commercially viable folding bicycles, which integrated technical problem-solving with market-driven design. Electrical Engineering students explored projects such as robotic vehicle design in microcontroller labs and HDTV system development within optoelectronics modules.

In Chemical Engineering, case studies focused on sustainable solutions such as recycling PET bottles to address plastic pollution, reinforcing concepts of reaction engineering and process sustainability. For Computer/Software Engineering, cases involved the use of Virtual and Augmented Reality to enhance experiential learning and simulate complex development environments, promoting innovation in user interaction and system integration.

Complexity and context were carefully considered to ensure the selected cases were appropriately challenging for the students' academic level—neither too simplistic nor overly complex. Each case was situated in a realistic professional context, allowing students to connect academic learning with the practical demands of their future careers.

Across all disciplines, the case studies used in PBL engaged students with ill-structured, open-ended problems that lacked a single correct solution. This approach encouraged inquiry-based learning, teamwork, and the development of transferable problem-solving skills, effectively preparing students to navigate the complexity and ambiguity of real-world engineering challenges.

The case studies were analysed using a set of comprehensive criteria, including institutional context and adoption, curriculum integration, the nature of the real-world problems addressed, PBL implementation strategies, student engagement and learning outcomes, teamwork and collaboration, design thinking and prototyping, assessment methods, instructor and student feedback, challenges encountered and mitigation strategies, as well as lessons learned, recommendations, and long-term impacts. While the core criteria remained consistent, marginal variations were applied across disciplines to reflect context-specific considerations.

10.3 Case Study 1: Civil Engineering

PBL has been increasingly adopted in Civil Engineering education, with case studies demonstrating its potential to enhance student engagement, learning outcomes, and career readiness. By integrating real-world challenges into the curriculum, PBL fosters critical thinking, problem-solving, and communication skills—core competencies for engineering professionals.

10.3.1 Institutional Context and Adoption

In various institutional contexts, PBL has been implemented to bridge the gap between theoretical instruction and practical application. Institutions such as the Singapore Institute of Technology and the University of Glasgow have used PBL to better align academic training with industry needs (Das et al., 2024). However, transitioning from traditional lecture-based models to student-centred approaches can be particularly challenging in non-Western contexts, where instructors are typically seen as the primary source of knowledge (Naji et al., 2020).

10.3.2 Design and Implementation of PBL Projects

PBL in Civil Engineering often centres on open-ended, real-world problems that require students to apply core concepts, such as in projects on fluid mechanics or soil liquefaction (Devi, 2017; Singh, 2023). Implementation typically blends theoretical instruction with hands-on activities like model design, frequency analysis, and IDF curve development (Das et al., 2024; Miranda et al., 2019). These projects encourage creative problem-solving and the exploration of alternative methods under resource constraints (Singh, 2023).

10.3.3 Student Engagement and Learning Outcomes

Student engagement is a key benefit of PBL, with increased autonomy, emotional investment, and collaborative learning contributing to deeper understanding and higher motivation (Kim et al., 2024; Naji et al., 2020). Students report enhanced critical thinking and improved problem-solving abilities, along with better preparation for real-world engineering challenges (Das et al., 2024; Kim et al., 2024).

10.3.4 Assessment Methods in PBL

Assessment in PBL is typically carried out using analytical rubrics that evaluate students' conceptual understanding, technical proficiency, communication, and teamwork (Devi, 2017). Tools such as graded reports, presentations, and surveys assess both academic progress and attitudes towards learning (Debnath & Pandey, 2011; Kim et al., 2024).

10.3.5 Challenges and Mitigation Strategies

Despite its benefits, PBL presents challenges. Students often struggle with teamwork dynamics, technical complexity, and confidence in problem-solving (Das et al., 2024; Kim et al., 2024). These issues can be mitigated through improved scaffolding, clearer project guidelines, and institutional support systems aimed at fostering a positive learning environment.

10.3.6 Lessons Learned and Recommendations

Case studies conclude that PBL significantly enhances applied learning and equips students for the complexities of professional engineering practice (Das et al., 2024). Successful implementation depends on thoughtful course design, transparent communication, and ongoing refinement to meet both student needs and industry expectations (Kim et al., 2024). By effectively integrating real-world contexts into the curriculum, PBL not only supports academic development but also ensures that graduates are well-prepared for the demands of the civil engineering profession.

10.4 Case Study 2: Mechanical Engineering

The integration of PBL into Mechanical Engineering education has demonstrated its potential to enhance student engagement, technical proficiency, and career readiness. By embedding real-world problems into the curriculum, PBL fosters both technical and transversal skills, aligning with modern educational paradigms that prioritise active, student-centred learning. This case study draws from multiple institutions to explore the institutional context, implementation strategies, student outcomes, assessment methods, challenges, and lessons learned.

10.4.1 Institutional Context

PBL has been integrated into Mechanical Engineering education by several institutions to better prepare students for industry challenges. At Universitat Politècnica de València (UPV), a two-year Educational Innovation and Improvement Project (EIIP) embeds PBL within the curriculum, emphasising transversal competences and Sustainable Development Goals (SDGs) (Llopis-Albert et al., 2024). Similarly, Oslo Metropolitan University incorporates PBL as part of its Education 4.0 strategy, which aligns student-centred learning with Industry 4.0 requirements (Keprate et al., 2023).

10.4.2 Curriculum Integration

Embedding real-world applications throughout the Mechanical Engineering curriculum is crucial for aligning academic learning with industry needs. For example, the University of Wisconsin–Madison offers an Interdisciplinary Experiential Design Projects course that highlights human-centred design and hands-on learning (Zhang, Jiang, Sun, & Cheadle, 2024). Jassim and Johnson (2019) developed a four-year hands-on design curriculum integrating design skills across all undergraduate years to prepare students for capstone projects and professional practice. The CDIO Initiative (Conceiving, Designing, Implementing, and Operating) further supports this by embedding design–build–test experiences, enhancing both technical and professional competencies (Hermon, 2016).

10.4.3 Nature of Real-World Problems

The real-world problems tackled in PBL are often client-driven and require innovative, practical solutions. For instance, students at North Carolina A&T State University designed machines to clear street-gutter debris in developing countries, addressing critical public health concerns (Layton & Owusu-Ofori, 1998). Another example includes designing and prototyping a commercially viable folding bicycle that blends technical problem-solving with market considerations (Hermon, 2016).

10.4.4 PBL Implementation Strategies

UPV implements PBL in a first-year master's course in Mechatronic Engineering, engaging students with authentic problems to increase motivation and relevance to industry needs (Llopis-Albert et al., 2024). At Oslo Metropolitan University, students undertake projects such as designing lightweight cranes, applying theoretical knowledge to practical challenges (Keprate et al., 2023). TU Dublin's module involves students designing, building, and testing actual machines, fostering hands-on skills within real-world constraints (Delaney & Nagle, 2019).

10.4.5 Student Engagement and Learning Outcomes

PBL bridges theory and practice, boosting student engagement and fostering collaboration and problem-solving skills (Debnath & Pandey, 2011; Keprate et al., 2023). Students at TU Dublin reported increased motivation and preparedness for professional careers, despite the heavier workload (Delaney & Nagle, 2019). Similarly, PBL at Jaipur Engineering College and Research Centre is linked to improved academic outcomes and job placements (Debnath & Pandey, 2011).

10.4.6 Teamwork, Design Thinking, and Prototyping

Collaborative teamwork is fundamental in PBL, with students often organised into vertically integrated teams that include undergraduates and graduate students to leverage diverse skills and perspectives (Palm et al., 2002). Design thinking is embedded through curricula combining theoretical instruction with practical lab work, promoting iterative design approaches, especially in first-year Mechanical Engineering courses (Naji et al., 2024). Rapid prototyping and mechanical performance testing are key activities enabling students to refine their designs effectively (Palm et al., 2002).

10.4.7 Assessment Methods

PBL assessment commonly includes graded reports, presentations, and project deliverables that evaluate both theoretical knowledge and practical skills (Debnath & Pandey, 2011). At UPV, additional evaluation through surveys and interviews measures students' understanding of transversal competences and SDGs, providing feedback to improve courses (Llopis-Albert et al., 2024).

10.4.8 Instructor and Student Feedback

Courses incorporate continuous feedback from instructors, peers, and industry partners to support students' technical and professional growth (Zhang, Zhang, Zhang, & Yu, 2024). Students generally favour open-ended, real-world assignments over traditional problem sets due to their relevance and challenge (Binu et al., 2024).

10.4.9 Challenges and Mitigation Strategies

Implementing PBL faces challenges such as time constraints, the complexity of real-world problems, and ensuring conceptual clarity (Chen, 2008). Effective mitigation includes modularising projects across courses, engaging industry professionals in project design, and leveraging advanced design tools (Chen, 2008; Wang et al., 2022). Additional barriers include limited financial resources, insufficient teaching support, and resistance from faculty (Jassim & Johnson, 2019). Balancing practical experience with theoretical rigour remains a priority.

10.4.10 Lessons Learned and Recommendations

Successful PBL requires strategic alignment with industry needs and adequate institutional support. Ongoing feedback and adaptability are essential for overcoming challenges and improving outcomes (Delaney & Nagle, 2019; Wang et al., 2022). Despite the resource demands, PBL offers significant benefits, including enhanced student engagement and the development of practical and transversal skills. Integrating PBL with broader institutional aims—such as sustainability and global competence—further strengthens its impact and relevance in mechanical engineering education.

10.4.11 Reflections and Impacts

Experiential learning through PBL has demonstrated positive effects, including improved student retention, stronger teamwork, and enhanced performance in complex design tasks (Jassim & Johnson, 2019). These approaches not only enhance employability but also provide industry partners with innovative solutions and recruitment opportunities (Zhang, Zhang, Zhang, & Yu, 2024). While challenges persist, integrating real-world design problems consistently proves effective in preparing engineering students for professional demands.

10.5 Case Study 3: Electrical/Electronic Engineering

PBL has become a widely adopted pedagogical approach in Electrical and Electronic Engineering (EEE) education. By embedding real-world problems into the curriculum, PBL aims to enhance student engagement, foster deeper learning, and develop essential skills such as teamwork, design thinking, and problem-solving. This report explores key dimensions of PBL implementation in EEE, drawing on insights from multiple case studies and academic research.

10.5.1 Institutional Context and Curriculum Integration

PBL is integrated across various levels of the EEE curriculum—from introductory courses to senior-level capstone projects. For example, institutions such as Prince Mohammad bin Fahd University and Griffith University have embedded PBL into core courses to align theory with practice (Hussain et al., 2024; Mohd-Yasin, 2021). At Western Carolina University, a structured PBL sequence extends from freshman through senior year, culminating in industry-sponsored capstone projects that bridge engineering and engineering technology disciplines (Stone & Jack, 2017).

10.5.2 Real-World Problems and Implementation Strategies

Central to PBL is the engagement with authentic engineering challenges. Examples include designing robotic vehicles in microcontroller labs or developing HDTV systems in optoelectronics modules (Hussain et al., 2024; Powell et al., 2007). Institutions have adopted various strategies to support successful implementation, including scaffolding, balanced assessment criteria, and explicit project scopes. Griffith University, for instance, has addressed common challenges such as lack of direction and student overload through structured planning and ongoing support (Mohd-Yasin, 2021).

10.5.3 Student Engagement and Learning Outcomes

Research consistently shows that PBL enhances student motivation, improves academic performance, and increases knowledge retention (Hussain et al., 2024; Kunicina et al., 2024). At the Technical University of Denmark, students participating in power converter design projects effectively translate theoretical knowledge into industrial applications (Zhang et al., 2016), demonstrating PBL's capacity to bridge academic and professional competencies.

10.5.4 Teamwork, Design Thinking, and Prototyping

PBL encourages collaboration and creative problem-solving, as students work in multidisciplinary teams to address complex engineering issues (Stone & Jack, 2017). Prototyping plays a critical role in the learning process, with students actively involved in designing, building, and testing electrical and electronic systems (Etchepareborda et al., 2018).

10.5.5 Assessment Methods and Feedback

Effective assessment in PBL includes formative feedback, peer evaluation, and integrated assessment tasks. These methods not only guide student progress but also enhance overall satisfaction and performance (Mohd-Yasin, 2021). At Griffith University, student feedback from an introductory electronics course has been instrumental in refining instructional approaches and improving learning experiences (Etchepareborda et al., 2018).

10.5.6 Challenges and Mitigation Strategies

Despite its benefits, PBL faces several implementation challenges, including resource limitations and the need for faculty training (Kunicina et al., 2024). Mitigation strategies include designing flexible scenarios, adapting to available resources, and cultivating faculty readiness. The University of Manchester, for example, successfully navigated these challenges in its robotics module through thoughtful design and sensitivity to student needs (Powell et al., 2007).

10.5.7 Lessons Learned and Recommendations

Key lessons from PBL initiatives highlight the importance of aligning projects with real-world complexities and promoting interdisciplinary collaboration (Lavado-Anguera et al., 2024). Moving forward, long-term evaluations of PBL's impact are needed, along with strategies to support its integration into online and hybrid learning environments (Kunicina et al., 2024).

10.5.8 Reflections and Impact

PBL has proven to be a transformative model in preparing students for the engineering profession. It equips them with not only technical knowledge but also essential soft skills, such as communication, collaboration, and adaptability (Kunicina et al., 2024). Although challenges remain—such as aligning projects with industry needs and ensuring sufficient support—ongoing refinement and institutional commitment can maximise PBL's effectiveness. Ultimately, the integration of PBL into EEE curricula represents a powerful approach to developing well-rounded engineers capable of meeting the demands of an evolving technological landscape.

10.6 Case Study 4: Chemical Engineering

PBL has been extensively explored in Chemical Engineering education through a range of case studies, each showcasing unique approaches to its implementation and impact. These studies highlight how PBL is integrated into curricula, the types of real-world challenges students address, and the pedagogical strategies employed to foster engagement and improve learning outcomes. The following sections examine these aspects in detail, drawing on insights from key research and case reports.

10.6.1 Institutional Context and Curriculum Integration

PBL is implemented across various stages of the chemical engineering curriculum. At the University of Sydney, it is embedded throughout all four years of study, aligning with professional accreditation requirements and ensuring consistent skill development across the programme (Gomes & Barton, 2005). Similarly, at the University of Toronto, PBL is introduced in second-year courses to strengthen research, communication, and problem-solving skills (Farmer & Wilkinson, 2018).

Collaborations with industry are a common feature of PBL integration. For instance, students have worked with virtual chemical plants related to pulp production processes as part of a distillation and absorption course, bridging academic learning with industrial application (Fregolente et al., 2018).

10.6.2 Nature of Real-World Problems

PBL initiatives often centre on pressing, real-world engineering challenges. Examples include projects addressing the recycling of PET bottles to combat plastic pollution (Rajan et al., 2019) and the design and testing of truss structures to apply theoretical principles in structural mechanics (Sarkawi et al., 2024). These projects not only enhance technical understanding but also promote environmental and societal awareness.

10.6.3 PBL Implementation Strategies

Effective implementation of PBL relies on structured methodologies. Strategies include introductory workshops, fieldwork opportunities, and active collaboration with industry professionals to contextualise learning (Noor & Nurhayati, 2024). Tools like Aspen Plus are frequently used for process simulation, allowing students to model and optimise real industrial systems (Fregolente et al., 2018). Additionally, e-learning modules and active learning exercises support student engagement and reinforce theoretical concepts (Farmer & Wilkinson, 2018).

10.6.4 Student Engagement and Learning Outcomes

PBL significantly enhances student motivation and involvement by immersing them in experiential learning. Engaging in complex, hands-on projects encourages critical thinking, innovation, and the practical application of knowledge (Lavado-Anguera et al., 2024). Students consistently report gains in communication, teamwork, and technical competencies, reflecting the holistic benefits of the PBL approach (Yusof et al., 2005).

10.6.5 Teamwork, Design Thinking, and Prototyping

Collaboration is a cornerstone of PBL. Students work in diverse teams to design and prototype engineering solutions, developing essential interpersonal and project management skills (Sarkawi et al., 2024). Design thinking is encouraged through tasks that require creativity, resourcefulness, and iterative problem-solving (dos Santos et al., 2021), often under real-world constraints.

10.6.6 Assessment Methods

Assessment in PBL is multifaceted and typically includes continuous feedback, oral presentations, and evaluations of project deliverables (Fregolente et al., 2018). Competency-based assessment frameworks are commonly used to track the development of both technical and soft skills (Gomes & Barton, 2005), aligning evaluation methods with learning objectives.

10.6.7 Instructor and Student Feedback

Both instructor and student feedback underline the effectiveness of PBL in fostering practical, career-relevant skills. Faculty observations highlight improvements in student preparedness and engagement, while student surveys reflect high satisfaction with the active learning environment (Cline & Powers, 1998; Sarkawi et al., 2024).

10.6.8 Challenges and Mitigation Strategies

PBL is not without its challenges. Issues such as unequal team participation, limited resources, and the unpredictable nature of real-world tasks can hinder implementation (Noor & Nurhayati, 2024). Successful programmes mitigate these issues through careful planning, strong industry partnerships, and iterative refinement of course design and project scope.

10.6.9 Lessons Learned and Recommendations

Case studies reveal that aligning PBL projects with authentic industry problems and fostering interdisciplinary collaboration enhances educational value (Lavado-Anguera et al., 2024). Key recommendations include expanding industry engagement, ensuring adequate institutional support, and integrating flexible resources to accommodate evolving student needs (Fregolente et al., 2018).

10.6.10 Reflections and Impact

PBL has a transformative impact on Chemical Engineering education. It equips students with the confidence, creativity, and technical skills needed to thrive in professional environments (Sarkawi et al., 2024). Despite the challenges involved, the evidence strongly supports PBL as a holistic and future-oriented pedagogical model. By bridging theory and practice, it not only enriches learning but also better prepares graduates for the dynamic demands of modern engineering careers (Lavado-Anguera et al., 2024).

10.7 Case Study 5: Computer/Software Engineering

PBL has emerged as a transformative pedagogical approach in Computer and Software Engineering education, aligning academic instruction with real-world industry needs. By emphasising active student engagement, critical thinking, and hands-on problem-solving, PBL bridges the gap between theoretical understanding and practical application. This educational model is increasingly adopted in curricula to equip students with the skills required to navigate the dynamic and interdisciplinary nature of modern engineering challenges. The following sections examine key dimensions of PBL implementation in computer and software engineering education.

10.7.1 *Institutional Context and Curriculum Integration*

PBL is often embedded throughout Computer and Software Engineering programmes to form a continuous "project spine" that supports the progressive development of technical and professional competencies (Shekar, 2014). At Massey University, for instance, the engineering curriculum has been comprehensively redesigned to incorporate PBL courses across all academic years, ensuring consistent application of theoretical knowledge in real-world contexts (Shekar, 2014).

This curricular integration is driven by the increasing demand for graduates with industry-relevant skills and the ability to work across disciplinary boundaries (Rehman, 2023). Institutions recognise that traditional lecture-based instruction alone is insufficient for preparing students for the complexities of contemporary software development environments.

10.7.2 *Real-World Problems and Implementation Strategies*

PBL engages students with authentic, complex engineering problems that promote a deeper understanding of core computing and engineering concepts (Eswaran, 2024). Implementation strategies vary but often include partnerships with industry, allowing students to work on actual projects that simulate real-world conditions and deliverables.

Innovative problems, such as Virtual and Augmented Reality, were investigated to enhance experiential learning and simulate complex development environments (Alves et al., 2014; Rehman, 2023). These tools not only increase engagement but also allow students to experiment and iterate in low-risk settings.

10.7.3 *Student Engagement and Learning Outcomes*

Research consistently demonstrates that PBL significantly improves student engagement and comprehension of computer science and software engineering principles (Li & Ma, 2024). The approach fosters the development of critical thinking, collaboration, and applied problem-solving skills—competencies essential for academic success and professional readiness (Li & Ma, 2024; Zhang, Zhang, Zhang, & Yu, 2024).

Students are more likely to retain knowledge and understand its practical value when they engage with meaningful, hands-on projects that reflect the demands of the software industry.

10.7.4 *Teamwork, Design Thinking, and Prototyping*

PBL inherently supports teamwork, helping students cultivate soft skills such as communication, leadership, and collaboration—all highly valued in software engineering (Garcés & Oliveira, 2024). Design thinking is also central to

PBL, encouraging students to approach problems creatively and iteratively as they develop innovative software solutions (Kirsten & Patel, 2018).

Prototyping activities are integrated into many PBL courses, allowing students to design, test, and refine their ideas in a structured yet flexible environment.

10.7.5 Assessment Methods and Feedback

Assessment in PBL settings relies on continuous, formative feedback and peer evaluation to support learning and reflect real-world project dynamics (Kirsten & Patel, 2018). Structured rubrics, client feedback, and partnerships with external stakeholders further enrich the evaluation process and help ensure alignment with professional expectations (Eswaran, 2024).

Such multidimensional assessment approaches allow instructors to evaluate both the process and the final product, including technical proficiency, teamwork, and communication.

10.7.6 Challenges and Mitigation Strategies

Despite its benefits, PBL presents several challenges. Common obstacles include faculty resistance to change, limited institutional resources, and difficulty developing robust and fair assessment methods (Rehman, 2023). To address these issues, institutions have implemented faculty development programmes, invested in infrastructure, and created streamlined evaluation frameworks that support both instructors and students (Rehman, 2023).

Clear project guidelines, effective team management practices, and regular feedback loops also help mitigate common issues such as unequal student participation and project misalignment.

10.7.7 Lessons Learned and Recommendations

Lessons from PBL implementations highlight the importance of aligning student projects with real-world complexity and promoting interdisciplinary collaboration (Lavado-Anguera et al., 2024). Successful programmes ensure that students work on projects relevant to current industry trends and provide opportunities to interact with real clients or stakeholders (Alves et al., 2014).

Recommendations include fostering ongoing industry partnerships, training faculty to manage PBL environments effectively, and allocating adequate time and resources to support student learning throughout the project lifecycle.

10.7.8 Reflections and Impact

PBL has a profound impact on preparing students for careers in computer and software engineering. It enhances both technical and soft skills, increases confidence, and fosters a deeper connection between academic study and professional practice (Garcés & Oliveira, 2024). By transforming traditional learning into a more interactive and relevant experience, PBL helps students develop the mindset and competencies needed to thrive in complex, collaborative, and rapidly evolving engineering environments (Shekar, 2014).

While challenges such as ensuring equal team participation and aligning academic projects with industry expectations persist, they can be addressed through thoughtful planning, institutional support, and continuous refinement of PBL practices. Overall, PBL represents a powerful approach for advancing computer and software engineering education in the 21st century.

10.8 Comparative Analysis across Disciplines

10.8.1 Commonalities in PBL Strategies and Results

Across engineering disciplines, the implementation of PBL reveals consistent strategies and outcomes. A defining commonality is the use of real-world problems to drive learning, enabling students to apply theoretical concepts in practical, often complex scenarios. This authenticity in task design enhances student engagement and motivation, fostering deeper learning. In all examined disciplines—Civil, Mechanical, Electrical, Chemical, and Software Engineering—PBL contributed to the development of key professional skills such as critical thinking, collaboration, communication, and self-directed learning. Furthermore, all disciplines adopted assessment frameworks that emphasise process and product, using rubrics, presentations, and reflective tasks to evaluate learning. Constructivist pedagogy underpinned these approaches, aligning with Bloom's taxonomy and promoting inquiry-based exploration as a cornerstone of student learning. Overall, the results demonstrate that, when properly implemented, PBL significantly enhances students' preparedness for real-world engineering challenges.

10.8.2 Discipline-Specific Adaptations and Innovations

While PBL principles are consistent across disciplines, each field exhibited distinct adaptations and pedagogical innovations to suit its unique technical and epistemic requirements. Table 10.1 presents a comparative analysis of PBL implementations across disciplines.

Civil Engineering: PBL projects frequently address infrastructural and environmental issues such as traffic/hydrological modelling and sustainable urban planning, incorporating contextual variables like local resource constraints.

Mechanical Engineering: PBL focused on hands-on prototyping and iterative design, often aligning with Sustainable Development Goals and emphasising transversal competences through design-build-test activities.

Table 10.1 Comparative Analysis of Discipline-Specific Challenges, Adaptations, and Innovations in PBL

Discipline	*Key Challenges*	*Adaptations*	*Innovations*
Civil Engineering	- Studentresistance to open-ended tasks - Integration with regulatory and environmental constraints	- Real-world infrastructure projects - Localised problems (e.g., hydrology, geotechnical issues)	- Use of IDF curve modelling - Hands-on design with structural trade-offs - Simulated field investigations
Mechanical Engineering	- High resource demands for prototyping - Ensuring consistent teamwork across diverse cohorts	- Modular project delivery - Scaffolded skill progression over curriculum years	- Design-build-test pedagogy (e.g., folding bicycles) - Integration with SDGs and transversal competences
Electrical Engineering	- Limited lab infrastructure - Difficulty balancing theory and hands-on work	- Layered complexity in projects from basic to advanced - Simulation-integrated lab modules	- IoT-basedcapstone projects - Power converter design linked to real industry needs - Early microcontroller use
Chemical Engineering	- Safety and complexity in real process modelling - Limited access to industrial equipment	- Use of digital simulation tools (e.g., Aspen Plus) - Emphasis on sustainability and process safety	- Projects on PET recycling - Virtual chemical plant modelling - Integrated SDGs in process design
Software Engineering	- Facultyresistance to agile teaching models - Ensuring fairness in team-based coding assignments	- Agile project management frameworks - Project spine across all years	- Use of AR/VR for virtual labs - Real-time coding sprints - Partnerships with industry on live software builds

Electrical Engineering: It integrated PBL into laboratory-based learning, with projects involving microcontrollers, robotics, and Internet of Things (IoT) systems.

Chemical Engineering: It placed greater emphasis on simulation tools like Aspen Plus and process modelling, while also integrating environmental themes such as recycling and green chemistry.

Software Engineering: PBL involved agile methodologies, real-time coding tasks, and the use of immersive technologies like Virtual Reality to replicate professional development environments.

These discipline-specific approaches ensured relevance and maximised the learning impact of PBL.

10.8.3 Institutional Enablers and Barriers

Several institutional factors influenced the success or difficulty of implementing PBL. Key enablers included strong alignment between curriculum goals and industry needs, support for curricular reform, and faculty engagement. Institutions that partnered with industry and provided faculty with professional development in PBL methodologies saw more sustainable implementation. Additionally, frameworks such as India's National Education Policy (NEP) 2020 encouraged interdisciplinary learning and supported PBL integration. However, significant barriers persisted. Faculty resistance—often rooted in the shift from traditional lecturing to facilitation—remained a challenge, as did the lack of institutional incentives for pedagogical innovation. Resource limitations, particularly in prototyping-heavy disciplines, posed logistical hurdles, and students sometimes resisted open-ended learning due to unfamiliarity or anxiety around ambiguous tasks. Mitigating these challenges required coordinated institutional strategies involving training, time allocation, and appropriate infrastructure.

10.8.4 Cross-Disciplinary Learning Opportunities

PBL opens rich opportunities for cross-disciplinary learning, especially in addressing complex, systems-level challenges that span traditional engineering boundaries. Projects like electric vehicle design naturally integrate Mechanical, Electrical, and Software Engineering, promoting systems thinking and collaborative problem-solving. Such integration mirrors industry practices and helps students understand the interconnectedness of engineering domains. Moreover, soft skills such as communication, leadership, and ethical awareness, cultivated through PBL, are highly transferable across disciplines. The collaborative nature of these projects fosters a culture of peer learning and interdisciplinarity. Policy environments such as NEP 2020 further bolster these opportunities by promoting curriculum designs that enable students to engage with multiple fields simultaneously, preparing them for a future where complex problems require integrated, interdisciplinary solutions.

10.9 Recommendations for Educators and Institutions

10.9.1 Guidelines for Effective PBL Design and Implementation

For PBL to be effectively integrated into engineering education, educators must focus on designing authentic, complex, and context-rich problems that reflect the challenges encountered in professional practice. These problems should encourage open-ended inquiry and collaborative exploration, meaningfully allowing students to apply theoretical knowledge. The design process must align with constructive alignment principles, ensuring coherence between learning outcomes, instructional methods, and assessment tasks. Incorporating iterative problem-solving processes, feedback loops, and opportunities for reflection enhances the learning experience. Additionally, scaffolding through structured milestones, formative feedback, and clear expectations can support students in navigating the complexities of PBL.

10.9.2 Faculty Development and Support

Successful PBL implementation hinges on faculty capacity and willingness to adopt new pedagogical roles. Institutions should provide continuous professional development opportunities that train faculty in facilitation techniques, collaborative learning strategies, and interdisciplinary project design. Establishing peer mentoring and teaching communities can foster a culture of pedagogical innovation and knowledge exchange. Furthermore, faculty contributions to PBL should be recognised and rewarded through institutional mechanisms such as teaching awards, career advancement criteria, and workload adjustments. Such measures not only motivate faculty but also build institutional capacity for sustaining PBL over the long term.

10.9.3 Resource Allocation and Stakeholder Engagement

Adequate resources and stakeholder involvement are critical for scaling and sustaining PBL. Institutions must invest in infrastructure such as maker spaces, simulation labs, and digital platforms to support hands-on, collaborative work. Financial resources should be allocated for tools, materials, and project management systems. Moreover, engaging external stakeholders—including industry partners, alumni, and community organisations—can enrich PBL projects with real-world relevance and provide mentorship and evaluation opportunities. Stakeholder involvement also ensures alignment with industry trends and enhances student employability. Institutional policies should promote such partnerships as integral to curriculum development and delivery.

10.9.4 Scalability and Sustainability of PBL Initiatives

To ensure the long-term viability of PBL, institutions should adopt a phased approach to implementation, beginning with pilot projects in selected modules and gradually scaling across the curriculum. Embedding a "project spine"—a series of progressively complex projects across academic years—can support curriculum integration and scaffold skill development. Digital collaboration, assessment, and project tracking tools can facilitate PBL in large-class or hybrid environments. Periodic evaluation of PBL outcomes, informed by student feedback and graduate performance, is essential for continuous improvement. Sustainable PBL requires strategic alignment with institutional goals, policy support, and a commitment to pedagogical excellence that evolves alongside technological and professional advancements.

10.10 Conclusion

This chapter has offered a comprehensive exploration of PBL through comparative case studies across five engineering disciplines—Civil, Mechanical, Electrical, Chemical, and Computer/Software Engineering. It has demonstrated that while the core principles of PBL remain consistent—namely, the use of real-world problems, student-centred inquiry, and collaborative learning—the implementation is shaped significantly by each discipline's unique epistemologies, technical demands, and contextual realities.

Commonalities across all fields underscore the transformative impact of PBL in fostering critical thinking, teamwork, communication, and applied problem-solving skills. These competencies align closely with the evolving expectations of engineering graduates and industry stakeholders. Furthermore, each discipline's adaptations reveal the pedagogical flexibility of PBL, whether through simulation tools in Chemical Engineering, prototyping in Mechanical Engineering, or agile development in Software Engineering.

Institutional enablers, such as supportive policy environments, interdisciplinary collaboration, and faculty development, have been pivotal in sustaining PBL initiatives. However, challenges persist, including resistance to pedagogical change, resource limitations, and the complexity of designing and assessing open-ended projects. Addressing these requires a strategic commitment from educators, administrators, and policymakers.

Eventually, the chapter illustrates that PBL, when contextually adapted and institutionally supported, serves as a method of instruction and a catalyst for curricular transformation. It nurtures a generation of engineers equipped with technical expertise and the ability to innovate, collaborate, and respond to the complex, multidisciplinary problems of the 21st century. As engineering education continues to evolve, PBL stands out as a robust and future-ready model that bridges the gap between academic knowledge and professional practice.

However, this chapter has limitations. It draws on existing case studies from various institutions rather than presenting a detailed analysis of one. While the examples drawn from various institutions are illustrative, they lack the granularity of a focused investigation. Consequently, future research centred on specific case studies from diverse engineering disciplines could provide more explicit and nuanced insights into the relevance, challenges, opportunities, and implications of implementing PBL.

References

Abu-Aisheh, A. A., Grant, L., Sumukadas, N., & Hadad, A. (2016). Fostering engineering students engagement using problem-based learning and course learner agent object portfolios. *International Journal of Engineering Pedagogy (IJEP)*, 6(4), 45–47. https://doi.org/10.3991/ijep.v6i4.6086

Alves, L. M., Ribeiro, P., & Machado, R. J. (2014). Project-based learning: An environment to prepare IT students for an industry career. In L. Yu (Ed.), *Overcoming Challenges in Software Engineering Education: Delivering Non-Technical Knowledge and Skills*, pp. 230–249. IGI Global Scientific Publishing. https://doi.org/10.4018/978-1-4666-5800-4.ch012

Ameta, D., Tiwari, S., & Singh, P. (2020). A preliminary study on case-based learning teaching pedagogy: Scope in SE education. In *ISEC'20: Proceedings of the 13th Innovations in Software Engineering Conference*, pp.1–12. https://doi.org/10.1145/3385032.3385045

Azam, R., Farooq, M. U., & Riaz, M. (2024). A case study of problem-based learning from a civil engineering structural analysis course. *Journal of Civil Engineering Education, 150*(3). https://doi.org/10.1061/jceecd.eieng-1861

Baluarte-Araya, C., & Bedregal-Alpaca, N. (2022). Influence of problem-based learning on the development of technical, methodological, participatory and personal competencies: Valuation by engineering students. In *2022 IEEE World Conference on Engineering Education (EDUNINE)* (pp. 1–6). https://doi.org/10.1109/EDUNINE53672.2022.9782362

Binu, G., Kumar, Y., D'Silva, R., Kaliveer, V., Shetty, A., Toney, G., Bhat, S., & Singh, A. B. (2024). Integrating real-world applications into the machine design course through an open-ended assignment—A case study. *Journal of Engineering Education Transformations, 37*(S2), 879–886. https://journaleet.in/index.php/jeet/article/view/2518

Bozic, M., Certic, J. D., Vukelić, M., & Čizmić, S. (2018). New instructional approach for fostering generic and professional competences: Case study of the project and problem-based learning engineering practice course. *International Journal of Engineering Education, 34*(5), 1581–1591. https://dialnet.unirioja.es/servlet/articulo?codigo=6880319

Butan, D. (2007). Case studies in problem-based learning in engineering. In *ASEE Annual Conference & Exposition*, Honolulu, HI.

Chen, X. (2008). Longitudinal integration of problem-based learning into undergraduate mechanical engineering courses. In *Proceedings of the ASME 2008 International Mechanical Engineering Congress and Exposition* (pp. 407–415). https://doi.org/10.1115/IMECE2008-68319

Cline, M. J., & Powers, G. J. (1998). Problem-based learning in a chemical engineering undergraduate laboratory [Paper presentation]. In *ASEE Annual Conference & Exposition*, Seattle, WA. https://peer.asee.org/7358.pdf

Das, S., Kim, Y., Shah, J. A., & Lim, L. H. I. (2024). Enhancing applied learning in civil engineering: Implementing project-based learning in hydrological studies. In *2024 IEEE International Conference on Teaching, Assessment, and Learning for Engineering (TALE)* (pp. 1–4). https://doi.org/10.1109/tale62452.2024.10834363

Daun, M., Salmon, A., Tenbergen, B., Weyer, T., & Pohl, K. (2014). Industrial case studies in graduate requirements engineering courses: The impact on student motivation. In *2014 IEEE 27th Conference on Software Engineering Education and Training (CSEE&T)* (pp. 3–12). https://doi.org/10.1109/CSEET.2014.6816775

Davis, C., & Yadav, A. (2014). Case studies in engineering. In *Cambridge handbook of engineering education research* (pp. 161–180). Cambridge University Press. https://doi.org/10.1017/CBO9781139013451.013

Debnath, M., & Pandey, M. (2011). Enhancing engineering education learning outcomes using project-based learning: A case study. *International Journal of Quality Assurance in Engineering and Technology Education*, *1*(2), 23–34. https://doi.org/10.4018/IJQAETE.2011070103

Delaney, K., & Nagle, G. (2019). Engineering education through problem-based learning: A case study of teaching mechanical engineering design in TU Dublin. [Conference paper]. In *30th Annual Conference of the Irish Association of Professional Applied Learning (IAPAL)*, Dublin, Ireland. https://doi.org/10.21427/0RSQ-4X20

Devi, T. V. R. (2017). Implementing project-based learning in civil engineering: A case study. *Journal of Engineering Education Transformations*, *30*(3), 272–277. https://journaleet.in/index.php/jeet/article/view/1424

dos Santos, F. R., Oliveira, I. M. de, Santos, L. C., Leal, R. E. G., & Cardoso, M. (2021). Pedagogical strategies for chemical engineering courses: Skills development through project-based learning (PBL). *Research, Society and Development*, *10*(7), e15545. https://doi.org/10.33448/RSD-V10I7.15545

Eswaran, U. (2024). Project-based learning: Fostering collaboration, creativity, and critical thinking. In M. Bhatia & M. Mushtaq (Eds.), *Enhancing Education With Intelligent Systems and Data-Driven Instruction* (pp. 23–43). IGI Global Scientific Publishing. DOI: 10.4018/979-8-3693-2169-0.ch002

Etchepareborda, P., Bierzychudek, M. E., Carducci, L., Veiras, F. E., Zacchigna, F. G., Corbellini, E., Marra, S. G., Iglesias, M., Teggia, M. M., Alvarez-Hamelin, J. I. (2018). A project-based learning method applied to an introductory course in electronics engineering. In *2018 IEEE World Engineering Education Conference (EDUNINE)*. Buenos Aires, Argentina (pp. 1–4). https://doi.org/10.1109/EDUNINE.2018.8450976

Farmer, J. L., & Wilkinson, L. (2018). Engineering success: Using problem-based learning to develop critical thinking and communication skills in a chemical engineering classroom. [Paper presentation]. In *Canadian Engineering Education Association (CEEA) Conference*, Vancouver, BC, Canada. https://doi.org/10.24908/PCEEA.V0I0.13057

Filho, W. M., Ribeiro, B. M. B., Ferreira, C. M., Feitosa, F., de Mauro, G.S., Alves, G. N., & Kawahara, H. Y. (2024). Fostering practical engineering skills through problem-based learning: A scissor lift design and automation case study. *Journal of Engineering Research*. Advance online publication. https://doi.org/10.22533/at.ed.3174724260210

Fregolente, L. V., Venturelli, H., Rodrigues, J., Da Silva, E., Diniz, I., & Maciel Maria, R. W. (2018). Project-based learning applied to distillation and absorption education: Integration between industry and a chemical engineering undergraduate course. *Chemical Engineering Transactions*, *69*, 427–432. https://doi.org/10.3303/CET1869072

Gandhi, S. J., Sauser, B., White, B. E., Gorod, A., & Ireland, V. (2012). Application of case studies to engineering management and systems engineering education. In *ASEE*

Annual Conference & Exposition, San Antonio, TX. https://peer.asee.org/application-of-case-studies-to-engineering-management-and-systems-engineering-education

Garcés, L., & Oliveira, B. R. N. (2024). Teaching software engineering with project-based learning: A four-year experience report. In *Anais do XXXVIII Simpósio Brasileiro de Engenharia de Software* (pp. 476–486). https://doi.org/10.5753/sbes.2024.3530

Gartner, G., Binn, A., Gabela, J., Retscher, G., Gikas, V., Schmidt, M., & Wang, W. (2022). From project-based to problem-based learning in engineering disciplines: Enhancing cartography and geomatics education. In *8th International Conference on Higher Education Advances (HEAd'22)*. https://doi.org/10.4995/head22.2022.14473

Gomes, V., & Barton, G. (2005). Problem-based learning in a new chemical engineering curriculum. In *Proceedings of the 7th World Congress of Chemical Engineering*, Glasgow, Scotland.

Hermon, J. (2016). Enhancing the manufacturing knowledge of undergraduate engineering students: A case study of a design-build-test challenge involving folding bicycles. [Paper presentation]. In *22nd International Conference on Engineering Design (ICED16)*, Belfast, United Kingdom. https://pureadmin.qub.ac.uk/ws/files/130043209/Enhancing_the_Manufac.uring_Knowledge_of_Undergraduate_Engineering_Students.pdf

Hussain, A. A., El-Nakla, S., Khan, A. H., Nayfeh, J., Ifurong, A., & Tayem, N. (2024). Project-based learning approach for undergraduate electrical engineering laboratories: A case study. In *2024 International Symposium on Applied Computing and Engineering Technology (ISAECT)* (pp. 1–7). https://doi.org/10.1109/isaect64333.2024.10799876

Jackson, H., Fleischmann, C., & Tarhini, K. (2015). An approach to improving student learning of civil engineering concepts using case studies. In *2015 IEEE Frontiers in Education Conference (FIE)* (pp. 1–5). https://doi.org/10.1109/FIE.2015.7344378

Jassim, E. W., & Johnson, B. E. (2019). An integrated four-year hands-on design curriculum: A case study. [Paper presentation]. In *ASEE Annual Conference & Exposition*, Tampa, FL. https://peer.asee.org/an-integrated-four-year-hands-on-design-curriculum-a-case-study

Keprate, A., Woodford, S., & Borrajo, R. (2023). From theory to practice: Leveraging project-based learning to cultivate student engagement in mechanical engineering education. In *2023 IEEE International Conference on Industrial Engineering and Engineering Management (IEEM)*. https://doi.org/10.1109/ieem58616.2023.10406362

Khalid, M., & Rahman, H. (2014). Problem-based learning to enhance students' understanding and motivation for learning mechanical science. *IIUM Journal of Educational Studies*, *2*(1), 56–66. https://doi.org/10.31436/ijes.v2i1.31

Kim, Y., Das, S., Shah, J. A., Lim, L. H. I., & Yeo, C. H. (2024). Analysis of project-based learning (PBL) characteristics in civil engineering education. In *2024 IEEE International Conference on Teaching, Assessment, and Learning for Engineering (TALE)* (pp. 1–7). https://doi.org/10.1109/tale62452.2024.10834330

Kirsten, R., & Patel, T. (2018). Systematic integration of project-based learning in an undergraduate human-computer interaction course: A case study. In *DS 92: Proceedings of the DESIGN 2018 15th International Design Conference* (pp. 656–661). The Design Society. https://www.designsociety.org/download-publication/40827/SYSTEMATIC+INTEGRATION+OF+PROJECT-BASED+LEARNING+IN+AN+UNDERGRADUATE+HUMAN-COMPUTER+INTERA+CTION+COURSE%3A+A+CASE+STUDY

Kolmos, A., Graaff, E., & Du, X. (2009). Diversity of PBL—PBL learning principles and models. In *Book: research on PBL practice in engineering education* (pp. 9–21). https://doi.org/10.1163/9789087909321_003

Kunicina, N., Uteševs, I., Chaiko, Y., & Grants, R. (2024). Enhancing electrical engineering education through project-based learning. In *2024, 65th Riga Technical University (RTUCON)* (pp. 1–6). https://doi.org/10.1109/rtucon62997.2024.10830810

Lavado-Anguera, S., Velasco-Quintana, P.-J., & Terrón-López, M.-J. (2024). Project-based learning (PBL) as an experiential pedagogical methodology in engineering education: A review of the literature. *Education Sciences*, *14*(6), 617. https://doi.org/10.3390/educsci14060617

Layton, R. A., & Owusu-Ofori, S. (1998). *A report on integrating design projects in mechanical engineering*, North Carolina A&T State University. https://www.scribd.com/document/7397403/2000witechpaper

Li, Y., & Ma, G. (2024). Application and effect evaluation of project-based learning in computer science teaching reform. *Region – Educational Research and Reviews*, *6*(9), 233. https://doi.org/10.32629/rerr.v6i9.3095

Llopis-Albert, C., Rubio, F., Mata-Amela, V., Devece, C., Torner-Feltrer, M. E., & Zeng, S. (2024). Project-based learning methodology (PBL) for the acquisition of transversal competences (TCs) and integration of sustainable development goals (SDGs) in mechanical engineering subjects. *Multidisciplinary Journal for Education, Social and Technological Sciences*. Advance online publication. https://doi.org/10.4995/muse.2024.21101

Loyens, S. M. M., Jones, S. H., Mikkers, J., & van Gog, T. (2015). Problem-based learning as a facilitator of conceptual change. *Learning and Instruction*, *38*, 34–42. https://doi.org/10.1016/j.learninstruc.2015.03.002

Miranda, M. A., da Costa, A., Lisboa de Castro, J. F., & Sáiz Linares, Á. (2019). A case study of a project-based learning activity implemented in a geotechnical engineering course. In *EDULEARN19 Proceedings* (pp. 2878–2883).

Mohd-Yasin, F. (2021). Effective strategies for project-based learning of practical electronics. *Electronics*, *10*(18), 2245. https://doi.org/10.3390/ELECTRONICS10182245

Naik, S. M., Bandi, S., & Reddy, C. (2024). Enhancing engineering education: Exploring the impact of problem-based learning on freshman students' skills and engagement – A case study. *Journal of Engineering Education Transformations*, *38*(S1), 133–139. https://journaleet.in/index.php/jeet/article/view/2264

Naji, K. K., Ebead, U., Al-Ali, A., & Du, X. (2020). Comparing models of problem and project-based learning (PBL) courses and student engagement in civil engineering in Qatar. *Eurasia Journal of Mathematics, Science and Technology Education*, *16*(8), em1867. https://doi.org/10.29333/EJMSTE/8291

Naji, M., Abdelhameed, W., & Awawdeh, S. (2024). Assessing student comprehension and implementation of design thinking in mechanical engineering education. In *2024 Design and Arts for STEAM Annual Conference (DASA)* (pp. 1–5). https://doi.org/10.1109/dasa63652.2024.10836335

Noor, A. H., & Nurhayati, S. (2024). Project-based learning implementation in a participative planning course: Strategies, outcomes, and challenges. *JCP (Jurnal Cahaya Pendidikan)*, *9*(2). https://doi.org/10.33373/chypend.v9i2.5660

Palm, W., Keggler, T., Taggart, D., Chelidze, D., & Stucker, B. (2002). Integration of mechanical design and prototyping activities. In *ASEE Annual Conference & Exposition*, Montreal, Quebec, Canada. https://peer.asee.org/integration-of-mechanical-design-and-prototyping-activities

Powell, N., Peaker, A. R., Truscott, W. S., Hicks, P. J., & Canavan, B. (2007). Seeding enquiry-based learning in electrical and electronic engineering: Case study 1—optoelectronics. [Conference paper]. In *International Conference on Engineering Education – ICEE 2007*, Coimbra, Portugal. http://icee2007.dei.uc.pt/proceedings/papers/297.pdf

Rajan, K. P., Gopanna, A., & Thomas, S. P. (2019). A project-based learning (PBL) approach involving PET recycling in chemical engineering education. *Recycling*, *4*(1), 10. https://doi.org/10.3390/RECYCLING4010010

Rehman, S. U. (2023). Trends and challenges of project-based learning in computer science and engineering education. In *Proceedings of the 2023 Conference on Innovation and Technology in Computer Science Education V. 2* (pp. 658–658). https://doi.org/10.1145/3629296.3629360

Riazifar, N., & Khajehsaeid, H. (2024). Enhancing engineering education through problem-based learning. In *Recent Advances in Technology Research and Education: Proceedings of the 1st International Conference on Technology Research and Education*, Tarumanagara University, 26–27 October 2023. https://doi.org/10.31273/10.31273/9781911675167/1636

Sarkawi, M. S., Mohamed, M., Shamjuddin, A., Alias, H., & Mohamad, Z. (2024). Project-based experiential learning in designing truss structure for first-year chemical engineering students. *Asean Journal of Engineering Education*, *8*(1), 49–55. https://doi.org/10.11113/ajee2024.8n1.149

Shekar, A. (2014). Project-based learning in engineering design education: Sharing best practices. [Paper presentation]. In *ASEE Annual Conference & Exposition*, Indianapolis, IN. https://peer.asee.org/project-based-learning-in-engineering-design-education-sharing-best-practices

Singh, V. (2023). A strategy for implementation of project-based learning in civil engineering curriculum. *Journal of Engineering Education Transformations*, *37*(3), 165–175. h

Stone, W. L., & Jack, H. (2017). Project-based learning: Integrating engineering technology and engineering. [Paper presentation]. In *ASEE Annual Conference & Exposition*, Columbus, OH. DOI: 10.18260/1-2--28770

Syed, K., Vijayalakshmi, M., Kandakatla, R., & Narayanan, S. R. (2024). Inculcating multidisciplinary learning in electrical engineering undergraduate curriculum through problem-based learning. *Journal of Engineering Education Transformations*, *38*(S1), 210–215. https://journaleet.in/index.php/jeet/article/view/2278

Taajamaa, V., Westerlund, T., Guo, X., Hupli, M., Salanterä, S., & Salakoski, T. (2014). Interdisciplinary engineering education—Practice-based case. In *2014 IEEE International Conference on Teaching, Assessment and Learning for Engineering (TALE)* (pp. 31–37). https://doi.org/10.1109/IEDEC.2014.6784677

Tham Yew Hua, E., & Breslow, L. (2017). A study on integrating case-based learning into the engineering curriculum. In *45th SEFI Annual Conference 2017*. https://search.informit.com.au/documentSummary;dn=384856779373583;res=IELENG;subject=Bibliographies%20%26%20catalogues

Wang, Y., Wang, X., Xin, Y., Zhao, J., Bai, J., & Yang, S. (2022). Implementation strategies of undergraduate GP for mechanical engineering major based on PBL. In *2022 7th International Conference on Educational and Information Technology (EEIT)* (pp. 1–5). https://doi.org/10.1109/EEIT56566.2022.00008

Warnock, J. N., & Mohammadi-Aragh, M. J. (2016). Case study: Use of problem-based learning to develop students' technical and professional skills. *European Journal of Engineering Education*, *41*(2), 142–153. https://doi.org/10.1080/03043797.2015.1040739

Yusof, K., Phang, F., Mohammad, S. (2016). Engineering education in Southeast Asia: Practice and research. In Abdulwahed, M., Hasna, M., Froyd, J. (Eds.), *Advances in engineering education in the Middle East and North Africa*. Springer. https://doi.org/10.1007/978-3-319-15323-0_3

Yusof, K. M., Kamaruddin Abd. Hamid, M., Abu Hassan, M. A., Hassim, M. H., Syed Hassan, S. A. H., & Tasir, Z. (2005). Outcomes of problem-based learning (PBL) implementation from students' perspectives [Paper presentation]. In *ASEE Annual Conference & Exposition*, Portland, OR. http://eprints.utm.my/id/eprint/951/

Zhang, J., Zhang, L., Zhang, G., & Yu, Y. (2024). From theory to practice: Project-based learning in computer science education. *Frontiers in Computing and Intelligent Systems*, *8*(1), 130–133. https://doi.org/10.54097/jv09dw34

Zhang, X., Jiang, C., Sun, J., & Cheadle, M. (2024). University of Wisconsin-Madison mechanical engineering experiential design project case study. *VTE*, *1*(4). https://doi.org/10.54844/vte.2024.0747

Zhang, Z., Hansen, C. T., & Andersen, M. A. E. (2016). Teaching power electronics with a design-oriented, project-based learning method at the Technical University of Denmark. *IEEE Transactions on Education*, *59*(1), 32–38. https://doi.org/10.1109/TE.2015.2426674

Chapter 11

Plausible Examples of Problem-Based Learning

11.1 Introduction

As mentioned previously (cf.: Chapter 1), Problem-Based Learning (PBL) is a student-centred instructional methodology that engages learners in exploring complex, real-world problems as the primary vehicle for acquiring and applying knowledge. Rooted in constructivist learning theory, PBL shifts the traditional paradigm from passive content absorption to active knowledge construction through inquiry, collaboration, and reflection. Originally developed in the context of medical education, PBL has since gained widespread adoption across disciplines, including engineering, due to its ability to foster critical thinking, problem-solving, and lifelong learning skills. At its core, PBL presents students with ill-structured problems that lack clear-cut solutions, requiring them to investigate, analyse, and propose solutions using interdisciplinary knowledge and teamwork. The role of the instructor in PBL transitions from content deliverer to facilitator, guiding students as they navigate the problem space and build competencies relevant to their professional context.

Engineering education increasingly faces the challenge of preparing graduates for a dynamic, complex, and interdisciplinary professional landscape. Traditional lecture-based approaches often fall short in equipping students with the practical skills, adaptive thinking, and collaborative abilities demanded by industry and society. PBL addresses this gap by providing a pedagogical framework that closely mirrors the processes engineers encounter in real-world settings—identifying needs, designing solutions under constraints, and iterating through feedback and testing. By engaging students in solving authentic engineering problems, PBL fosters not only technical proficiency but also the development of transferable skills such as communication, leadership, ethical reasoning, and systems thinking. Furthermore, PBL supports curriculum alignment with global competency frameworks such as Accreditation Board for Engineering and Technology (ABET) and Conceive–Design–Implement–Operate (CDIO), which emphasise integrated knowledge, personal and interpersonal skills, and engineering reasoning. In the context of

DOI: 10.1201/9781003529743-15

engineering education reform, PBL stands out as a transformative approach that cultivates innovation, student agency, and responsiveness to global challenges, including sustainability and technological change.

This chapter aims to illustrate the practical application of PBL in engineering education through diverse, discipline-specific examples. Drawing hypothetical examples from Civil, Mechanical, Electrical, Chemical, and Computer/Software Engineering, as well as interdisciplinary programmes involving Management Sciences, the chapter highlights how PBL can be effectively designed and implemented to meet both pedagogical and professional objectives. harnessed to enhance engineering education in the 21st century.

11.2 Pedagogical Foundations and Principles Adopted for the Setting Up of the Examples

The examples discussed in this chapter on PBL in different engineering disciplines are grounded in a set of pedagogical principles that support meaningful and transferable learning. The key principles include student-centred learning, problem contextualisation, collaborative knowledge construction, and facilitator-guided inquiry. Students are placed at the centre of the learning process, actively engaging with real-world problems that lack predetermined solutions. These problems are designed to simulate the complexity and ambiguity of professional engineering practice, requiring learners to apply multidisciplinary knowledge, conduct independent research, and iteratively refine their solutions.

Another critical principle adopted is self-directed learning, where students take ownership of their learning goals, monitor their progress, and evaluate their outcomes. The learning process is inherently collaborative, encouraging peer-to-peer interaction and team-based problem-solving—skills that are essential in engineering contexts. The role of the educator in PBL shifts from knowledge provider to learning facilitator, providing scaffolding, prompting critical thinking, and guiding reflective practices. In engineering, PBL often involves open-ended projects such as sustainable infrastructure design, energy system optimisation, or software development cycles, which inherently demand iterative problem solving and systems thinking.

The theoretical foundations adopted are primarily constructivist and experiential learning theories, both of which are highly relevant to engineering education. According to constructivism, learners actively construct their own understanding based on experiences and prior knowledge, rather than passively receiving information. PBL creates authentic contexts in which students must engage in inquiry, hypothesis testing, and knowledge integration—hallmarks of constructivist learning. In engineering classrooms, this may involve analysing real data, designing solutions within constraints, or navigating trade-offs in technical systems.

Experiential learning, as advanced by scholars like David Kolb, complements constructivism by emphasising learning through reflection on concrete experience. The experiential cycle—comprising concrete experience, reflective observation, abstract conceptualisation, and active experimentation—is particularly well suited to engineering tasks that involve prototyping, simulation, testing, and iteration. For example, a PBL project requiring students to design a microgrid or optimise a water distribution network allows them to experience engineering practice, reflect on their processes, apply theory, and test their designs—all within an integrated learning framework.

The integration of PBL into engineering education aligns closely with global engineering competency frameworks such as ABET and CDIO. These frameworks emphasise not only technical knowledge but also the development of professional, interpersonal, and system-level competencies.

ABET's student learning outcomes include abilities such as applying knowledge of science and engineering, designing systems within constraints, functioning effectively on multidisciplinary teams, and engaging in lifelong learning—all of which are directly supported through PBL activities. For instance, when students work on projects such as retrofitting a building for energy efficiency, they must apply thermodynamics and control theory, manage teamwork, and navigate economic and environmental considerations.

Similarly, the CDIO framework provides a curriculum model where students learn through projects that mirror the engineering product lifecycle. PBL is naturally suited to CDIO environments, especially in courses where students must conceive design ideas, prototype them, and assess their societal impact. PBL thus becomes an operational mechanism for implementing CDIO principles, particularly in capstone design courses, interdisciplinary innovation labs, and entrepreneurship-oriented engineering programmes.

Similarly, engineering education is guided by the demonstration of competencies or graduate attributes defined by professional bodies in various countries, such as the Engineering Council of South Africa (ECSA). These attributes include problem-solving, design, application of scientific and engineering principles, sustainability, teamwork, and independent learning. Such competencies are carefully considered in the development of the PBL examples presented in this chapter. Furthermore, although specific examples and attributes vary across disciplines, generic assessment rubrics and the peer evaluation process (Appendix A; Tables 11.A1–11.A3), as well as a student feedback questionnaire for collecting responses on the PBL (Appendix B), are provided for reference.

11.3 PBL Examples across Engineering Disciplines

11.3.1 Civil Engineering

Title: Designing a Sustainable Urban Mobility Plan in South Africa

11.3.1.1 Institutional Context

Engineering Institutes/Universities.

Discipline: Civil Engineering
Course: Transportation Engineering and Planning (Final Year Undergraduate)
Class Size: ~60 Students
Purpose: Aligns with ECSA graduate attributes and national smart city and sustainability *agendas*.

11.3.1.2 Curriculum Integration

This PBL case can be embedded within a final-year Transportation Engineering Module on Urban Transport Planning. It was designed to address:

- Graduate Attribute 1: Problem-Solving
- Graduate Attribute 7: Environmental and Sustainability Impact (The engineer and the world)
- SDG 11: Sustainable Cities and Communities

Duration: The project spans eight weeks, integrating both theoretical instruction and field-based analysis.

11.3.1.3 Nature of Real-World Problems

Problem Scenario: Design a sustainable, inclusive, and multimodal urban mobility plan for a high-density, low-income township facing:

- Severe congestion
- Unsafe pedestrian environments
- Poor access to formal public transport

Constraints: Limited funding, poor infrastructure, limited formal data, and sociopolitical sensitivity.

11.3.1.4 PBL Implementation Strategies

The PBL cycle can follow the steps as mentioned in Table 11.1. Alternatively, the generic steps and rubrics suggested in the Appendices may be followed.

11.3.1.5 Student Engagement and Learning Outcomes

- Application of transportation engineering principles to real urban challenges.
- Development of systems-thinking and critical evaluation skills.

Table 11.1 Steps in PBL in Civil Engineering

Step No	*Steps*	*Activities*
1	Problem Introduction	Municipal planning brief with GIS maps and traffic incident data.
2	Analysis	Site assessment, stakeholder analysis, and traffic data collection.
3	Research & Ideation	Case studies, literature review, and stakeholder role-play
4	Design	Development of multimodal transport plans including cycle lanes, improved pedestrian zones, and minibus taxi integration.
5	Simulation	Use of VISSIM/PTV Visum for traffic modelling
6	Presentation	Final presentations and submission of a detailed report

- Enhanced ability to design for socio-economic diversity.
- Improved presentation and report-writing abilities.

11.3.1.6 Teamwork, Design Thinking, and Prototyping

Teams of four or five students used design thinking steps:

- Empathise: Stakeholder personas and journey mapping
- Define: Key urban mobility issues
- Ideate: Brainstorm multimodal strategies
- Prototype: Models and GIS-based visualisations
- Test: Peer-reviewed designs and software simulations

Simulation models of traffic flow under a multimodal system can be developed.

11.3.1.7 Assessment Methods

Formative

- Weekly group progress reviews
- Peer feedback and interim expert critique
- Pitching and Dragons' Den

Summative

Table 11.2 presents the assessment and marking rubrics in Civil Engineering.

***Challenges and Mitigation Strategies*:** Table 11.3 presents the Challenges and Mitigation Strategies in Civil Engineering.

Table 11.2 Marking Rubrics in Civil Engineering

Assessment No	*Types of Assessment/Rubrics*	*Marks (%)*
1	Group Technical Report: Engineering design, sustainability analysis, stakeholder considerations	40
2	Oral Presentation to Expert Panel: Clarity, technical justification, stakeholder engagement	20
3	Simulation & Modelling Output: Examples of simulation scenarios	15
4	Peer Evaluation of Contribution: Contribution to teamwork and problem-solving	15
5	Individual Reflection Essay: Lessons learned, challenges, and personal growth	10

Note: Alternatively, the generic rubrics (Appendices) may be followed.

11.3.1.8 Important Pointers

- Real-world problem framing drives student motivation.
- Interdisciplinary approaches are essential in transport design.
- Faculty support in software and team management is critical.
- Scaffold early stages to reduce cognitive overload.

11.3.1.9 Impacts

- Students can become more confident in applying knowledge to complex systems.
- Some proposals can contribute to local community advocacy efforts.
- Fosters industry partnerships for mentorship and assessment.
- More dynamic and engaging classroom environments.

11.3.2 Mechanical Engineering

Title: Developing an Automated Sorting System—A Mechatronics PBL Case

Table 11.3 Challenges and Mitigation Strategies in Civil Engineering

Challenge	*Mitigation Strategy*
Uneven Digital Skills	Introductory sessions on transport software
Incomplete Data	Open-source and crowdsourced data usage
Time Constraints	Structured milestones and progress tracking
Role Conflict in Teams	Rotating leadership and peer evaluation

11.3.2.1 Institutional Context

Institution: Engineering Institutes/Universities
Discipline: Mechanical and Mechatronic Engineering
Course: Mechatronics Systems Design (3rd/4th Year Undergraduate)
Class Size: ~40 students
Purpose: To integrate mechanical design, electronics, sensors, actuators, and programming through a practical, real-world automation challenge.

11.3.2.2 Curriculum Integration

This project can be embedded in a core third- or fourth-year course focusing on integrated system design. It addresses:

- Graduate Attribute 3: Engineering Design
- Graduate Attribute 5: Use of Engineering Tools
- Graduate Attribute 9: Independent Learning
- SDG 9: Industry, Innovation and Infrastructure

Duration: The project spans eight weeks, integrating both theoretical instruction and field-based analysis.

11.3.2.3 Nature of Real-World Problems

Problem Scenario: "Design and prototype a low-cost automated sorting system for a small-scale recycling facility to separate plastic, metal, and glass items."

Constraints
- Cost-effective for deployment in informal settlements
- Operable via solar energy
- Low maintenance
- Simple user interface

11.3.2.4 PBL Implementation Strategies

Steps that should be followed in the PBL cycle (Table 11.4).

11.3.2.5 Student Engagement and Learning Outcomes

- High level of engagement through hands-on iterative prototyping
- Better grasp of system integration
- Practical understanding of sensors and actuators
- Ability to work within real-world constraints

Table 11.4 Steps and Activities in PBL in Mechanical Engineering

Step no	*Steps*	*Activities*
1	Problem Introduction	Video briefing from a local recycling NGO.
2	Analysis	Decomposition of the problem into sensor selection, mechanical design, control logic, and user interface.
3	Research & Ideation	Investigation into detection technologies (e.g., capacitive, inductive, IR sensors), microcontroller options, and sorting mechanisms
4	Design/Prototyping	CAD modelling, circuit design, and rapid prototyping.
5	Testing & Iteration/ Simulation	Debugging, performance evaluation, and refinement
6	Presentation	Final presentations and submission of a detailed report/Public demonstration day with feedback from industry partners.

11.3.2.6 Teamwork, Design Thinking, and Prototyping

- Teams of three to five students employed agile development cycles
- Design thinking phases include:
  - Empathise: Understanding the needs of informal recyclers
  - Define: Clear system requirements and constraints
  - Ideate: Concept brainstorming and benchmarking
  - Prototype: Arduino-controlled conveyor with sorting gates
  - Test: Field trials and debugging

11.3.2.7 Assessment Methods

Formative
- Weekly team progress updates
- Peer evaluations
- Instructor coaching feedback sessions

Summative

Table 11.5 presents the marking rubrics in Mechanical Engineering.

Challenges and Mitigation Strategies: The challenges and mitigation strategies are presented in Table 11.6.

11.3.2.8 Important Pointers

- Early-stage prototyping accelerates engagement
- Peer learning is crucial in hardware-intensive projects

Table 11.5 Marking Rubrics in Mechanical Engineering

Assessment No	*Types of Assessment/Rubrics*	*Marks (%)*
1	Final Group Technical Report: Engineering design, sustainability analysis, stakeholder considerations.	40
2	Oral Presentation to Expert Panel: Clarity, technical justification, stakeholder engagement	20
3	Functional Demonstration: Prototype	20
4	Peer Evaluation of Contribution: Contribution to teamwork and problem-solving	10
5	Individual Reflection Essay: Lessons learned, challenges, and personal growth.	10

Note: Alternatively, the generic rubrics (Appendices) may be followed.

Table 11.6 Challenges and Mitigation Strategies in Mechanical Engineering

Challenge	*Mitigation Strategy*
Technical skill disparities	Provided an optional Arduino/sensor bootcamp
Hardware availability	Created component-sharing library
Debugging complexity	Introduced structured testing protocols

- Real-world clients enhance accountability
- Partner with NGOs or industry for authentic challenges
- Incorporate PBL earlier in the program
- Allocate dedicated lab time and flexible deadlines

11.3.2.9 Impacts

- Students can demonstrate real-world readiness
- Development of prototypes and solutions to real-world problems
- Cross-disciplinary collaboration with Electrical Engineering

11.3.3 Electrical/Electronic Engineering

Title: Designing a Smart Energy Monitoring and Control System for Campus Buildings

11.3.3.1 Institutional Context

Institution:
Department: Electrical, Electronic and Computer Engineering
Course: Embedded Systems and IoT (4th Year Undergraduate)
Class Size: ~50 students
Purpose: Enhance systems thinking, sustainability awareness, and embedded design skills.

11.3.3.2 Curriculum Integration

This PBL case was implemented in a senior-level course integrating power electronics, IoT, and embedded systems. It aligns with:

- Graduate Attribute 1: Problem-Solving
- Graduate Attribute 3: Engineering Design
- Graduate Attribute 6: Communication
- SDG 7: Affordable and Clean Energy

Duration: The project can run for 10 weeks alongside lectures and lab sessions.

11.3.3.3 Nature of Real-World Problems

Problem Scenario: Design and prototype a smart system to monitor, analyse, and optimise energy usage in university campus buildings.

Constraints

- Must integrate multiple sensors (current, temperature, light)
- Wireless data transmission and real-time dashboard
- User-configurable alerts for abnormal consumption
- Limited budget and off-the-shelf components

11.3.3.4 PBL Implementation Strategies

Structured PBL phases: Table 11.7 presents the structured PBL phases in Electrical and Electronic Engineering.

Table 11.7 Structured PBL Phases in Electrical and Electronic Engineering

Step no	*Steps*	*Activities*
1	Problem Briefing	Provided by the university's energy management office
2	Investigation	Energy audit of pilot buildings
3	Research, Ideation & Conceptualisation	Sensor and platform evaluation (e.g., ESP32, Raspberry Pi)
4	Design/ Development	Hardware interfacing, software programming, and dashboard design
5	Testing/Simulation	Load testing and data validation
6	Deployment/ Presentation	Final presentations and submission of a detailed report/Pilot system installed in one lab classroom

11.3.3.5 Student Engagement and Learning Outcomes

Students will have high motivation and collaboration. They would

- Apply theory to meaningful sustainability challenges
- Gain practical skills in circuit design, PCB development, and cloud integration
- Develop a deeper understanding of smart grid technologies

11.3.3.6 Teamwork, Design Thinking, and Prototyping

Multidisciplinary teams of 4–6 students each engaged in:

- Empathise: Interviewed facility managers and users
- Define: Created user needs and system requirement specs
- Ideate: Explored energy metering and visualisation solutions
- Prototype: Developed working models using Internet of Things (IoT) kits
- Test: Validated the accuracy and response time of systems

Final prototypes included power monitors with Wi-Fi modules and web dashboards.

11.3.3.7 Assessment Methods

Formative

- Biweekly team presentations and code reviews
- Peer feedback cycles

Summative: Table 11.8 presents the assessment of rubrics of Electrical/ Electronic Engineering.

Table 11.8 Assessment of Rubrics of Electrical/Electronic Engineering

Assessment No	*Types of Assessment/Rubrics*	*Marks (%)*
1	Final Group Technical Report: Engineering design, sustainability analysis, stakeholder considerations.	40
2	Oral Presentation to Expert Panel: Clarity, technical justification, stakeholder engagement	20
3	System Demonstration	20
4	Peer Evaluation of Contribution: Contribution to teamwork and problem-solving	10
5	Individual Reflection Essay: Lessons learned, challenges, and personal growth.	10

Table 11.9 Challenges and Mitigation Strategies in Electrical/Electronic Engineering

Challenge	*Mitigation Strategy*
Hardware integration delays	Early procurement and kits prepared ahead
Limited programming skills	Offered weekend coding tutorials
Group conflicts	Facilitated conflict resolution sessions
Data security issues	Introduced encryption and cloud authentication labs

***Challenges and Mitigation Strategies*:** Table 11.9 presents the Challenges and Mitigation Strategies in Electrical/Electronic Engineering.

11.3.3.8 Important Pointers

- Real-world problems foster ownership and accountability.
- Embedded systems require careful scheduling due to testing cycles.
- Integration of cloud platforms significantly boosts student interest.
- Include structured version control training early.
- Engage campus facilities staff as problem owners.
- Offer flexible deadlines for hardware-intensive phases.

11.3.3.9 Impacts

- Expression of a new interest in energy tech and IoT careers
- Improved critical thinking and technical documentation
- Prototype can be adopted for an extended pilot
- Introduction of an Energy Tech Innovation Lab on campus

11.3.4 Chemical Engineering

Title: Designing a Sustainable Bioethanol Production Process from Agricultural Waste

11.3.4.1 Institutional Context

Institution: Engineering Institutions/Universities
Department: Chemical Engineering
Course: Process Design and Simulation (Final-Year Undergraduate)
Class Size: ~60 students
Objective: To integrate biochemical process design, sustainability, and chemical engineering principles in a real-world project.

11.3.4.2 Curriculum Integration

This PBL project is embedded in a capstone design course and supports:

- Graduate Attribute 1: Problem-Solving
- Graduate Attribute 3: Engineering Design
- Graduate Attribute 7: Sustainability and Environmental Impact (The engineer and the world)
- SDG 12: Responsible Consumption and Production

Duration: 8 to 10 weeks

11.3.4.3 Nature of Real-World Problems

Problem Scenario: Design a scalable, cost-effective, and environmentally sustainable process to produce bioethanol from sugarcane bagasse, considering local South African agricultural conditions.

Constraints

- Low carbon footprint
- Local sourcing of raw materials and enzymes
- Compliance with national environmental standards
- Scalable to rural industrial clusters

11.3.4.4 PBL Implementation Strategies

Table 11.10 presents the structured phases in PBL in Chemical Engineering.

11.3.4.5 Student Engagement and Learning Outcomes

High engagement may be observed through:

Table 11.10 Structured Phases in PBL in Chemical Engineering

Step No	*Steps*	*Activities*
1	Problem Introduction	Industry-sponsored webinar on biofuels
2	Research/Investigation	Literature review on biomass pretreatment and fermentation
3	Ideation & Concept Development	Process flow diagram creation and reactor selection
4	Modelling and Simulation	Using Aspen Plus and MATLAB
5	Economic Evaluation	CAPEX/OPEX analysis
6	Sustainability Assessment	Lifecycle and carbon footprint analysis
7	Presentation	Final presentations and submission of a detailed report

- Active participation in design reviews
- Use of peer teaching in simulation tools
- Student-led webinars on enzyme kinetics and renewable energy policy
- Improved ability to synthesise interdisciplinary knowledge
- Developed proficiency in process simulation
- Enhanced communication and teamwork skills

11.3.4.6 Teamwork, Design Thinking, and Prototyping

Teams of four or five students employed the design thinking methodology:

- Empathise: Engaged with rural cooperatives and NGOs
- Define: Identified key local challenges (e.g., water scarcity, cost)
- Ideate: Brainstormed pre-treatment and fermentation alternatives
- Prototype: Simulated process configurations in Aspen Plus
- Test: Sensitivity analysis on process parameters and yields
- No physical prototyping, but virtual testing and optimisation.

11.3.4.7 Assessment Methods

Formative

- Supervisor consultations and design milestone reviews
- Peer and self-assessment of contributions

Summative

Table 11.11 presents the assessment types/rubrics for PBL in Chemical Engineering.

***Challenges and Mitigation Strategies*:** The challenges and mitigation strategies in PBL in Chemical Engineering are presented in Table 11.12.

Table 11.11 Assessment Types/Rubrics for PBL in Chemical Engineering

Assessment No	*Types of Assessment/Rubrics*	*Marks (%)*
1	Process Design Report: Engineering design, sustainability analysis	40
2	Oral Presentation to Expert Panel: Clarity, technical justification, stakeholder engagement	15
3	Simulation Files and Code Review	25
4	Peer Evaluation of Contribution: Contribution to teamwork and problem-solving	10
5	Individual Reflection Essay: Lessons learned, challenges, and personal growth.	10

Table 11.12 Challenges and Mitigation Strategies for PBL in Chemical Engineering

Challenge	*Mitigation Strategy*
Simulation learning curve	Offered step-by-step video tutorials and support hours
Scope creep	Set clear project boundaries with deliverable milestones
Access to real data	Partnered with local distilleries for case data

11.3.4.8 Important Pointers

- Problem authenticity drives engagement and innovation
- Process simulation tools should be scaffolded early
- Industry partnerships improve realism and student accountability
- Begin simulation training in the third year
- Integrate community engagement in project framing
- Offer mini-workshops on sustainability metrics and lifecycle analysis

11.3.4.9 Impacts

- Strengthened ties with the biofuel industry and sustainability NGOs
- Student-led research into circular bioeconomy opportunities

11.3.5 Computer/Software Engineering

Title: Developing a Campus-Wide Mobile App for Smart Navigation and Resource Booking

11.3.5.1 Institutional Context

Institution: Engineering Institutions/Universities
Department: Computer Science and Software Engineering
Course: Software Engineering Project (Final-Year Undergraduate)
Class Size: 40–50 students
Objective: To enhance full-stack development, agile project management, and user-centred design skills through a real-world, multi-featured software solution.

11.3.5.2 Curriculum Integration

This PBL project is part of a capstone course aligned with:

- Graduate Attribute 1: Problem-Solving
- Graduate Attribute 5: Software System Design and Implementation (Use of engineering tools)

- Graduate Attributes 6 & 8: Communication and Teamwork
- SDG 9: Industry, Innovation and Infrastructure

The project builds on prior coursework in data structures, databases, UI/UX design, and mobile app development.

11.3.5.3 Nature of Real-World Problems

Problem Scenario: Develop an integrated mobile application to help students navigate campus, book shared resources (e.g., study rooms, computer labs), and receive real-time campus updates.

Constraints

- Android platform (Kotlin/Java)
- Integration with the university's backend systems (APIs)
- Accessibility features and data privacy compliance
- Limited development time (10 weeks)

11.3.5.4 PBL Implementation Strategies

Phased Agile approach: The steps in PBL in Computer/Software Engineering is presented in Computer/Software Engineering in Table 11.13.

11.3.5.5 Student Engagement and Learning Outcomes

Students will likely show high levels of participation and ownership:

- Formed cross-functional Scrum teams
- Practised CI/CD pipelines and Git workflows
- Conducted campus-wide user testing
- Enhanced coding proficiency and architectural design

Table 11.13 The Steps in PBL in Computer/Software Engineering

Step No	*Steps*	*Activities*
1	Sprint 0	Project scoping, user persona creation, and system requirements
2	Sprint 1–3	Incremental app development (map, booking system, notifications)
3	Sprint Reviews	Stakeholder demos every two weeks
4	Sprint Retrospectives	Reflection on teamwork and technical blockers
5	Final Deployment	Deployment of the Alpha version on test devices for feedback
7	Presentation	Final presentations and submission of a detailed report

- Improved agile documentation and collaboration skills
- Learned real-world constraints of multi-user systems

11.3.5.6 Teamwork, Design Thinking, and Prototyping

Teams of five or six students applied design thinking:

- Empathise: Interviewed students and admin staff
- Define: Identified usability and access pain points
- Ideate: Brainstormed feature sets and wireframes
- Prototype: Built interactive mockups using Figma
- Test: Conducted usability testing sessions and adjusted based on feedback

Prototypes progressed from wireframes to functional mobile apps.

11.3.5.7 Assessment Methods

Formative

- Sprint reviews and peer evaluations
- Technical code walkthroughs
- Team retrospectives

Summative: The assessment rubrics in Computer/Software Engineering are presented in Table 11.14.

Challenges and Mitigation Strategies: The challenges and mitigation strategies in Computer/Software Engineering are presented in Table 11.15.

11.3.5.8 Important Pointers

- Agile PBL fosters adaptability and professional readiness
- Real user involvement motivates iterative improvement
- Continuous integration prevents last-minute failures
- Include testing automation and DevOps basics
- Involve the IT department for real deployment opportunities

Table 11.14 Assessment Rubrics in Computer/Software Engineering

Assessment No	*Types of Assessment/Rubrics*	*Marks (%)*
1	Final App Demonstration and Codebase	40
2	Project Documentation and UML Diagrams	20
3	Presentation and demonstration	20
4	Client Feedback and Usability Metrics	10
5	Individual Reflection and Contribution Report	10

Table 11.15 Challenges and Mitigation Strategies in Computer/Software Engineering

Challenge	*Mitigation Strategy*
Backend integration issues	Scheduled API walkthroughs and mock servers
Uneven workload distribution	Introduced structured peer evaluations weekly
Device compatibility bugs	Provided emulator/simulator labs and test devices

11.3.5.9 Impacts

- Solution for local tech incubator
- Strengthened alumni–industry links through guest mentorships
- Increased interest in product-based startups

11.3.6 Interdisciplinary Programmes: Engineering + Management Sciences

Title: Designing and Deploying a Smart Modular Water Purification System for Rural Communities

11.3.6.1 Project Brief

A national development agency has called for innovative, scalable solutions to improve access to clean drinking water in underserved rural regions. Student teams must design, plan, and simulate the deployment of a smart modular water purification system that is both technically feasible and economically sustainable, with potential for public–private partnership.

11.3.6.2 Nature of Real-World Problems

Problem Scenario: Design a scalable, solar-powered, sensor-enabled modular water purification system for off-grid rural communities in Southern Africa. Ensure technical, financial, and social feasibility.

Constraints

- Local manufacturing and maintainability
- Water quality data must be monitored and managed via IoT sensors
- Budget cap: ZAR 500,000 per unit
- 5-year sustainability plan
- Must comply with national water and environmental standards
- Must use renewable energy sources (e.g., solar-powered pumps)
- Stakeholder engagement with rural community cooperatives
- Project must be financially viable with a 5-year break-even plan

11.3.6.3 Institutional Context

Institution: Engineering Universities/Institutions
Departments: School of Engineering & School of Management Sciences
Course: Interdisciplinary Design and Innovation Project (Final-Year Undergraduate)
Class Size: 60–80 students (Mixed Engineering and Management students)
Objective: To foster interdisciplinary collaboration, systems thinking, and entrepreneurial approaches to solving complex societal challenges.

11.3.6.4 Curriculum Integration

This PBL project is embedded in a joint final-year capstone course and aligned with:

- Graduate Attributes 1: Problem-Solving
- Graduate Attributes 3 & 5: Engineering Design and Innovation
- Graduate Attributes 6 & 8: Teamwork and Interdisciplinary Communication
- SDG 6: Clean Water and Sanitation
- SDG 9: Industry, Innovation and Infrastructure

It integrates knowledge from systems engineering, water treatment, cost analysis, project management, stakeholder engagement, and entrepreneurship.

11.3.6.5 Cross-Functional Team Roles & Role-Playing

Each team (five or six students) includes the following designated roles:

- Systems Engineer: Oversees technical system integration (IoT + purification tech)
- Civil/Mechanical Engineer: Designs the structural and mechanical components
- Project Manager: Coordinates timelines, milestones, and stakeholder inputs
- Financial Analyst: Conducts cost–benefit analysis, ROI, and funding strategy
- Community Liaison (Role-played): Represents rural user perspectives
- Policy & Compliance Officer/ regulatory advisor (Role-played): Reviews regulatory impact

Role-playing is introduced via simulation sessions (town hall meetings, pitch presentations, and stakeholder briefings).

11.3.6.6 Core Skills and Methodologies

- Systems Thinking: Mapping water demand, technology integration, community impact, and supply chain logistics

- Cost–Benefit Analysis: Capital Expenditure (CAPEX)/Operating Expenditure (OPEX) projection, social Return on Investment (ROI), lifecycle costing
- Risk Management: Identification of technical, financial, operational risks; mitigation planning using Failure Mode and Effects Analysis (FMEA) or Monte Carlo simulations
- Design Thinking: Empathy interviews, ideation, prototyping (using CAD/IoT platforms)

11.3.6.7 PBL Implementation Strategies

The key phases in PBL in Interdisciplinary PBL projects are presented in Table 11.16.

11.3.6.8 Expected Outcomes

- Communication
  - Technical and non-technical reporting
  - Stakeholder negotiations and team decision documentation
- Entrepreneurial Thinking
  - Business model development (e.g., pay-per-use or community co-ownership)
  - Pitch to simulated investors and development agencies

Table 11.16 The Steps in PBL in Interdisciplinary PBL Projects

Step No	*Steps*	*Activities*
1	Problem Definition & Team Formation	Mixed-discipline teams created with defined roles
2	Design Thinking Sprint	Empathise, Define, Ideate, Prototype, Prototyping (using CAD/IoT platforms), Test
3	Systems Mapping	Mapping water demand, technology integration, community impact, supply chain logistics, stakeholders, technical subsystems, and financial models
4	Design and Development	Design, simulate, and validate technical and business models
5	Cost–Benefit Analysis	CAPEX/OPEX projection, social ROI, lifecycle costing
6	Risk Management	Identification of technical, financial, and operational risks; mitigation planning using FMEA or Monte Carlo simulations
7	Presentation	Formal pitch, impact reflection, role-based critique

- Real-World Decision-Making
  - Trade-offs between cost, quality, and scalability
  - Data-driven strategy revisions using scenario modelling

11.3.6.9 Learning Benefits

- Engineering Students learn about budgeting, regulatory impacts, and stakeholder alignment.
- Management Students engage with technical design trade-offs, feasibility, and infrastructure finance.
- All students collaborate across domains, gaining exposure to start-up-style execution in a socially relevant context.

11.3.6.10 Student Engagement and Learning Outcomes

- Field visits to rural communities and water boards
- Interviews with NGOs and government water agencies
- Hackathons and solution workshops
- Integrated problem-solving under technical, social, and financial constraints
- Confidence in public speaking and investor engagement
- Understanding of sustainable development and scalable innovation

11.3.6.11 Design Thinking Steps

- Empathise: Conducted interviews in rural communities
- Define: Identified key water access and quality challenges
- Ideate: Brainstormed multiple purification and sensor technologies
- Prototype: Created CAD models and financial simulation tools
- Test: Conducted user testing with a scaled mock-up in simulation labs

11.3.6.12 Application of Systems Thinking, Cost–Benefit Analysis, and Risk Management

Systems Thinking

- Developed causal loop diagrams to visualise water access barriers
- Used system dynamics modelling for long-term impact estimation

Cost–Benefit Analysis

- CAPEX and OPEX projections
- ROI estimation with social return on investment (SROI) metrics
- Comparative analysis of centralised vs decentralised models

Risk Management

- Risk register with technical, financial, and operational categories
- Monte Carlo simulations for cost sensitivity
- SWOT and FMEA applied during design reviews

11.3.6.13 Assessment Strategy

Formative

- Sprint reviews with instructors and external mentors
- Peer assessment of role contribution and conflict resolution

Summative

The assessment types and rubrics in interdisciplinary PBL projects are presented in Table 11.17.

Challenges and Mitigation Strategies: Table 11.18 presents the challenges and mitigation strategies in interdisciplinary PBL projects.

11.3.6.14 Important Pointers

- Interdisciplinary learning enhances innovation and holistic thinking
- Role-play deepens understanding of real-world stakeholder dynamics
- Systems thinking tools foster long-term impact orientation
- Extend the timeline for prototyping and stakeholder iterations
- Include professional mentors from industry/government
- Develop a repository of tools (CAD, IoT platforms, financial models)

11.3.6.15 Impacts

- Possible patent filing for novel sensor integration methods
- Local NGOs can adopt elements of one business model for a pilot deployment
- Interest in tech-for-development start-ups

Table 11.17 Assessment Types and Rubrics in Interdisciplinary PBL Projects

Assessment No	*Component*	*Weight (%)*
1	Technical Design Report	40
	Product/Design Demonstration	15
2	Cost–Benefit and Risk Report	15
3	Team Presentation (Investor Pitch)	10
4	Individual Role Reflection	10
5	Peer & Role Evaluation	10

Table 11.18 Challenges and Mitigation Strategies in Interdisciplinary PBL Projects

Challenge	*Mitigation Strategy*
Uneven technical/business knowledge	Pre-project bootcamp for management and engineering concepts
Difficulty in defining shared goals	Weekly inter-role alignment meetings
Prototype testing limitations	Used digital twins and VR-based simulations

11.4 Conclusion

The diverse PBL examples presented in this chapter illustrate the transformative potential of this pedagogy across engineering and interdisciplinary domains. From sustainable urban transport systems to smart energy monitoring, bioethanol production, and integrated sociotechnical water purification solutions, PBL demonstrates its capacity to engage students in authentic, complex challenges that mirror real-world professional practice.

Rooted in constructivist and experiential learning principles, PBL facilitates deep learning by situating knowledge acquisition within the context of purposeful inquiry, collaboration, and reflection. It supports curriculum alignment with globally recognised competency frameworks such as ABET, CDIO, and ECSA, reinforcing critical graduate attributes like problem-solving, systems thinking, innovation, teamwork, and ethical responsibility.

Across the examples, several key themes emerge:

- Interdisciplinary collaboration and role-based learning significantly enhance the realism and relevance of engineering education.
- Design thinking and prototyping, whether physical or virtual, support iterative learning and creativity.
- Systems thinking, cost–benefit analysis, and risk assessment empower students to make informed decisions under constraints.
- Cross-functional team dynamics and role-playing (e.g., as project managers, engineers, analysts) foster professional communication and an entrepreneurial mindset.

Challenges such as uneven skill levels, resource limitations, and project scope management are best addressed through structured scaffolding, formative feedback loops, and institutional support systems.

The chapter affirms that well-designed PBL experiences prepare engineering graduates not only to solve technical problems but also to navigate complexity, lead interdisciplinary teams, and contribute to sustainable and inclusive innovation. As engineering education continues to evolve, PBL offers a robust framework for developing agile, competent, and socially responsive professionals equipped for the demands of the 21st century.

Appendices (Marking Rubrics, Peer Evaluation and Student Feedback)

Appendix A

Scale:

5 = Exceptional
4 = Above Average
3 = Meets Expectations
2 = Below Expectations
1 = Poor

Table 11.A1 PBL Assessment Rubric (Representative)

Criteria	*Excellent (5)*	*Good (4)*	*Fair (3)*	*Needs Significant Improvement (2)*	*Poor/ Unacceptable (1)*
1. Problem Understanding & Definition	Demonstrates exceptional insight; clearly articulates the problem, scope, and constraints.	Clearly defines the problem with relevant context and boundaries.	Adequately defines the problem with some context.	The problem definition lacks clarity or depth.	Misinterprets or fails to define the problem.
2. Research & Information Gathering	It uses high-quality, diverse, and credible sources; it is thoroughly researched.	Gathers relevant and mostly credible information; well-researched.	Adequate research with some relevant sources.	Limited research; credibility or relevance of sources is questionable.	Inadequate or no research conducted.
3. Application of Engineering Concepts	Applies advanced, accurate, and creative use of engineering concepts.	Applies concepts accurately and appropriately.	Applies fundamental concepts with minor errors.	Shows misunderstanding or gaps in the application.	Incorrect or irrelevant application of engineering concepts.
4. Innovation & Creativity	Develops original, highly creative, and feasible solutions.	Provides creative and practical solutions.	Solutions are somewhat creative and workable.	Minimal creativity; standard or partially workable solutions.	No creativity; impractical or copied solution.
5. Collaboration & Teamwork	Actively supports team success; leads, listens, respects others.	Works well in a team; participates and communicates effectively.	Participates in team activities; minor issues in collaboration.	Limited participation; struggles to work with others.	Disengaged or disruptive in teamwork.
6. Problem-Solving Process	Follows a structured, iterative, and reflective process with clear logic.	Logical and organised approach with minor gaps.	Reasonable structure but with inconsistencies or missing steps.	Unstructured or simplistic process.	No evident problem-solving strategy.

(Continued)

Table 11.A1 (Continued)

Criteria	*Excellent (5)*	*Good (4)*	*Fair (3)*	*Needs Significant Improvement (2)*	*Poor/ Unacceptable (1)*
7. Use of Tools & Technology	Skillfully integrates appropriate tools/ software with high proficiency.	Uses tools effectively with few errors.	Basic use of tools with some guidance.	Limited or inaccurate use of tools.	No appropriate use of engineering tools.
8. Project Outcome/ Design/Model/ Prototype	The final product is functional, well-designed, innovative, and exceeds requirements.	Product meets requirements; well-executed and functional.	Functional but with minor flaws or limited depth/ detail.	Incomplete or partially functional; basic design.	Product not functional, missing, or irrelevant.
9. Communication & Presentation	Exceptionally clear, professional, and engaging; excellent visuals/ documentation.	Clear, well-organised communication; minor presentation flaws.	Adequate presentation; some issues with clarity or organisation.	Weak presentation; lacks clarity or structure.	Poor or incomplete presentation.
10. Ethical, Social & Environmental Considerations	Insightful analysis of ethics, sustainability, and social impact.	Addresses key ethical/social/ environmental issues.	Mentions some broader impacts, but lacks depth.	Minimal or superficial awareness of issues.	No evidence of broader consideration.
11. Reflection & Continuous Learning	Deep, honest reflection; strong insights and clear plans for growth.	Reflects thoughtfully; identifies strengths and areas to improve.	Basic reflection; some insights offered.	Superficial reflection with little depth.	No meaningful reflection.

Table 11.A2 Suggested Weighting (Representative and Can Be Adjusted)

Criterion No	*Criterion*	*Weightage (%)*
1	Problem Understanding & Definition/ Formulation	10
2	Research & Information Gathering	10
3	Application of Engineering Concepts	10
4	Innovation & Creativity	10
5	Collaboration & Teamwork	5
6	Problem-Solving Process	10
7	Use of Tools & Technology	5
8	Project Outcome/Design/Prototype	20
9	Communication & Presentation	10
10	Ethical & Environmental Considerations	5
11	Reflection & Continuous Learning	5

Table 11.A3 Peer Evaluation Component (Separate—Optional 10% of Total Grade)

Peer Evaluation Criteria	*Description*	*Rating*
Contribution to Teamwork	Level of effort, quality of work, and initiative.	
Communication	Actively listens, shares ideas, and provides constructive feedback.	
Dependability & Responsibility	Meets deadlines, attends meetings, and fulfils assigned tasks.	
Respect & Professionalism	Treats others respectfully, supports team dynamics.	
Leadership (if applicable)	Takes initiative, facilitates discussion, resolves conflict.	

Note: Each team member evaluates their peers using the following scale (1–5), across key dimensions.

Appendix B: Student Feedback Questions on PBL Experience

Student feedback questionnaires may be developed as relevant, premised upon the following aspects.

I. **Learning Experience & Outcomes**

1 What were the most valuable skills or concepts you learned during this PBL activity?
2 How well did the project help you understand the engineering concepts covered in the course?
3 Did the PBL experience encourage you to think critically and solve real-world problems?

4 To what extent did this project enhance your ability to apply theoretical knowledge in practice?

II. **Teamwork & Collaboration**

5 How effective was your team's collaboration and communication?
6 Did all team members contribute fairly to the project? Why or why not?
7 What challenges (if any) did your team face in working together?
8 What would you do differently in future team projects?

III. **Problem-Solving & Project Development**

9 How clear and engaging was the problem you worked on?
10 Were the problem-solving steps (research, brainstorming, prototyping) meaningful and logical?
11 What was the most difficult part of developing your solution/design/prototype?
12 Did you feel adequately supported (resources, tools, time) to complete the project?

IV. **Use of Tools, Technology, & Resources**

13 Did you have access to the tools and technologies needed for this project?
14 Did the project give you opportunities to practice using real engineering tools or software?

V. **Assessment & Feedback**

15 Was the assessment rubric clear and fair?
16 Did you receive useful feedback from your peers or the instructor during the project?
17 How would you rate the balance between individual accountability and group work in grading?

VI. **Ethics, Society, & Reflection**

18 Did the project encourage you to consider ethical, social, or environmental impacts?
19 Did you reflect on your learning and performance during or after the project?
20 How has this project influenced your view of what it means to be an engineer?

VII. **Overall Satisfaction & Suggestions**

21 What did you enjoy most about the PBL experience?
22 What are your suggestions for improving the PBL experience in future courses?
23 Would you recommend this type of learning activity in other courses? Why or why not?
24 On a scale of 1 to 5, how would you rate your overall experience with this project?

VIII. **Final Open-Ended Question**

25 Is there anything else you'd like to share about your experience?

Chapter 12

Reflections and Future Directions in Problem-Based Learning for Engineering Education

12.1 Introduction

This book has explored the transformative potential of Problem-Based Learning (PBL) in engineering education by comprehensively examining its theoretical foundations, pedagogical practices, implementation challenges, and empirical outcomes. Through interdisciplinary case studies, conceptual frameworks, and evidence-based discussions, the book has highlighted how PBL addresses the demanding need to equip engineering graduates with not only technical skills but also the critical thinking, collaboration, and adaptability required to thrive in a rapidly evolving world.

As engineering disciplines continue to grapple with complex global challenges—including sustainability, digital transformation, and social equity—PBL has emerged as a student-centred, inquiry-driven model that bridges the gap between academic preparation and professional competence. By engaging learners in real-world problem-solving, PBL cultivates essential 21st-century skills such as lifelong learning, ethical reasoning, and systems thinking. These attributes are increasingly recognised as vital for engineers expected to contribute meaningfully to sustainable development and societal progress.

This concluding chapter offers a reflective synthesis of the key insights presented throughout the book, while also charting future directions for PBL in engineering education. It begins by revisiting the key insights and takeaways from the PBL approach, particularly regarding PBL philosophy and pedagogy, curriculum integration, assessment strategies, faculty roles and training, case studies, or institutional examples. It then outlines the Impact of PBL on Engineering Students, Institutional and Faculty Perspectives, Challenges and Limitations, Future Directions and Recommendations, Research Agenda and Final Reflections, including a call to action for educators and institutions.

12.2 Key Insights and Takeaways

Several important themes and lessons have emerged that underscore the transformative potential of PBL in engineering education. At its core, PBL is presented as a student-centred and inquiry-driven pedagogical approach,

DOI: 10.1201/9781003529743-16

grounded in constructivist, experiential, and sociocultural theories of learning. Rather than relying on traditional lectures, PBL engages students in solving complex, real-world problems that require critical thinking, self-directed learning, and collaboration. This approach enhances cognitive engagement and nurtures important professional attributes such as communication, adaptability, and ethical awareness. The shift from passive knowledge transmission to active problem-solving marks a fundamental philosophical change in how engineering education is conceptualised and delivered.

Effective curriculum integration emerges as another central theme. Rather than treating PBL as a discrete instructional strategy, successful programmes embed it across modules and disciplines, using authentic problems as the core organising units. This interdisciplinary model allows for the seamless blending of technical knowledge with transversal competencies such as teamwork, systems thinking, and social responsibility. The integration of sustainability challenges, digital technology, and community-based issues into engineering projects reflects the increasing relevance of PBL to real-world applications. Examples drawn from international case studies illustrate how institutions have redesigned courses to align with broader competency frameworks, such as Outcomes-Based Education (OBE), Conceive–Design–Implement–Operate (CDIO) standards, and professional accreditation criteria, thereby making PBL a vehicle for institutional transformation.

Assessment strategies within PBL environments are also extensively addressed. Traditional assessment tools—often focused on rote memorisation and individual performance—are shown to be insufficient in capturing the multifaceted learning outcomes that PBL cultivates. Instead, it is advocated for authentic, formative, and performance-based assessments. These include portfolios, peer and self-assessments, reflective journals, and structured rubrics that evaluate not only technical competence but also group collaboration, problem-solving process, and communication skills. Aligning assessment methods with learning goals remains a critical component in ensuring the credibility and impact of PBL implementation.

The evolving role of faculty in PBL is another primary focus. As illustrated, the success of PBL depends heavily on faculty members' ability to act as facilitators rather than content deliverers. This transformation requires targeted faculty development programmes that equip instructors with skills in guiding inquiry, managing group dynamics, and offering constructive feedback. Institutions that have adopted PBL effectively often invest in sustained professional development, peer learning communities, and institutional support systems to enable this pedagogical shift. Faculty willingness to embrace ambiguity, encourage student autonomy, and assess both content and process is key to cultivating a culture of active learning.

In addition, a range of case studies and examples that offer practical insights into PBL implementation are presented. From the pioneering work at McMaster University to newer initiatives in developing regions, the cases illustrate how PBL can be adapted to diverse educational and cultural contexts. These examples highlight both the opportunities and challenges of transitioning to a PBL model, especially in resource-constrained environments. Despite differences in institutional capacity, a common thread across successful implementations is aligning curriculum, faculty development, and assessment practices with the overarching philosophy of student-centred, real-world learning.

Drawing from these insights, several critical success factors are identified for the effective implementation of PBL. These include strong institutional commitment, strategic alignment with educational goals, and leadership support for curriculum reform. Faculty training is equally essential, with emphasis on facilitating rather than instructing, and on developing skills to support reflective and collaborative learning. Curriculum coherence, particularly in the design of integrative and scaffolded problem scenarios, ensures that students encounter progressively complex challenges that build both depth and breadth of understanding.

Furthermore, robust assessment systems that go beyond summative testing are necessary to evaluate the full range of student competencies. Access to supportive learning environments—both physical and digital—is another enabling factor, along with careful orientation of students to the expectations of active, self-regulated learning. Finally, the importance of reflective practice is underscored and continuous improvement is emphasised, suggesting that successful PBL implementation is an iterative process grounded in evidence, feedback, and adaptive thinking.

The cumulative insights suggest that PBL is not merely an instructional technique but a holistic educational model capable of bridging the gap between academic learning and professional competence. When implemented thoughtfully and carefully, with systemic alignment and institutional support, PBL can serve as a powerful lever for innovation and transformation in engineering education.

12.3 Impact of PBL on Engineering Students

PBL has emerged as a transformative approach in engineering education, offering a pedagogical shift from traditional, lecture-based instruction to a student-centred model that actively engages learners in real-world problem-solving. It has gained considerable traction in engineering education due to its ability to enhance critical skills, foster deeper engagement, and lead to more meaningful learning outcomes. Substantial evidence was found of PBL's multifaceted impact on engineering students, particularly in areas such as skill development, motivation, academic performance, professional readiness, and student adaptability.

One of the most significant outcomes of PBL is the development of transferable and professional skills. By engaging with complex, open-ended problems, students are encouraged to think critically, apply interdisciplinary reasoning, and make iterative decisions (Lalwani, 2024; Meifitri, 2020). Through this process, they build robust problem-solving abilities, learn to identify knowledge gaps, research independently, and translate theory into practice. The inherently collaborative nature of PBL also fosters essential teamwork and communication skills, as students negotiate group roles, provide and receive feedback, and manage interpersonal dynamics within diverse teams—mirroring the realities of professional engineering practice (Kunicina et al., 2024; Prasad & O'Malley, 2022; Yusof et al., 2016).

In addition, PBL supports the development of entrepreneurial thinking and leadership, particularly through active learning strategies such as brainstorming, design thinking, and role-playing (Naik et al., 2024; Warke & Kukker, 2023). Students take ownership of their learning processes, enhancing their capacity for self-directed inquiry and lifelong learning. The iterative nature of PBL promotes habits of reflection and adaptability—skills essential for navigating the rapidly evolving demands of modern engineering and technology fields (Dolmans & Schmidt, 2006; Duch et al., 2001; Neves & Ribeiro, 2024; Riazifar & Khajehsaeid, 2024).

Another key area of impact is student motivation and engagement. PBL's use of real-world, socially relevant problems enhances the perceived relevance of academic content, fostering intrinsic motivation and deeper emotional investment in learning (Caesar et al., 2016; Garanina, 2020). Engineering students consistently report higher levels of interest, participation, and satisfaction in PBL contexts than in traditional lecture-based environments (Naik et al., 2024; Rendón & Martínez, 2024). Autonomy in learning and meaningful contributions to team-based projects lead to a stronger sense of ownership and accountability, reinforcing students' engagement and commitment to their educational journey (Neves & Ribeiro, 2024).

Evidence of improved learning outcomes associated with PBL was also observed. Academically, students demonstrate greater understanding and retention of theoretical concepts, as well as improved performance in assessments, project deliverables, and presentations (Neves & Ribeiro, 2024). These outcomes are reinforced by students' enhanced ability to apply knowledge across disciplinary boundaries and in varied real-life contexts (Rendón & Martínez, 2024; Riazifar & Khajehsaeid, 2024). From a professional standpoint, PBL prepares students for industry by developing competencies in project planning, time management, interdisciplinary collaboration, and the use of engineering tools and technologies (Bercial et al., 2024; Du et al., 2013; Guerra et al., 2017).

Despite these benefits, students often face initial challenges in adapting to the demands of PBL. Those accustomed to passive learning environments may find the shift to self-directed, collaborative inquiry disorienting and stressful. Difficulties such as unclear learning objectives, uneven participation in group work, and time management issues are commonly reported (Hunter et al., 2025).

However, with appropriate scaffolding—such as structured timelines, reflective journals, and facilitator guidance—students gradually develop strategies that enable them to thrive in PBL settings (Dolmans et al., 2005; Yusof et al., 2016). Mentorship and regular feedback also play a crucial role in helping students transition from dependent to independent learners.

PBL thus has a substantial and positive impact on engineering students by cultivating a wide array of cognitive, interpersonal, and professional skills. This approach: enhances critical thinking, collaboration, and communication; promotes lifelong learning; increases student motivation and engagement; and leads to demonstrable improvements in academic and practical outcomes. While some students may initially struggle with the openness and demands of PBL, most adapt over time, especially when supported by thoughtful facilitation and a conducive learning environment. Eventually, PBL equips engineering graduates not only with technical knowledge but also with the adaptive, integrative, and reflective capabilities needed to address the complex challenges of the contemporary world.

12.4 Institutional and Faculty Perspectives

The successful implementation demands significant adaptation by faculty and sustained institutional support to overcome common challenges related to infrastructure, resources, and policy frameworks. This discussion draws on institutional and faculty perspectives to explore the core elements necessary for effective PBL integration, focusing on faculty adaptation, institutional backing, and enabling infrastructure.

Faculty members often encounter difficulties when transitioning to a PBL framework, particularly due to unfamiliarity with the method and its underlying pedagogy. Faculty development programmes are, therefore, essential in equipping educators with the necessary skills to facilitate PBL effectively. These programmes typically unfold in stages—curriculum transition, implementation, and advancement—and emphasise continuous learning and reflective practice (Farmer, 2004). A key shift in this process involves moving from the role of content deliverer to that of a facilitator, guiding students in their exploration without providing direct instruction. This transformation demands a solid understanding of PBL dynamics and the ability to support students in becoming autonomous learners (Chan, 2009; Ribas, 2004). Resistance from faculty is not uncommon, often stemming from scepticism about the effectiveness of PBL or concerns that students may leave with gaps or misconceptions in foundational knowledge. Addressing these concerns through evidence-based faculty training and sharing successful PBL implementations can help mitigate resistance and build greater confidence in the approach (Schwartz et al., 2006).

Beyond faculty development, institutional support is critical to embedding PBL within engineering and other professional programmes. Institutions play a central role in providing the resources, encouragement, and policy flexibility needed for curriculum innovation. This includes supporting faculty development initiatives, allocating time for collaborative planning, and fostering a culture that values pedagogical experimentation and student-centred learning (Müller, 2011). Cultural change within institutions is often necessary to fully embrace PBL, requiring a re-evaluation of traditional educational paradigms. Engaging faculty, students, and administrative leaders in open dialogue can help create shared ownership of the learning process and foster an environment more conducive to innovation (Heuchemer et al., 2020). Moreover, institutions must work to bridge the gap between intended curriculum reforms and actual classroom practices by aligning institutional policies with the broader goals of PBL. Without this alignment, PBL risks remaining a surface-level intervention rather than becoming a transformative practice that supports holistic learning (Müller, 2011).

Infrastructure and resources also play a crucial role in enabling PBL. Effective implementation requires physical and digital environments that support group work, collaboration, and student autonomy. Access to learning technologies, flexible classroom layouts, and relevant instructional materials are all essential components (Donnelly & Massa, 2015). At the policy level, institutions must create frameworks that support the integration of PBL into existing curricula. This includes revising assessment strategies to measure not only knowledge acquisition but also the learning process, such as critical thinking, teamwork, and communication skills. Traditional assessment models may need to be redesigned to reflect PBL's process-oriented nature (Schwartz et al., 2006). Institutions should also consider the resource intensity of PBL, particularly the increased demand for faculty tutors and the potential workload implications. Addressing these concerns through strategic planning, workload management, and targeted resource allocation is essential to sustaining PBL initiatives (Schwartz et al., 2006).

Thus, institutions must be willing to invest in robust faculty development, foster a culture that values innovative teaching methods, and address the logistical and policy-related demands of PBL.

12.5 Challenges and Limitations

While PBL offers significant potential to enhance engineering education by fostering problem-solving, teamwork, communication, and lifelong learning, its implementation is not without challenges (Kjellberg, Adawi & Brolin, 2015). The book highlights several persistent limitations that institutions, educators, and students must address in order to unlock the full transformative capacity of PBL. These challenges span pedagogical, structural, cultural, and resource-related domains.

One of the foremost issues is the uneven development of key student competencies. While PBL is designed to cultivate autonomy, time management, and collaborative learning, many students struggle to adapt to the shift from traditional, instructor-led methods to self-directed inquiry. Skills such as reflective thinking, effective communication, and peer negotiation are not always intuitive and often require gradual, scaffolded development (Dolmans & Schmidt, 2006). Without structured support and well-designed facilitation, students may fail to internalise these skills, particularly in group settings where participation is unequal or where dominant voices overshadow quieter learners (Prasad & O'Malley, 2022).

Student motivation and engagement also shift significantly in a PBL context. On the one hand, real-world problems and authentic tasks increase intrinsic motivation and enhance learner engagement (Caesar et al., 2016; Garanina, 2020). Students frequently report that PBL makes learning more meaningful and personally relevant. On the other hand, for those accustomed to lecture-based instruction, the open-ended, ambiguous nature of PBL can lead to frustration, anxiety, or disengagement—especially when there is a lack of clear expectations or adequate orientation to the approach (Henderson et al., 2021; Kjellberg, Adawi & Brolin, 2015). This transition can be particularly challenging for students from educational systems that prioritise memorisation and teacher authority.

Empirical evidence from the book suggests that PBL can lead to improved academic and professional learning outcomes, such as enhanced analytical reasoning, interdisciplinary application, and long-term knowledge retention (Du et al., 2013; Lalwani, 2024). Students engaged in PBL environments also tend to demonstrate stronger metacognitive awareness and a capacity for lifelong learning, skills increasingly demanded by contemporary engineering practice. However, assessing these outcomes remains complex. Conventional metrics, such as exams or content recall, often fail to capture the full range of cognitive and interpersonal skills that PBL develops. Moreover, the lack of longitudinal data makes it difficult to track the sustainability of PBL-acquired competencies beyond the academic setting.

The adaptation process for students is another area of concern. Common obstacles include confusion about learning objectives, ineffective group communication, and reliance on more vocal or experienced team members. In the early stages, students often experience disorientation due to the lack of structured content delivery. Over time, many adapt by developing coping strategies such as self-organised planning, peer support networks, and proactive feedback-seeking. However, this transition is highly dependent on the presence of supportive facilitation, reflective practices, and institutional mechanisms that guide and monitor student development (Chen et al., 2023; Eboi et al., 2024).

At an institutional level, resistance to change from both faculty and students remains a significant barrier. Faculty members may find it difficult to transition from being content experts to learning facilitators, a shift that challenges their professional identity and entrenched pedagogical beliefs (Jowitt, 2009; Li, 2022). Similarly, students may resist the demands of active participation and autonomous learning, particularly when these expectations are not clearly communicated or supported (Henderson et al., 2021). This dual resistance can impede the diffusion of PBL and limit its impact unless addressed through effective change management and faculty development initiatives.

Assessment design is another major limitation. Traditional methods such as examinations and written assignments often do not align with the learning goals of PBL, which emphasise process-oriented skills like collaboration, critical thinking, and applied problem-solving. Designing assessments that capture these complex outcomes requires innovative tools and significant time investments—resources that are frequently unavailable or undervalued in conventional educational systems (Epler & Jacobs, 2022; Ni'mah et al., 2024).

Also, logistical and resource-related challenges complicate the effective scaling of PBL. Implementing PBL at the programmatic level demands considerable investments in faculty training, curriculum design, and physical learning environments conducive to group work and inquiry (Mansor et al., 2015). In resource-constrained or overcrowded institutions, these requirements may prove particularly difficult to meet. Additionally, many academic programmes have rigid, content-heavy curricula that leave little room for flexible, inquiry-driven methodologies, thereby necessitating comprehensive curriculum restructuring.

A recurring issue is the lack of comprehensive faculty training and support. Without adequate preparation, educators struggle to implement PBL effectively, leading to superficial applications that fall short of the approach's intended outcomes (Chen et al., 2023; Eboi et al., 2024).

Perhaps one of the most important concerns is the disconnect between PBL theory and practice. Although PBL is conceptually grounded in student autonomy and deep engagement, its implementation is often inconsistent and superficial. A lack of shared understanding among faculty about what constitutes "authentic" PBL contributes to fragmented or diluted practices (Li, 2022). This misalignment can reduce PBL to a procedural exercise rather than a transformative learning experience. Institutions also struggle to balance the demand for pedagogical innovation with the need to maintain structured, measurable, and scalable educational outcomes.

Thus, while PBL remains a powerful educational strategy capable of reshaping engineering education, its success is contingent upon thoughtful implementation and systemic support. Overcoming these challenges requires strategic investment in faculty development, curriculum redesign, and institutional culture change. A commitment to continuous improvement, grounded in reflective practice and supported by evidence-based policy, is essential to ensure that PBL realises its potential in preparing engineering graduates for the complexities of the 21st-century world.

12.6 Future Directions and Recommendations

Innovations in PBL are essential for keeping pace with the evolving demands of higher education, particularly in the context of engineering education.

As the need for student-centred, real-world learning intensifies, the integration of digital tools, online and hybrid learning formats, and interdisciplinary approaches has become central to enhancing the effectiveness of PBL. These innovations not only help overcome some of the limitations associated with traditional pedagogical methods but also better prepare students for the complex, collaborative problem-solving required in contemporary professional environments. Addressing these opportunities and challenges requires a concerted effort from curriculum designers, educators, and administrators to rethink and sustain the implementation of PBL across diverse educational contexts.

One of the most impactful innovations in PBL is its integration with digital tools. Digital PBL leverages technologies such as simulations, collaborative platforms, and virtual labs to create more interactive and accessible learning environments. In engineering education in particular, this integration supports the development of both technical competencies and professional skills (Patnawar, 2023). Additionally, the rise of online and hybrid learning, accelerated by the COVID-19 pandemic, has compelled institutions to adapt PBL to digital formats. These formats offer increased flexibility and support self-directed learning, allowing students to progress at their own pace and engage with content in more personalised ways (Epler & Jacobs, 2022). Another key innovation involves interdisciplinary applications of PBL, which encourage collaboration among students from various academic backgrounds. This approach reflects the complexity of real-world problems, which often require input from multiple domains of expertise (Kolmos & Graaff, 2014). By working across disciplines, students develop a broader perspective and enhance their ability to collaborate, communicate, and think critically.

Scaling and sustaining PBL, however, requires strategic planning and investment. Resource allocation is a significant factor; effective PBL depends on well-trained facilitators, access to appropriate materials, and time for thoughtful planning and implementation. Institutions must prioritise the provision of these resources to ensure the quality and continuity of PBL at scale (Tse & Bona, 2019). Institutional support is equally crucial. Administrative backing, supportive policies, and embedding PBL into the core curriculum help legitimise and normalise the approach, fostering a culture that values active, student-centred learning (LeCesne et al., 2007). Continuous evaluation and feedback mechanisms are also vital to sustaining PBL. Regular assessment of both learning outcomes and pedagogical processes enables institutions to identify areas for improvement and adapt practices accordingly. Feedback from students and educators plays a key role in refining PBL implementation and ensuring alignment with intended learning outcomes (Hung, 2015).

For curriculum designers, educators, and administrators, specific recommendations can guide the effective integration of PBL. Curriculum designers should aim to create flexible curricula that embed PBL as a central feature, with carefully constructed problems that reflect real-world challenges and promote critical thinking and collaboration (Madden et al., 2007). Educators must be equipped with the skills to facilitate PBL effectively, which involves understanding the dynamics of group work and fostering an environment where students are encouraged to take ownership of their learning while receiving appropriate guidance (Roth et al., 2024). Administrators have a responsibility to ensure that the necessary infrastructure is in place to support PBL in both physical and digital learning environments. This includes investing in technological tools and platforms that enable effective collaboration and knowledge sharing (Walecka, 2023).

However, as mentioned previously, the transition from traditional to PBL-based instruction is often resource-intensive and may face institutional resistance. Moreover, the success of PBL can vary depending on the discipline, institutional context, and fidelity of implementation. Therefore, it is essential for all stakeholders to engage in continuous evaluation, adapt strategies as needed, and remain responsive to the changing needs of learners and the broader demands of the professional world.

12.7 Research Agenda

12.7.1 Emerging Areas for Future Research

As PBL continues to gain traction as a transformative pedagogical approach, several key areas for future research are emerging. One of the most pressing needs is the development of longitudinal studies that assess the sustained impact of PBL on students' academic and professional trajectories. Such research is vital to understanding how PBL influences long-term knowledge retention, skill acquisition, and career adaptability. By tracking students over extended periods, these studies can offer valuable insights into the effectiveness of PBL in cultivating lifelong learning, critical thinking, and self-directed problem-solving (Othman et al., 2024; Roopashree, 2014).

In addition to long-term evaluations, comparative studies are essential to position PBL within the broader spectrum of pedagogical methods. Research that systematically compares PBL with traditional lecture-based learning, as well as other student-centred approaches like project-based and inquiry-based learning, can illuminate its unique advantages and limitations (Epler & Jacobs, 2022). These studies should be conducted across varied educational settings and disciplines to identify contexts in which PBL is most effective and to explore how it may be integrated synergistically with other teaching strategies.

Another significant area for investigation lies in the cultural and contextual adaptation of PBL. Educational outcomes associated with PBL may differ based on cultural values, institutional norms, and learner expectations. Research should therefore explore how PBL can be customised to suit diverse cultural settings while maintaining its core principles. Such studies would address the need to make PBL more inclusive and responsive to localised educational challenges (Susanty et al., 2023; Xiang, 2024). Understanding these cultural dimensions is crucial for scaling PBL effectively and ensuring its relevance in global educational landscapes.

12.7.2 Need for Evidence-Based Practices and Metrics

While the pedagogical potential of PBL is widely acknowledged, its implementation and evaluation still lack consistent and rigorous evidence-based practices and metrics. One of the central challenges lies in developing standardised tools for assessing learning outcomes that align with the goals of PBL. Traditional evaluation methods often fail to capture the depth of learning and the development of transversal skills such as collaboration, creativity, and critical thinking. Therefore, there is an urgent need to design and validate new assessment instruments that can measure both cognitive and non-cognitive outcomes effectively (Woodward, 1996).

In addition to assessment tools, research should aim to identify key indicators of successful PBL implementation, including student engagement, facilitator effectiveness, and institutional readiness. Developing evaluation frameworks that encompass these indicators can support educators and administrators in monitoring the quality of PBL practices and making data-informed decisions about curriculum design and instructional strategies (Zhang et al., 2022). These frameworks should also consider the perspectives of various stakeholders—students, educators, and industry partners—to ensure a holistic evaluation of PBL's impact.

Despite its promise, PBL's widespread adoption is hindered by a lack of empirical clarity regarding what constitutes effective practice. Consequently, there is a strong imperative for educational researchers to close this gap by producing actionable, evidence-based insights. Moreover, exploring hybrid pedagogical models that integrate PBL with other methods may provide new pathways for innovation in teaching and learning. Thus, advancing research in these areas will equip institutions with the tools and knowledge needed to refine PBL, making it a more adaptable, scalable, and impactful educational strategy across disciplines and cultures.

12.8 Final Reflections

12.8.1 The Evolving Role of Engineers in a Complex, Globalised World

The role of engineers is undergoing a significant transformation in response to the growing complexity of the globalised society. Today's engineers are expected to go beyond technical competence—they must also confront and contribute to solving broad societal challenges such as sustainability, climate change, global equity, and the integration of rapidly advancing technologies.

These demands are reshaping the expectations of engineering professionals and, by extension, the educational systems that prepare them.

One of the most significant areas of change is the increasing responsibility engineers bear in addressing sustainability and environmental issues. As the world confronts climate change, resource depletion, and ecological degradation, engineers are tasked with developing innovative and sustainable solutions that balance technological advancement with environmental protection. This requires a long-term, systems-oriented approach that incorporates ecological, social, and economic considerations (Cruickshank & Fenner, 2007; Foley & Leahy, 2010).

In parallel, globalisation has transformed the nature of engineering work. Engineers frequently operate in multicultural, geographically dispersed teams, making cross-cultural competence and international collaboration essential. The ability to navigate diverse cultural norms, communicate effectively across boundaries, and understand global perspectives is critical for solving complex, transnational problems (Ilyas, 2021).

Engineers are also increasingly involved in the design and management of complex sociotechnical systems—integrated networks of people, technology, and institutions. This requires not only technical acumen but also the ability to assess societal impacts, engage with non-technical stakeholders, and incorporate ethical and policy considerations into decision-making processes (Aslaksen, 2017; El Maraghy, 2011). These evolving demands highlight the urgent need for educational reform.

Engineering education must expand beyond traditional disciplinary boundaries to include the social sciences, humanities, and environmental studies. A multidisciplinary approach equips future engineers with the tools needed to understand and address complex, real-world challenges (Cruickshank & Fenner, 2007; Foley & Leahy, 2010). Equally important is the cultivation of global competencies. Institutions should emphasise local relevance, cultural awareness, international collaboration, and a nuanced understanding of global systems to prepare students for work in a connected and diverse world (Bourn & Neal, 2008; Ilyas, 2021).

12.8.2 The Transformative Potential of PBL in Preparing Future-Ready Engineers

In the context of rapid technological change and the growing complexity of global challenges, PBL has emerged as a transformative approach in engineering education. By shifting the focus from passive content delivery to active, problem-centred inquiry, PBL empowers students to engage more deeply with engineering concepts while developing the critical skills needed to thrive in an evolving professional landscape. Particularly in light of Industry 5.0—which emphasises digital transformation, environmental sustainability, and human-centred innovation—PBL offers a powerful framework for preparing engineers who are not only technically competent but also adaptable, creative, and future-ready.

One of the most evident impacts of PBL is its ability to enhance student engagement and understanding. Unlike traditional lecture-based instruction, PBL invites students to explore real-world problems, encouraging them to take ownership of their learning. This approach fosters greater motivation, satisfaction, and retention of core engineering principles. In a recent study, Riazifar and Khajehsaeid (2024) found that students engaged in PBL reported a deeper grasp of fundamental concepts, largely due to the active, contextualised nature of their learning. Similarly, Delaney & Nagle (2019) observed that students who participated in the hands-on design, construction, and testing of real machines under practical constraints reported feeling more confident and better prepared for professional work.

Beyond content mastery, PBL is instrumental in cultivating future-ready skills essential for Industry 5.0. The ability to solve complex problems, lead diverse teams, and think creatively is increasingly vital in a world where engineers are expected to drive innovation while addressing sustainability and ethical considerations. Studies have shown that PBL fosters these competencies by challenging students to work collaboratively, manage projects, and make decisions in uncertain, real-world contexts (Ciolacu et al., 2023). For instance, PBL was explicitly used to target professional skills such as project management and systems thinking, demonstrating that the method can be effectively tailored to meet diverse educational goals (Fernandez & Duarte, 2013).

Another significant contribution of PBL is its capacity to bridge the gap between academia and industry. One of the persistent criticisms of traditional engineering education is its disconnect from practical applications. PBL addresses this by simulating the complexities of workplace environments and aligning student projects with actual industry needs. As Grolinger (2011) noted, this alignment ensures that graduates are better prepared to transition smoothly into the workforce, equipped not only with technical knowledge but also with the soft skills and professional judgement required in real-world settings. In African universities, where educational reform is often linked with broader development goals, the adoption of PBL has also been identified as a strategy for modernising curricula and making engineering education more inclusive and relevant. Abbott et al. (2020) highlight how PBL not only engages students more effectively but also supports SDGs by embedding social and environmental relevance into technical education.

Of course, the implementation of PBL is not without challenges. A consistent theme across studies and institutional experiences is the need for significant faculty training and support. Teaching in a PBL environment requires a shift in mindset—from delivering information to facilitating learning—and many educators may feel unprepared for this transition. Additionally, effective PBL often requires interdisciplinary collaboration, posing logistical and curricular hurdles. Lavi et al. (2023) emphasise the importance of strategic planning and institutional commitment in overcoming these barriers. Moreover, some studies suggest that when PBL is used in isolation, students may show weaker mastery of specific disciplinary content (Fernandez & Duarte, 2013). This points to the need for a balanced approach that integrates PBL with structured, traditional methods to ensure both depth and breadth of learning.

Therefore, PBL offers immense potential to transform engineering education, aligning it with the complex realities of Industry 5.0. Through its emphasis on real-world relevance, active engagement, and skill development, PBL prepares students to become not only engineers but also leaders, innovators, and problem-solvers in a rapidly changing world. However, to fully realise this potential, institutions must address implementation challenges thoughtfully, support faculty development, and remain committed to continuous improvement. As the future of engineering education is imagined, PBL stands out as a critical driver in shaping professionals who are equipped to lead both technological innovation and sustainable societal progress.

12.8.3 A Call to Action for Educators and Institutions

Amidst the escalating complexity and transformative pressures on the engineering profession—fuelled by globalisation, digitalisation, and sustainability imperatives—PBL has increasingly emerged as a critical educational approach for cultivating future-ready engineers. This method, grounded in real-world problem-solving, emphasises active, student-centred learning and fosters essential competencies such as critical thinking, creativity, collaboration, and adaptability. These are precisely the kinds of transversal skills demanded by Industry 5.0, where engineering is not only about solving technical problems but also about navigating ambiguity, leading innovation, and addressing societal challenges.

The call to action for educators and institutions is clear: to align more closely with the evolving expectations of the profession, there must be a concerted effort to integrate PBL meaningfully and systematically across engineering curricula.

The rationale for adopting PBL is well supported by both research and practice. One of its most compelling advantages is its ability to bridge the gap between academic learning and industry needs. By immersing students in real-world challenges, PBL ensures they are not just absorbing theoretical knowledge but are also learning how to apply it in practical, professional settings. This alignment with industry expectations enhances graduates' readiness and employability. Moreover, PBL supports the development of 21st-century skills—such as problem-solving, communication, leadership, and teamwork—which are increasingly essential for working across disciplines and geographies in a globalised world. Through its emphasis on student-led inquiry, PBL also cultivates greater motivation and engagement. Students are encouraged to take ownership of their learning, leading to deeper understanding and improved retention of core concepts.

The benefits of PBL extend beyond skill development. Studies consistently show that students involved in PBL report higher levels of satisfaction and stronger connections to their field of study. They feel more prepared to confront complex, interdisciplinary challenges and are better equipped to navigate uncertainty. In addition, the integration of technology—such as simulations, digital collaboration tools, and virtual labs—within the PBL framework enhances the relevance and accessibility of engineering education, particularly in today's digitally driven world. These tools not only enrich learning experiences but also prepare students for modern workplaces where such technologies are ubiquitous.

However, while PBL offers clear advantages, its implementation is not without challenges. Faculty may require substantial training to shift from traditional lecturing to facilitation roles, and large class sizes or rigid curriculum structures may initially hinder the adoption of this approach. Furthermore, concerns persist about maintaining sufficient depth in technical knowledge when using PBL as the primary method of instruction. Addressing these challenges requires thoughtful and careful curriculum design, ongoing professional development, and institutional support. A blended approach—combining PBL with more structured content delivery—can offer a balanced model that preserves both rigour and relevance.

In moving forward, educators and institutions must recognise that integrating PBL is more than a pedagogical change; it is a strategic decision that reflects a broader commitment to educational innovation and social responsibility. Preparing engineers to lead in an era of complex, interconnected challenges requires more than technical competence—it demands critical inquiry, ethical awareness, and the ability to work collaboratively in diverse contexts. By embracing PBL, institutions have the opportunity to reshape engineering education into a more engaging, responsive, and future-focused endeavour.

References

Abbott, D., Chipika, S., & Wilson, G. (2020). The potential of problem-based learning to enhance engineering education in African universities. *Journal of International Development*, *32*(1), 44–61. https://doi.org/10.1002/JID.3444

Aslaksen, E. W. (2017). *Engineers and the evolution of society* (pp. 113–124). Springer. https://doi.org/10.1007/978-3-319-45193-0_9

Bercial, P. J., Gaya-López, M. C., Martínez-Orozco, J.-M., & Lavado-Anguera, S. (2024). PBL impact on learning outcomes in computer engineering: A 12-year analysis. *Education Sciences*, *14*(6), 653. https://doi.org/10.3390/educsci14060653

Bourn, D., & Neal, I. (2008). *The Global Engineer: Incorporating Global Skills within UK Higher Education of Engineers. Engineers against Poverty*, Institute of Education, University of London, pp. 1–30.

Caesar, M. I. M., Jawawi, R., Matzin, R., Shahrill, M., Jaidin, J. H., & Mundia, L. (2016). The benefits of adopting a problem-based learning approach on students' learning development in secondary geography lessons. *International Education Studies*, *9*(2), 51–65. https://doi.org/10.5539/ies.v9n2p51

Chan, A. (2009). Reflecting on the essence of our problem-based learning discussions: The importance of faculty development and our continuous quest for applications of problem-based learning. *Kaohsiung Journal of Medical Sciences*, *25*(5), 276–281. https://doi.org/10.1016/S1607-551X(09)70074-9

Chen, F., Zakaria, M. I., & Muslim, N. (2023). A systematic review: Challenges in implementing problem-based learning in mathematics education. *International Journal of Academic Research in Progressive Education and Development*, *12*(3). https://doi.org/10.6007/ijarped/v12-i3/19158

Ciolacu, M., Alves, G. R., Terkowsky, C., Alzoubi, A., Boettcher, K., Pozzo, M. I., & Kist, A. A. (2023). Developing future skills in engineering education for Industry 5.0: Enabling technologies in the age of digital transformation and green transition. In *Proceedings of the International Conference on Engineering Education* (pp. 1019–1031). Springer. https://doi.org/10.1007/978-3-031-42467-0_94

Cruickshank, H., & Fenner, R. (2007). The evolving role of engineers: Towards sustainable development of the built environment. *Journal of International Development*, *19*(1), 111–121. https://doi.org/10.1002/JID.1352

Delaney, K., & Nagle, G. (2019). Engineering education through problem-based learning: A case study of teaching mechanical engineering design in TU Dublin. https://doi.org/10.21427/0RSQ-4X20

Dolmans, D. H. J. M., De Grave, W., Wolfhagen, I. H. A. P., & van der Vleuten, C. P. M. (2005). Problem-based learning: Future challenges for educational practice and research. *Medical Education*, *39*(7), 732–741. https://doi.org/10.1111/j.1365-2929.2005.02205.x

Dolmans, D. H. J. M., & Schmidt, H. G. (2006). What do we know about cognitive and motivational effects of small group tutorials in problem-based learning? *Advances in Health Sciences Education*, *11*(4), 321–336. https://doi.org/10.1007/s10459-006-9012-8

Donnelly, J., & Massa, N. (2015). The PBL projects: Where we've been and where we are going. *Proceedings of SPIE*, *9793*, 452–468. https://doi.org/10.1117/12.2223219

Du, X., Emmersen, J., Toft, E., & Sun, B. (2013). PBL and critical thinking disposition in Chinese medical students—A randomised cross-sectional study. *Journal of Problem-Based Learning in Higher Education*, *1*(1), 72–83. https://doi.org/10.5278/ojs.jpblhe.v1i1.275

Duch, B. J., Groh, S. E., & Allen, D. E. (Eds.). (2001). *The power of problem-based learning*. Stylus Publishing.

Eboi, A. A., Obonyo, L., Kowuor, R., Aswani, D. R., & Radoli, L. (2024). Advancements and challenges of problem-based learning. In *Advances in educational marketing, administration, and leadership* (pp. 175–214). IGI Global. https://doi.org/10.4018/979-8-3693-6930-2.ch007

El Maraghy, W. H. (2011). *Future trends in engineering education and research* (pp. 11–16). Springer. https://doi.org/10.1007/978-3-642-20183-7_2

Epler, P., & Jacobs, J. (2022). Challenges with designing, implementing, and teaching PBL in higher education. In *Problem-based learning in higher education* (Ch. 5). IGI Global. https://doi.org/10.4018/978-1-7998-8177-3.ch005

Farmer, E. A. (2004). Faculty development for problem-based learning. *European Journal of Dental Education*, *8*(2), 59–66. https://doi.org/10.1111/J.1600-0579.2003.00337.X

Fernandez, F. H., & Duarte, J. E. (2013). El aprendizaje basado en problemas como estrategia para el desarrollo de competencias específicas en estudiantes de ingeniería. *Formación Universitaria*, *6*(5), 29–38. https://doi.org/10.4067/S0718-50062013000500005

Foley, A., & Leahy, P. (2010). The role of the professional engineer in the 21st century. https://pure.qub.ac.uk/en/publications/the-role-of-the-professional-engineer-in-the-21st-century

Garanina, R. M. (2020). Problem-based learning as a means of chemistry education effectiveness improvement at a medical university. *Samara Journal of Science*, *9*(4), 282–289. https://doi.org/10.17816/SNV202094303

Grolinger, K. (2011). Problem-based learning in engineering education: Meeting the needs of industry. *Teaching Innovation Projects*, *1*(2). https://ir.lib.uwo.ca/cgi/viewcontent.cgi?article=1011&context=tips

Guerra, A., Ulseth, R., Johnson, B., & Kolmos, A. (2017). Engineering grand challenges and the attributes of the global engineer: A literature review. In J. C. Quadrado, J. Bernardino, & J. Rocha (Eds.), *Proceedings of the 45th SEFI Annual Conference 2017* (pp. 1222–1235). European Association for Engineering Education.

Henderson, K. J., Coppens, E. R., & Burns, S. (2021). Addressing barriers to implementing problem-based learning. *AANA Journal*, *89*(2), 117–124. https://pubmed.ncbi.nlm.nih.gov/33832571/

Heuchemer, S., Martins, E., & Szczyrba, B. (2020). Problem-based learning at a "learning university": A view from the field. *Interdisciplinary Journal of Problem-Based Learning*, *14*(2). https://doi.org/10.14434/IJPBL.V14I2.28791

Hung, W. (2015). *Problem-based learning: Conception, practice, and future* (pp. 75–92). Springer. https://doi.org/10.1007/978-981-287-521-1_5

Hunter, C., Archer, S., & Illingworth, C. (2025). Problem-based learning. In R. Papa & S. H. K. Tan (Eds.), *Bloomsbury handbook of educational leadership and management* (pp. 302–305). Bloomsbury Publishing. https://doi.org/10.4337/9781035302932.ch83

Ilyas, M. (2021). Reflections on engineering education in the age of globalisation. In *Ideas to overcome and emerge from the pandemic crisis: Proceedings of the 1st International Multiconference on Entrepreneurship, Innovation and Regional Development*. https://doi.org/10.18687/leird2021.1.1.51

Jowitt, A. (2009). Issues in making the transition from traditional teaching to project-based learning. In *Proceedings of the 2009 Annual Conference of the Australasian Association for Engineering Education* (p. 417).

Kjellberg, M., Adawi, T., & Brolin, K. (2015). Challenges in implementing PBL: Chalmers Formula Student as a case. https://research.chalmers.se/en/publication/226614

Kolmos, A., & de Graaff, E. (2014). Problem-based and project-based learning in engineering education: Merging models. In A. Johri & B. M. Olds (Eds.), *Cambridge handbook of engineering education research* (pp. 141–161). Cambridge University Press. https://doi.org/10.1017/CBO9781139013451.012

Kunicina, N., Uteševs, I., Chaiko, Y., & Grants, R. (2024). Enhancing electrical engineering education through project-based learning. In *Proceedings of the RTUCON 2024 Conference* (pp. 1–6). https://doi.org/10.1109/rtucon62997.2024.10830810

Lalwani, K. (2024). How to design and moderate an effective problem-based learning discussion. In Lalwani, K., Cohen, I.D., Choi, E.Y., Robins, B., Kirsch, J. et al. (eds), *Professional, Ethical, Legal, and Educational Lessons in Medicine: A Problem-Based Learning Approach*, pp. 301–306. online edn., Oxford Academic, New York. https://doi.org/10.1093/med/9780197655979.003.0052

Lavi, R., Bertel, L. B., & Du, X. (2023). Guest editorial special issue on transforming engineering education. *IEEE Transactions on Education*, *66*, 404–409. https://doi.org/10.1109/te.2023.3308668

LeCesne, T. M., Katalanos, N., White, T., & Taslitz, N. (2007). Future directions in problem-based learning. *The Journal of Physician Assistant Education*, *18*(1), 67–70. https://doi.org/10.1097/01367895-200718010-00014

Li, H. (2022). *Educational change towards problem-based learning*. CRC Press. https://doi.org/10.1201/9781003357230

Madden, D. L., Black, M. E., Naylor, C. J., & Hecker, R. (2007). Future directions for the Bulletin. *New South Wales Public Health Bulletin*, *18*(2), 1–2. https://doi.org/10.1071/NB07018

Mansor, A. N., Abdullah, N., Wahab, J. A., Rasul, M. S., Nor, M. Y. M., Nor, N. M., & Raof, R. A. (2015). Managing problem-based learning: Challenges and solutions for educational practice. *Asian Social Science*, *11*(4), 259–268. https://doi.org/10.5539/ASS.V11N4P259

Meifitri. (2020). Problem-Solving Ability and Communication Skills Through a Problem-Based Learning Approach. *Proceedings of the Eighth International Conference on Languages and Arts (ICLA-2019)*. https://doi.org/10.2991/assehr.k.200819.028

Müller, C. (2011). Implementation von problem-based learning—institutionelle Bedingungen und Anforderungen. *Zeitschrift für Hochschulentwicklung*, *6*(3). https://doi.org/10.3217/zfhe-6-03/10

Naik, S. M., Bandi, S., & Reddy, C. (2024). Enhancing engineering education: Exploring the impact of problem-based learning on first-year students' skills and engagement—A case study. *Journal of Engineering Education Transformations*, *38*(IS1), 133–139. https://doi.org/10.16920/jeet/2024/v38is1/24222

Neves, A. J. R., & Ribeiro, F. (2024). Enhancing engineering education through project-based learning: A case study of biometry and computer vision courses. In *2024 IEEE Global Engineering Education Conference (EDUCON)* (pp. 1–5). https://doi.org/10.1109/EDUCON60312.2024.10578896

Ni'mah, A., Arianti, E. S., Suyanto, S., Putera, S. H. P., & Nashrudin, A. (2024). Problem-based learning (PBL) methods within an independent curriculum: A literature review. *Sintaksis*, *2*(4), 165–174. https://doi.org/10.61132/sintaksis.v2i4.859

Othman, N., Khalid, N., & Hamid, N. A. (2024). Mapping the intersection of problem-based learning and lifelong learning: A bibliometric analysis. *Attarbawiy*, *8*(2), 153–162. https://doi.org/10.53840/attarbawiy.v8i2.258

Patnawar, S. T. (2023). A comprehensive review on PBL and digital PBL in engineering education—Status, challenges and future prospects. *Journal of Engineering Education Transformations*, *37*(2), 142–157. https://doi.org/10.16920/jeet/2023/v37i2/23157

Prasad, S., & O'Malley, C. B. (2022). An introductory framework of problem-based learning (PBL) and perspectives on enhancing facilitation approaches. *HAPS Educator*, 26 (3), 52–58. https://doi.org/10.21692/haps.2022.016

Rendón, A., & Martínez, F. (2024). Impact of project-based learning on conceptual achievements and motivation of technology students in electricity. *World Journal of Advanced Engineering Technology and Sciences*, *12*(2), 001–008. https://doi.org/10.30574/wjaets.2024.12.2.0262

Riazifar, N., & Khajehsaeid, H. (2024). Enhancing engineering education through problem-based learning. *The Future of Engineering Education Series*. https://doi.org/10.31273/10.31273/9781911675167/1636

Ribas, A. F. (2004). Líneas maestras del Aprendizaje por Problemas. *Revista Interuniversitaria de Formación del Profesorado*, *18*(1), 79–95. https://dialnet.unirioja.es/servlet/articulo?codigo=1057106

Roopashree, B. J. (2014). PBL: Future challenges for educational practice and research. *Journal on School Educational Technology*, *10*(2), 9–16. https://doi.org/10.26634/JSCH.10.2.2969

Roth, A. M., Manocha, D., Sriram, R. D., & Tabassi, E. (2024). Opportunities, challenges, and future directions. In *Trustworthy engineering computing* (pp. 77–88). Morgan & Claypool Publishers. https://doi.org/10.1007/978-3-031-47518-4_5

Schwartz, P., Mennin, S., & Webb, G. (2006). *Problem-based learning: Case studies, experience and practice*. Routledge. http://ci.nii.ac.jp/ncid/BB08534054

Susanty, A. I., Artadita, S., Neo, T.-K., Pradana, M., Neo, M., & Amphawan, A. (2023). Problem-based learning research mapping (2013–2022): A bibliometric analysis. In *2023 International Conference on Advances in Digital Education and Intelligent Systems (ICADEIS)* (pp. 1–6). https://doi.org/10.1109/icadeis58666.2023.10270960

Tse, N. M. K., & di Bona, R. (2019). Large-scale vertically integrated PBL. In *Proceedings of the 2019 IEEE Global Engineering Education Conference (EDUCON)*, 1–8. https://researchers.mq.edu.au/en/publications/large-scale-vertically-integrated-pbl

Walecka, J. D. (2023). Future directions. In *Electron scattering for nuclear and nucleon structure* (pp. 278–287). Cambridge University Press. https://doi.org/10.1017/9781009290616

Warke, A., & Kukker, A. (2023). Role of engineering projects in prospective of teaching-learning process. In *2023 IEEE PuneCon* (pp. 1–5). https://doi.org/10.1109/punecon58714.2023.10450085

Woodward, C. A. (1996). Problem-based learning in medical education: Developing a research agenda. *Advances in Health Sciences Education*, *1*(1), 83–94. https://doi.org/10.1007/BF00596231

Xiang, C. (2024). Research on applying problem-based learning (PBL) teaching mode to forestry translation classrooms. *International Journal of New Developments in Education*, *6*(9). https://doi.org/10.25236/ijnde.2024.060903

Yusof, K. M., Sadikin, A. N., Phang, F. A., & Aziz, A. A. (2016). Instilling professional skills and sustainable development through problem-based learning (PBL) among first-year engineering students. *International Journal of Engineering Education*, *32*(1), 333–347. http://eprints.utm.my/id/eprint/73506/

Zhang, F., Wang, H., Bai, Y., & Zhang, H. (2022). A bibliometric analysis of the landscape of problem-based learning research (1981–2021). *Frontiers in Psychology*, *13*, Article 828390. https://doi.org/10.3389/fpsyg.2022.828390

Index

Pages in *italics* refer to figures, pages in **bold** refer to tables.

For Product Safety Concerns and Information please contact our
EU representative GPSR@taylorandfrancis.com Taylor & Francis
Verlag GmbH, Kaufingerstraße 24, 80331 München, Germany